Advances in Mathematical Fluid Mechanics

Advances in Mathematical Fluid Mechanics is a forum for the publication of high quality monographs, or collections of works, on the mathematical theory of fluid mechanics, with special regards to the Navier-Stokes equations. Its mathematical aims and scope are similar to those of the *Journal of Mathematical Fluid Mechanics*. In particular, mathematical aspects of computational methods and of applications to science and engineering are welcome as an important part of the theory. So also are works in related areas of mathematics that have a direct bearing on fluid mechanics.

The monographs and collections of works published here may be written in a more expository style than is usual for research journals, with the intention of reaching a wide audience. Collections of review articles will also be sought from time to time.

Conformal and Potential Analysis in Hele-Shaw Cells

Björn Gustafsson
Alexander Vasil'ev

Birkhäuser Verlag
Basel · Boston · Berlin

Authors:

Björn Gustafsson
Department of Mathematics
Royal Technology Institute
100 44 Stockholm
Sweden
e-mail: gbjorn@kth.se

Alexander Vasil'ev
Department of Mathematics
University of Bergen
Johannes Brunsgate 12
5008 Bergen
Norway
e-mail: alexander.vasiliev@uib.no

2000 Mathematical Subject Classification 76D27, 76M40, 65E05

A CIP catalogue record for this book is available from the
Library of Congress, Washington D.C., USA

Bibliographic information published by Die Deutsche Bibliothek
Die Deutsche Bibliothek lists this publication in the Deutsche Nationalbibliografie;
detailed bibliographic data is available in the Internet at <http://dnb.ddb.de>

ISBN 3-7643-7703-8 Birkhäuser Verlag, Basel – Boston – Berlin

Printed on acid-free paper produced from chlorine-free pulp. TCF ∞
Printed in Germany
ISBN-10: 3-7643-7703-8
ISBN-13: 978-3-7643-7703-8
e-ISBN-10: 3-7643-7704-6

9 8 7 6 5 4 3 2 1

www.birkhauser.ch

Contents

Preface

One of the most influential works in Fluid Dynamics at the edge of the 19-th century was a short paper [166] written by **Henry Selby Hele-Shaw** (1854 – 1941). There Hele-Shaw first described his famous cell that only became a subject of deep investigation more than 50 years later. A Hele-Shaw cell is a device for investigating two-dimensional flow of a viscous fluid in a narrow gap between two parallel plates. This cell is the simplest system in which multi-dimensional convection is present. Probably the most important characteristic of flows in such a cell is that when the Reynolds number based on gap width is sufficiently small, the Navier-Stokes equations averaged over the gap reduce to a linear relation similar to Darcy's law and then to a Laplace equation for pressure. Different driving mechanisms can be considered, such as surface tension or external forces (suction, injection). Through the similarity in the governing equations, Hele-Shaw flows are particularly useful for visualization of saturated flows in porous media, assuming they are slow enough to be governed by Darcy's law. Nowadays, the Hele-Shaw cell is used as a powerful tool in several fields of natural sciences and engineering, in particular, matter physics, material science, crystal growth and, of course, fluid mechanics.

The next important step after Hele-Shaw's work was made by **Pelageya Yakovlevna Polubarinova-Kochina** (1899-1999) and **Lev Aleksandrovich Galin** (1912 – 1981) in 1945 [117], [262], [263], who developed a complex variable method to deal with non-gravity Hele-Shaw flows neglecting surface tension. The main idea was to apply the Riemann mapping from an appropriate canonical domain (the unit disk in most situations) onto the phase domain to parameterize the free boundary. The equation for this map, named after its creators, allows one to construct many explicit solutions and to apply methods of conformal analysis and geometric function theory to investigate Hele-Shaw flows. In particular, solutions to this equation in the case of advancing fluid give subordination chains of simply connected domains which have been studied for a long time in the theory of univalent functions. The Löwner-Kufarev equation [212], [224] plays a central role in this study (**Charles Loewner** or **Karel Löwner** originally in Czech, 1893 – 1968; **Pavel Parfen'evich Kufarev**, 1909 – 1968). The Polubarinova-Galin equation and the Löwner-Kufarev one, having some evident geometric connections, are not closely related analytically. The Polubarinova-Galin equation is essentially non linear and the corresponding subordination chains are of rather complicated nature.

Among other remarkable contributions we distinguish the discovery of the viscous fingering phenomenon by **Sir Geoffrey Ingram Taylor** (1886 – 1975) and **Philip Geoffrey Saffman** [293], [294], and the first modern description of the complex variable approach and the study of the complex moments made by **Stanley Richardson** [283]. Contributions made by scientists from Great Britain (L. J. Cummings, D. Crowdy, C. M. Elliott, S. D. Howison, J. R. King, J. R. Ockendon, S. Richardson) are to be emphasized. They have substantially developed the complex variable approach and actually converted the Hele-Shaw problem into a modern challenging branch of applied mathematics.

During the last couple of decades, interest in Hele-Shaw flows has increased considerably and problems related to such flows are now studied from different aspects all over the world.

In the present monograph, we aim at giving a presentation of recent and new ideas that arise from the problems of planar fluid dynamics and which are interesting from the point of view of geometric function theory and potential theory. In particular, we are concerned with geometric problems for Hele-Shaw flows. We also view Hele-Shaw flows on parameter spaces (e.g., the Teichmüller space) and reveal connections with string theory. Ultimately, we see the interaction between several branches of complex and potential analysis, and planar fluid mechanics.

For most parts of this book we assume the background provided by graduate courses in real and complex analysis, in particular, the theory of conformal mappings and in fluid mechanics. We also try to make some historical remarks concerning the persons that have contributed to the topic. We have tried to keep the book as self-contained as possible.

Acknowledgements. We would like to acknowledge many conversations with J. R. Arteaga, J. Becker, L. Cummings, V. Goryainov, V. Gutlyanskiĭ, R. A. Hidalgo, Yu. Hohlov, S. Howison, J. King, K. Kornev, J. Ockendon, S. Ruscheweyh, Ch. Pommerenke, D. Prokhorov, H. Shahgholian, H. S. Shapiro. Both authors especially want to thank their wives Eva Odelman and Irina Markina. They always inspire our work. Irina Markina is, moreover, a colleague and co-author of the second author. The project has been supported by the Swedish Research Council, the Göran Gustafsson Foundation (Sweden), by the projects FONDECYT (Chile), grants #1030373, # 7030011 and # 1040333; Project UTFSM # 12.03.23, by the grant of the University of Bergen.

Björn Gustafsson & Alexander Vasil'ev
Bergen-Stockhom-Valparaíso, 2003-2006

Chapter 1

Introduction and Background

1.1 Newtonian fluids

A *fluid* is a substance which continues to change shape as long as there is a small shear stress (dependent on the velocity of deformation) present. If the force F acts over an area A, then the ratio between the tangential component of F and A gives a shear stress across the liquid. The liquid's response to this applied shear stress is to flow. In contrast, a solid body undergoes a definite displacement or breaks completely when subjected to a shear stress. Viscous stresses are linked to the velocity of deformation. In the simplest model, this relation is just linear, and a fluid possessing this property is known as a *Newtonian fluid.* The constant of the proportionality between the viscous stress and the deformation velocity is known as the *coefficient of viscosity* and it is an intrinsic property of a fluid.

Certain fluids undergo very little change in density despite the existence of large pressures. Such a fluid is called *incompressible* (modelled by taking the density to be constant). In fluid dynamics we speak of incompressible flows, rather than incompressible fluids. A laminar flow, that is a flow in which fluid particles move microscopically in straight parallel lines without macroscopic velocity fluctuations, satisfies Newton's Viscosity Law (or is said do be Newtonian) if the shear stress in the direction x of flow is proportional to the change of velocity V in the orthogonal direction y as

$$\sigma := \frac{dF}{dA} = \mu \frac{\partial V}{\partial y}.$$

The coefficient of proportionality μ is called the coefficient of viscosity or dynamic viscosity. Many common fluids such as water, all gases, and petroleum products are Newtonian. A non-Newtonian fluid is a fluid in which shear stress is not simply proportional solely to the velocity gradient, perpendicular to the plane of shear. Non-Newtonian fluids may not have a well-defined viscosity. Pastes, slurries, high polymers etc. are not Newtonian. Pressure has only a small effect on viscosity and this effect is usually neglected. The kinematic viscosity is defined as the quotient

$$\nu = \frac{\mu}{\rho},$$

where ρ stands for density of the fluid. All these considerations can be made with dimensions and their units taken into account or else be made dimensionless.

1.2 The Navier-Stokes equations

Important quantities that characterize the flow of a fluid are

- m – mass;
- p – pressure;
- $\mathbf{V}$ – velocity field;
- Θ – temperature;
- ρ – density;
- μ – viscosity.

Various approaches to the equations of fluid motion can be summarized in the so-called *Reynolds' Transport Theorem* (**Osborne Reynolds** 1842–1912). From a mathematical point of view this simply means a formula for the derivative of an integral with respect to a parameter (e.g., time) in the case that both integrand and the domain of integration depend on the parameter.

We always assume that a fluid system is composed of the same fluid particles. Let us consider a fluid that occupies a *control volume* $V(t)$ bounded by a *control surface* $S(t)$. Let $N(t)$ be an extensive property of the system, such as mass, momentum, or energy. Let $\mathbf{x} = (x_1, x_2, x_3)$ be the spatial variable and let t be time. We denote by $\eta(\mathbf{x}, t)$ the corresponding intensive property which is equal to the extensive property per unit of mass, $\eta = dN/dm$,

$$N(t) = \int_{V(t)} \eta\rho \, dv, \qquad dv = dx_1 dx_2 dx_3.$$

Reynolds' Transport Theorem states that the rate of change of N for a system at time t is equal to the rate of change of N inside the control volume V plus the rate of flux of N across the control surface S at time t:

$$\left(\frac{dN}{dt}\right)_{sys} = \int_{V(t)} \frac{\partial}{\partial t}(\eta\rho) \, dv + \oint_{S(t)} \eta\rho\mathbf{V} \cdot \mathbf{n} \, dS. \tag{1.1}$$

Here $\mathbf{V} = (V_1, V_2, V_3)$, and $\mathbf{n}$ is the unit normal vector in the outward direction. The Gauss theorem implies

$$\left(\frac{dN}{dt}\right)_{sys} = \int_{V(t)} \left[\frac{\partial}{\partial t}(\eta\rho) + \nabla\cdot(\eta\rho\mathbf{V})\right] dv.$$

Let us introduce a derivative $\frac{D}{Dt}$ which is called the *convective derivative*, or Eulerian derivative, and which is defined as

$$\frac{D}{Dt} = \frac{\partial}{\partial t} + \mathbf{V}\cdot\nabla,$$

or in coordinates

$$\frac{D}{Dt} = \frac{\partial}{\partial t} + V_1\frac{\partial}{\partial x_1} + V_2\frac{\partial}{\partial x_2} + V_3\frac{\partial}{\partial x_3}.$$

Then we have

$$\left(\frac{dN}{dt}\right)_{sys} = \int_{V(t)} \left(\frac{D(\eta\rho)}{Dt} + \eta\rho(\nabla\cdot\mathbf{V})\right) dv.$$

1.2.1 The continuity equation

If we take the mass as the extensive property, then $N \equiv m$, $\eta \equiv 1$ and Reynolds' Transport Theorem (1.1) becomes

$$\left(\frac{dm}{dt}\right)_{sys} = \int_{V(t)} \frac{\partial\rho}{\partial t} dv + \oint_{S(t)} \rho\mathbf{V}\cdot\mathbf{n}\, dA.$$

The law of conservation of mass states that $\left(\frac{dm}{dt}\right)_{sys} = 0$. Therefore,

$$\int_{V(t)} \left(\frac{\partial\rho}{\partial t} + \nabla\cdot(\rho\mathbf{V})\right) dv = 0.$$

The latter equation is known as the *continuity equation.* Since this equation holds for any control volume, we get

$$\frac{\partial\rho}{\partial t} + \nabla\cdot(\rho\mathbf{V}) = 0.$$

When ρ is constant, the fluid is said to be *incompressible* and the above equation reduces to

$$\nabla\cdot\mathbf{V} = 0. \tag{1.2}$$

1.2.2 The Euler equation

Let us consider only incompressible fluids. Linear momentum of an element of mass dm is a vector quantity defined as $d\mathbf{P} = \mathbf{V}\, dm$, or for the whole control volume,

$$\mathbf{P} = \int\limits_{V(t)} \rho \mathbf{V}\, dv.$$

Applying Reynolds' Transport Theorem we get

$$\left(\frac{d\,\mathbf{P}}{d\,t}\right)_{sys} = \int\limits_{V(t)} \rho \frac{D\mathbf{V}}{Dt} dv = \int\limits_{V(t)} \frac{D\mathbf{V}}{Dt} dm,$$

which infinitesimally is $\frac{D\mathbf{V}}{Dt} dm$, i.e., just the product of the mass element and acceleration.

Newton's second law for an inertial reference frame states that the rate of change of the momentum $\mathbf{P}$ equals the force exerted on the fluid in $V(t)$:

$$d\mathbf{F} = \frac{D\,\mathbf{V}}{D\,t} dm = \left(\frac{\partial}{\partial t}\mathbf{V} + (\mathbf{V}\cdot\nabla)\mathbf{V}\right) dm, \tag{1.3}$$

where $\mathbf{F}$ is the vector resultant of forces. Suppose for a moment that there are no shear stresses (inviscid fluid). If the surface forces $\mathbf{F}_s$ on a fluid element are due to pressure p and the body forces are due to gravity in the x_3-direction, then we have $d\mathbf{F} = d\mathbf{F}_s + d\mathbf{F}_b$, or

$$d\mathbf{F} = -(\nabla p)\, dv - g(\nabla x_3)(\rho\, dv), \tag{1.4}$$

where $\mathbf{F}_b$ is the gravity force per unit of mass and g is the gravity constant. Substituting (1.4) into (1.3) we obtain

$$-\frac{1}{\rho}\nabla p - g\nabla x_3 = \frac{\partial \mathbf{V}}{\partial t} + (\mathbf{V}\cdot\nabla)\mathbf{V}\,,$$

or

$$-\nabla p - \rho g \nabla x_3 = \rho \frac{D\,\mathbf{V}}{D\,t}. \tag{1.5}$$

The equation (1.5) is known as *the Euler equation.*

In terms of control volume we have

$$\left(\frac{d}{dt}\right)_{sys} \int\limits_{V(t)} \rho \mathbf{V}\, dv = -\int\limits_{V(t)} (\nabla p + \rho g \nabla x_3) dv,$$

or

$$\left(\frac{d}{dt}\right)_{sys} \int\limits_{V(t)} \rho \mathbf{V}\, dv = \oint\limits_{S(t)} \sigma \cdot \mathbf{n}\, dA - \int\limits_{V(t)} \rho g \nabla x_3 dv, \tag{1.6}$$

where $\sigma = (\sigma_{ij})_{i,j=1}^3$, $\sigma_{jj} = -p$, $\sigma_{ij} = 0$, $i \neq j$, is the stress tensor. In general, the *stress tensor* $(\sigma_{ij})_{i,j=1}^3$ is defined by the relationship $dF_i = \sum_{j=1}^3 \sigma_{ij} n_j \, dA$ between the surface force $d\mathbf{F}$ on an infinitesimal area element dA and the normal vector $\mathbf{n}$ of it ($\mathbf{F} = (F_1, F_2, F_3)$, $\mathbf{n} = (n_1, n_2, n_3)$).

1.2.3 The Navier-Stokes equation

The first term in the right-hand side of the Euler equation (1.6) is due the surface forces and the second one is due to body forces (or forces per unit mass in (1.5)). Let us consider the shear and normal stresses σ_{ij} in a mass element $dm = \rho\, dv = \rho\, dx_1 dx_2 dx_3$ that occupies a volume bounded by a parallelepiped such that its principal diagonal joins the points $\mathbf{x} = (x_1, x_2, x_3)$ and $\mathbf{x} + d\mathbf{x} = (x_1 + dx_1, x_2 + dx_2, x_3 + dx_3)$. We call the *$x_i$-surface*, that surface with one of the vertices at the point $\mathbf{x}$ and with the normal vector parallel to the x_i axis. The *surface parallel to the x_i-surface* is the one with a vertex at $\mathbf{x} + d\mathbf{x}$. We denote by σ_{jj} the normal stress on the x_j-surface in the outward direction. The normal stress on the parallel surface is $\sigma_{jj} + \frac{\partial \sigma_{jj}}{\partial x_j} dx_j$. By σ_{ij}, $i \neq j$, we denote the shear stress on the x_i-surface in the direction x_j and similarly for the parallel surface. The shear and normal stresses are given by a stress-velocity relation which is more general than Newton's law and which is known as Stokes' viscosity law for incompressible fluids. It states that the stress tensor $(\sigma_{ij})_{i,j=1}^3$ is given by

$$\sigma_{ii} = -p + 2\mu \frac{\partial V_i}{\partial x_i}, \qquad \sigma_{ij} = \mu \left(\frac{\partial V_i}{\partial x_j} + \frac{\partial V_j}{\partial x_i} \right), \qquad i \neq j,$$

where μ is the viscosity coefficient. The *Navier-Stokes equation* is just a generalization of the Euler equation when allowing both normal and shear stresses for surface forces. Replacing the stress tensor in (1.6) by the above expression we obtain the Navier-Stokes equation. The Gauss theorem leads to a point-wise equation in vector form for a Newtonian incompressible fluid with constant viscosity

$$\frac{D\mathbf{V}}{Dt} = \mathbf{F}_b + \frac{1}{\rho}(-\nabla p + \mu \Delta \mathbf{V}). \tag{1.7}$$

If body forces are negligible, then we can put $\mathbf{F}_b = 0$. The equations (1.2) and (1.7) are called the Navier-Stokes equations for incompressible fluids.

1.2.4 Dynamical similarity and the Reynolds number

Letting L be a representative scale (that can be thought of as the distance between enclosing boundaries), and U be a representative velocity (that can be thought of as the steady speed of a rigid boundary), we change variables

$$\mathbf{x} \to L\mathbf{x}, \quad \mathbf{V} \to U\mathbf{V}, \quad t \to \frac{L}{U} t.$$

Let us choose a scaling for the pressure as

$$p \to \rho U^2 p.$$

Substituting these new values into the Navier-Stokes equation (with $\mathbf{F}_b = 0$) we have

$$\frac{D\mathbf{V}}{Dt} = -\nabla p + \frac{1}{\mathrm{R}}\Delta \mathbf{V}, \tag{1.8}$$

where $\mathrm{R} = \rho UL/\mu$ is the *Reynolds number.* This equation is just the Navier-Stokes equation in dimensionless variables. Taking into account units

$$\rho = \frac{kg}{m^3}, \quad U = \frac{m}{s}, \quad L = m, \quad \mu = \frac{kg}{ms},$$

we reach the conclusion that the Reynolds number is a non-dimensional number.

Nondimensionalization, being a seemingly superficial step, becomes important when considering different flows with the same Reynolds number. A three-parameter family of solutions for a specific flow is equivalent to a one-parameter family for some modelling flow. Two flows with the same Reynolds number and the same geometry are called *dynamically similar.*

There are two different types of real fluid flow: laminar and turbulent. A well-ordered flow, free of macroscopic velocity fluctuations, is said to be *laminar.* Fluid layers are assumed to slide over one another without fluid being exchanged between the layers. In *turbulent flow*, secondary random motions are superimposed on the principal flow and there is an exchange of fluid from one adjacent segment to another. More important, there is an exchange of momentum such that slowly moving fluid particles speed up and fast moving particles give up their momentum to the slower moving particles and slow down themselves.

In an experiment in 1883, Reynolds demonstrated that, under certain circumstances, the flow in a tube changes from laminar to turbulent over a given region of the tube. He used a large water tank that had a long tube outlet with a tap at the end of the tube to control the flow speed. The tube went smoothly into the tank. A thin filament of coloured fluid was injected into the flow at the mouth as is shown in Figure 1.1. When the speed of the water flowing through the tube was low, the filament of colored fluid maintained its identity for the entire length of the tube. However, when the flow speed was high, the filament broke up into the turbulent flow that existed throughout the cross section. Thus, laminar flow occurs when the Reynolds number R is not too large. When R is sufficiently large, then turbulence comes into consideration. It is observed empirically that the flow becomes turbulent whenever the Reynolds number exceeds a certain value R^* which is critical. The Landau theory of the transition from steady laminar flow to turbulence suggests another limiting critical number $\mathrm{R}^{**} > \mathrm{R}^*$. Passing R^* the flow becomes unstable and bifurcations occur until it arrives at turbulence passing R^{**}. For the water flow $\mathrm{R}^* = 2,300$ and $\mathrm{R}^{**} = 40,000$ in Reynolds' experiment.

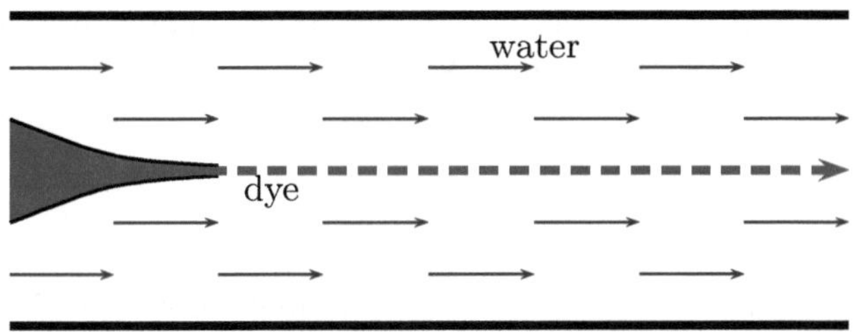

Figure 1.1: Reynolds' experiment

1.2.5 Vorticity, two-dimensional flows

When the Reynolds number is rather large, the distribution of vorticity proves to be an important entity to be taken into account. Let us consider two-dimensional flow with the velocity field $\mathbf{V} = (V_1, V_2, 0)$, subject to the restriction of incompressibility $\nabla \cdot \mathbf{V} = 0$, from which it follows that $V_1 dx_2 - V_2 dx_1$ is (locally) an exact differential $d\psi$. Then $V_1 = \partial\psi/\partial x_2$ and $V_2 = -\partial\psi/\partial x_1$ or $\mathbf{V} = \nabla \times (\psi \nabla x_3)$. If γ is a curve in the (x_1, x_2)-plane with the rightward normal vector $\mathbf{n} = (n_1, n_2, 0)$, then

$$\int_\gamma d\psi = \int_\gamma V_1 dx_2 - V_2 dx_1 = \int_\gamma \mathbf{V} \cdot \mathbf{n}\, ds.$$

Hence the flux of volume across any curve joining two points is equal to the difference between the values of ψ at these points, the function ψ is constant along a streamline, and it is called the *stream function.* The curl $\nabla \times \mathbf{V} = \boldsymbol{\omega}$ is called the *vorticity* of the fluid. In terms of the stream function, $\boldsymbol{\omega} = -\nabla x_3 \Delta\psi$. Taking the curl of Navier-Stokes equation ((1.8)) the term ∇p disappears and one gets an equation in $\boldsymbol{\omega}$ alone

$$\frac{D\boldsymbol{\omega}}{Dt} = \frac{1}{\mathrm{R}}\Delta\boldsymbol{\omega},$$

or for the stream function

$$\frac{D\,\Delta\psi}{D\,t} = \frac{1}{\mathrm{R}}\Delta(\Delta\psi). \tag{1.9}$$

Equation (1.9) has several benefits. For example, it is a scalar equation rather than a vector one.

As we have remarked, the flow is laminar until the Reynolds number reaches its first critical value, or it can be thought of as a "slow" flow. When the Reynolds number passes its second critical value the flow becomes turbulent and it can be either steady or unsteady. Even though it may be generated by a globally steady process, such as a steady volume flow through a pipe, turbulent flow is never a locally steady flow. We can see that $\mathbf{V}$ can be considered to be the sum of a time-averaged value $\tilde{\mathbf{V}}$ and a time variable increment $\mathbf{V}'$ that is usually significantly smaller than the time-averaged value: $\mathbf{V} = \tilde{\mathbf{V}} + \mathbf{V}'$,

$$\tilde{\mathbf{V}} = \frac{1}{T}\int_{t}^{t+T} \mathbf{V}(\mathbf{x}, \tau)d\tau.$$

Note that the time-averaged value of $\mathbf{V}'$ is automatically zero. The random component $\mathbf{V}'$ of the velocity has some of the characteristics of random noise signals, such as electrical noise in electronic circuits. Obviously, there are small amplitude, high frequency, random motions involved in turbulent flow, the details of which are very difficult to calculate or to predict.

Adding the so-called Reynolds turbulent stress into the Navier-Stokes equation gives the equation of turbulent flow

$$\frac{D\mathbf{V}}{Dt} = -\nabla p + \frac{1}{\mathrm{R}}\Delta \mathbf{V} - \mathbf{V}' \cdot \nabla \mathbf{V}',$$

where $\mathbf{V}' \cdot \nabla \mathbf{V}'$ means the vector with k-th coordinate $\mathbf{V}' \cdot \nabla V_k'$, $k = 1, 2, 3$.

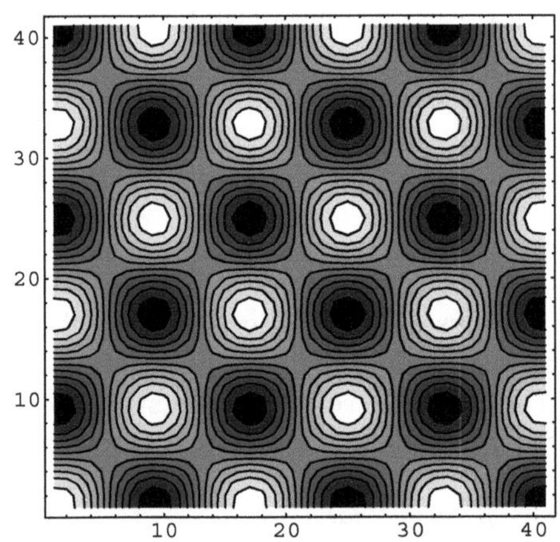

Figure 1.2: Kolmogorov's flow

An external force $\mathbf{F}_{ext}$ added to equation (1.8) or (1.9) in different forms can generate interesting flows. For example, **Andrei Nikolaevich Kolmogorov** (1903–1987) presented in 1959 a seminar in which he suggested a toy problem with which theorists might explore the transition to fluid turbulence in two dimensions. The flow is conceptually simple, and exhibits several shear instabilities before becoming fully turbulent. This flow is governed by the incompressible Navier-Stokes equation (1.8) in two dimensions with a forcing term that is periodic in one spatial direction and steady in time: $\mathbf{F}_{ext} = F_0 \sin(2\pi x_2)\nabla x_1$. Periodic boundary conditions are assumed in both directions of a rectangular box $[0,1] \times [0,1]$. Equation (1.9) with the term corresponding to F_{ext} added becomes

$$\frac{D\,\Delta\psi}{Dt} = \frac{1}{\mathrm{R}}\Delta(\Delta\psi) + F_0\frac{8\pi^3}{\mathrm{R}}\cos(2\pi x_2).$$

The stationary solution is just $\psi_{st} = -\frac{1}{2\pi}\cos(2\pi x_2)$. For small values of the forcing parameter F_0 the fluid develops a steady state spatial profile corresponding to the spatial profile of the forcing. This flow was named the *Kolmogorov flow.* Above a critical value of the forcing parameter F_0, the flow becomes unstable to small velocity perturbations perpendicular to the direction of forcing. The resulting flow is a steady cellular pattern of vorticies. More generally, the external force can be chosen to be

$$\mathbf{F} = F_0 \begin{pmatrix} \sin(2\pi n x_1)\cos(2\pi m x_2) \\ -\cos(2\pi n x_1)\sin(2\pi m x_2) \end{pmatrix}.$$

For a weak forcing, i.e., for a small value of F_0, the $2n \times 2n$ array of counterrotating vortices (for the case $n = m$ see Figure 1.2) is the only time-asymptotic state.

1.3 Riemann map and Carathéodory kernel convergence

In this section we present some background on conformal maps, in particular, two basic instruments that we shall use throughout this monograph: the Riemann mapping theorem and the Carathéodory kernel convergence. A map of one domain (or surface) onto another is said to be *conformal* if it preserves angles between curves. The unit sphere S^2 without its north pole admits stereographic projection onto the complex plane $\mathbb{C}$ which is conformal. Adding the north pole we obtain a compactification of S^2, and consequently, a compactification $\overline{\mathbb{C}}$ of $\mathbb{C}$ which is called the *Riemann sphere* or the extended complex plane. Any analytic map from $\mathbb{C}$ to $\mathbb{C}$ is conformal at a point where the derivative is non zero. Let D be a domain in $\overline{\mathbb{C}}$. A map f is called univalent in D if it is injective (one-to-one) in D. A meromorphic function $f(\zeta)$ is univalent in D if and only if it is analytic in D except for at most one pole and $f(\zeta_1) \neq f(\zeta_2)$ whenever $\zeta_1 \neq \zeta_2$ in D. Univalence in D implies univalence in every subdomain in D. A univalent map is a conformal homeomorphism. The starting point of many considerations in this monograph is the *Riemann Mapping Theorem* (**Georg Friedrich Bernhard Riemann**, 1826–1866). Riemann had formulated his mapping theorem already in 1851, but his proof was incomplete. Carathéodory and Koebe (**Paul Koebe**, 1882–1945) proved the mapping theorem around 1909.

Theorem 1.3.1. *Let Ω be a simply connected domain in $\overline{\mathbb{C}}$ whose boundary contains at least two points, and let $a \in \Omega$, $|a| < \infty$. Then there exists a real number R and a unique conformal univalent map $\zeta = f(z)$ that maps Ω onto $U_R = \{\zeta : |\zeta| < R\}$ and satisfies $f(a) = 0$, $f'(a) = 1$.*

Remark 1.3.2. *Generally, a domain whose universal covering is conformally equivalent to the unit disk is called hyperbolic. So the domain in the above theorem is hyperbolic.*

If $f : \Omega \to U_R$ is the map in Theorem 1.3.1 (or the Riemann map), then the number $R = R(\Omega, a)$ is called the *conformal radius* of the domain Ω with respect to the point a.

In the case $a = \infty$ it is more natural to let f map Ω onto the exterior of a disk $|\zeta| > R$. Then $R = R(\Omega, \infty)$ is uniquely determined by taking the expansion at infinity as $f(z) = z + a_0 + a_1/z + \dots$.

One of the principal tools to study evolution of domains is the Carathéodory kernel convergence. **Constantin Carathéodory** (1873–1950) gave in 1912 [51] a complete characterization of convergence of univalent maps in terms of convergence of the images of a canonical domain under these maps. Its formulation is found also in [13], [92], [270].

Let $\{\Omega_n\}_{n=1}^{\infty}$ be a sequence of domains in the Riemann sphere $\overline{\mathbb{C}}$ such that a fixed point z_0 belongs to all Ω_n excluding possibly a finite number of them. A domain Ω is said to be the *kernel* of $\{\Omega_n\}_{n=1}^{\infty}$,

$$\Omega = \operatorname{Ker}_{z=z_0}\{\Omega_n\},$$

if Ω satisfies the following three conditions:

- $z_0 \in \Omega$;
- any compact set of Ω belongs to all Ω_n starting with a certain number N;
- any domain $\tilde{\Omega}$ satisfying the preceding conditions is a subset of Ω.

If the point z_0 belongs to all Ω_n, starting with a certain number $N(z_0)$, but there is no neighbourhood of z_0 that is contained in all Ω_n for $n > N$, then $\operatorname{Ker}_{z=z_0}\{\Omega_n\} = z_0$ and the kernel degenerates. For the kernel with respect to the origin we write simply $\Omega = \operatorname{Ker}\{\Omega_n\}$.

A sequence $\{\Omega_n\}_{n=1}^{\infty}$ is said to converge to the kernel Ω with respect to z_0 if every subsequence $\{\Omega_{n_k}\}_{k=1}^{\infty}$ has Ω as its kernel. This type of convergence is called *kernel convergence.* If Ω_n is decreasing and Ω^0 is the set of interior points of $\Omega = \bigcap_{n=1}^{\infty} \Omega_n$, $0 \in \Omega$, then Ω_n converges to the component of Ω^0 that contains 0 if $0 \in \Omega^0$, or to $\{0\}$ if $0 \notin \Omega^0$.

Theorem 1.3.3. (Carathéodory kernel theorem) *Let the functions $f_n(\zeta)$ be analytic and univalent in $U \equiv U_1$, $f_n(0) = 0$, $f_n'(0) > 0$, and let $\Omega_n = f_n(U)$. Then the sequence f_n converges locally uniformly in U if and only if Ω_n converges to its kernel Ω, $\Omega \neq \mathbb{C}$, with respect to the origin. If* $\operatorname{Ker}\Omega_n \neq \{0\}$, *then the limiting function is a univalent map of U onto Ω. If* $\operatorname{Ker}\Omega_n = \{0\}$, *then* $\lim_{n\to\infty} f_n(z) \equiv 0$.

The kernel convergence can be generalized to continuous intervals as follows. Let $\{\Omega(t)\}$, $t \in [a, b]$ be a one-parameter family of domains in the Riemann sphere $\overline{\mathbb{C}}$ such that a fixed point z_0 belongs to all $\Omega(t)$. Consider first the case $t_0 \in [a, b]$, and let there be a neighbourhood of z_0 that belongs to all $\Omega(t)$, $t \neq t_0$. A domain Ω is said to be the *kernel* of $\{\Omega(t)\}$ with respect to z_0, if Ω satisfies the following three conditions:

- $z_0 \in \Omega$;
- for any compact set D of Ω there is a small positive number ε, such that $D \subset \Omega(t)$ for all $0 < |t - t_0| < \varepsilon$;
- any domain satisfying the preceding conditions is a subset of Ω.

If there is no such neighbourhood, then we say that the kernel degenerates and $\mathrm{Ker}_{z=z_0}\{\Omega(t)\} = \{z_0\}$.

A generalized Carathéodory kernel theorem states that if the functions $f(\zeta, t)$ are analytic and univalent in U, $f(0,t) = 0$, $f'(0,t) > 0$, $\Omega(t) = f(U,t)$, then the family $f(\zeta, t)$ converges locally uniformly in U if and only if $\Omega(t)$ converges to its kernel Ω, $\Omega \neq \mathbb{C}$, as $t \to t_0$ with respect to the origin. If $\mathrm{Ker}\,\Omega(t) \neq \{0\}$, then the limiting function is a univalent map of U onto Ω. If $\mathrm{Ker}\,\Omega(t) = \{0\}$, then $\lim_{t\to t_0} f(z,t) \equiv 0$.

1.4 Hele-Shaw flows

First, let us give some historical remarks. Around 1770 **Charles Augustin Coulomb** (1736–1806) studied the motion of a disk suspended by a torsion wire to oscillate in a vessel of liquid. He observed that the resistance of the liquid under a slow motion is proportional to the velocity. Later Beaufoy [23] in 1834[1] and **William Froude** (1810–1879) found that at higher velocities the resistance varied as the square of the velocity. **Colonel Mark Beaufoy** (1764–1827) (who founded the Society for the Improvement of Naval Architecture in 1791) described in [23] his Nautical Experiments on the resistance to propulsion through water of variously shaped solids, carried out in Greenland Dock, Rotherhithe, in 1793–1798, under the direction of the Society for the Improvement of Naval Architecture. Reynolds, about 1883, investigated the critical velocity at which the change of state occurred and a liquid flowed quite steadily until a certain velocity was reached.

Henry Selby Hele-Shaw (1854–1941), an English mechanical and naval engineer, was working during the period 1885–1904 at the Engineering Department of the University of Liverpool. He was a fellow of The Royal Society (see his biography in [158]). In 1898 he published in Nature [166], see also [167], a short note where he started to study the following situation. For a liquid flow in a tube or in a channel with wetted sides, the velocity reaches its maximum in the middle and vanishes at the sides. Thus, the transition from laminar flow to turbulent can be observed somewhere between. To make the separation interface visible Hele-Shaw proposed to inject a gas (an inviscid fluid) into the system. This injection can be interpreted as suction of the original viscous fluid. To avoid gravity effect he

[1] The book was published by his son after Beaufoy's death.

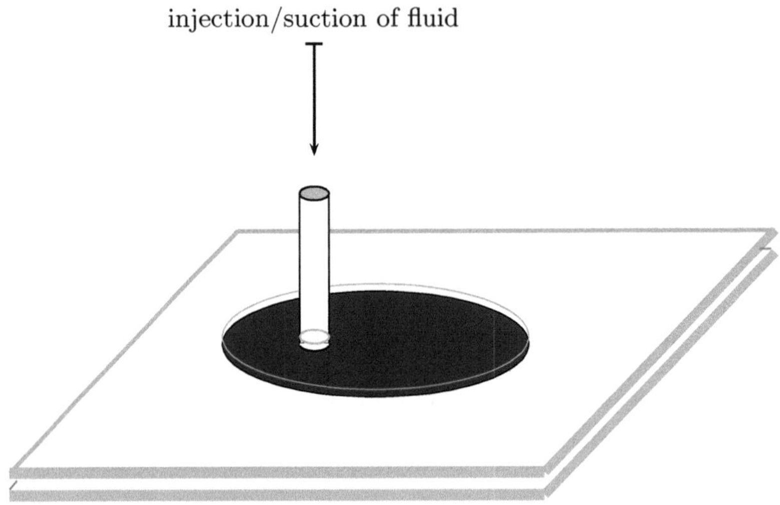

Figure 1.3: A Hele-Shaw cell

suggested considering a flow between two parallel horizontal plates with a narrow gap between them.

Later a model with slightly different geometry appeared in [117], [262], [263], and [283], see Figure 1.3. In this model the viscous fluid occupies a bounded phase domain with free boundary and more fluid is injected or removed through a point well. The free boundary starts moving due to injection/suction. Similar problems appear in metallurgy in the description of the motion of phase boundaries by capillarity and diffusion [241]; in the dissolution of an anode under electrolysis [114]; in the melting of a solid in a one-phase Stefan problem with zero specific heat [66], etc.

This book will expose some of the developments in two-dimensional Hele-Shaw theory that have taken place over the last few decades. Several other models, methods, and applications exceed the scope of our work, such as the treatment of the rectangular dam by Polubarinova-Kochina [264] who gave solutions in terms of the Riemann P-function [67], [181]; mathematical treatment of rotating Hele-Shaw cells [63], [105]; some nice analytical and numerical results found in [54], [55], [56], [247]; a study of Hele-Shaw flows on hyperbolic surfaces [164] [165]; applications to electromagnetic problems [69], [114]; models of diffusion-limited aggregation

[52], [345], [346]; Hele-Shaw flows with multiply connected phase domains [285]; development of singularities in non-smooth free boundary problems [170], [197], [198]; connections between Stokes and Hele-Shaw flows [68] (a large collection of references on Hele-Shaw and Stokes flows is found in [125]), two-phase Muskat problems [1], [180], [311]; some applications of quasiconformal maps are found in [41], [232]. Recently, it was shown [3], that the semiclassical dynamics of an electronic droplet confined in the plane in a quantizing inhomogeneous magnetic field in the regime when the electrostatic interaction is negligible is similar to the Hele-Shaw problem in the plane. Further development of these ideas and applications to the complex moments are found in [206], [231], [344].

1.4.1 The Stokes-Leibenzon model

(**Leonid Samuilovich Leibenzon**, 1879-1951, see [223]). We consider a slow parallel flow of an incompressible fluid between two parallel flat plates which are fixed at a small distance h. The reference velocity $\mathbf{V}$ is generated by some external pumping mechanism. A vertical section is given in Figure 1.4. We agree that the flow attains its maximal velocity at the middle of the cell and the velocity vanishes at the sides. We follow Lamb's method [217] of deriving the Hele-Shaw equation starting from

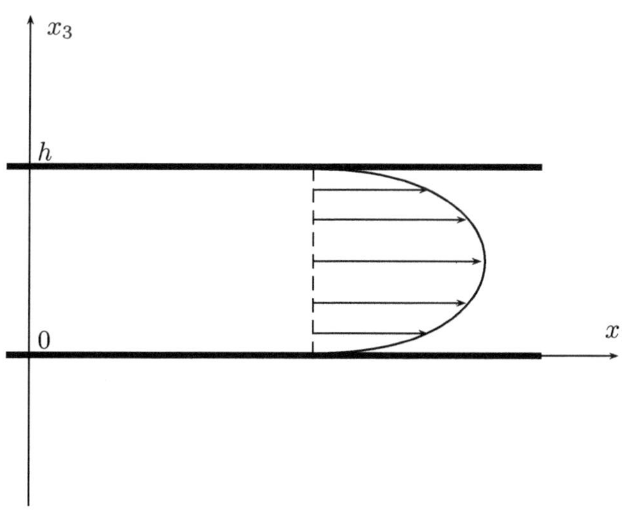

Figure 1.4: The section of a Hele-Shaw cell in the x_1-direction

the Navier-Stokes equations (1.2), (1.7), which, neglecting gravity, become

$$\frac{\partial \mathbf{V}}{\partial t} + (\mathbf{V} \cdot \nabla)\mathbf{V} = \frac{1}{\rho}(-\nabla p + \mu \Delta \mathbf{V}), \quad \nabla \cdot \mathbf{V} = 0. \tag{1.10}$$

We assume that the injection of fluid is slow enough for the flow to be approximately steady and parallel. This means that

$$\frac{\partial \mathbf{V}}{\partial t} = 0, \quad V_3 = 0.$$

These assumptions reduce (1.10) to

$$\begin{aligned} \left(V_1 \tfrac{\partial}{\partial x_1} + V_2 \tfrac{\partial}{\partial x_2}\right) V_1 &= -\tfrac{1}{\rho}\tfrac{\partial p}{\partial x_1} + \tfrac{\mu}{\rho}\Delta V_1, \\ \left(V_1 \tfrac{\partial}{\partial x_1} + V_2 \tfrac{\partial}{\partial x_2}\right) V_2 &= -\tfrac{1}{\rho}\tfrac{\partial p}{\partial x_2} + \tfrac{\mu}{\rho}\Delta V_2, \\ 0 &= -\tfrac{1}{\rho}\tfrac{\partial p}{\partial x_3}, \end{aligned}$$

with boundary conditions

$$V_1\Big|_{x_3=0,h} = V_2\Big|_{x_3=0,h} = 0.$$

If h is sufficiently small and the flow is slow, then we can assume that the derivatives of V_1 and V_2 with respect to x_1 and x_2 are negligible compared to the derivatives with respect to x_3. Therefore, we can simplify the system by putting

$$\frac{\partial V_1}{\partial x_j} = \frac{\partial V_2}{\partial x_j} = \frac{\partial^2 V_1}{\partial x_j^2} = \frac{\partial^2 V_2}{\partial x_j^2} = 0, \quad j = 1, 2,$$

which gives the system

$$\begin{aligned} \frac{\partial p}{\partial x_1} &= \mu \frac{\partial^2 V_1}{\partial x_3^2}, \\ \frac{\partial p}{\partial x_2} &= \mu \frac{\partial^2 V_2}{\partial x_3^2}, \\ 0 &= \frac{\partial p}{\partial x_3}. \end{aligned}$$

The last equation in the system shows that p does not depend on x_3, whence V_1, V_2 are polynomials of degree at most 2 as functions of x_3. The boundary conditions then imply

$$V_1 = \frac{1}{2}\frac{\partial p}{\partial x_1}\left(\frac{x_3^2}{\mu} - \frac{hx_3}{\mu}\right), \quad V_2 = \frac{1}{2}\frac{\partial p}{\partial x_2}\left(\frac{x_3^2}{\mu} - \frac{hx_3}{\mu}\right).$$

The integral means $\tilde{V_1}$ and $\tilde{V_2}$ of V_1 and V_2 across the gap are

$$\tilde{V_1} = \frac{1}{h}\int_0^h V_1\, dx_3 = -\frac{h^2}{12\mu}\frac{\partial p}{\partial x_1}, \quad \tilde{V_2} = \frac{1}{h}\int_0^h V_2\, dx_3 = -\frac{h^2}{12\mu}\frac{\partial p}{\partial x_2},$$

so the integral mean $\tilde{\mathbf{V}}$ of $\mathbf{V}$ satisfies

$$\tilde{\mathbf{V}} = -\frac{h^2}{12\mu}\nabla p. \tag{1.11}$$

Here $\tilde{\mathbf{V}}$ and p depend only on x_1 and x_2, so we may consider (1.11) as a purely two-dimensional equation. Thus equation (1.11) describes a two-dimensional potential flow for which the potential function is proportional to the pressure. By incompressibility (1.2) the pressure is a harmonic function. Equation (1.11) is called the *Hele-Shaw equation.* It is of the same form as Darcy's law, which governs flows in porous media.

In the sequel we write just $\mathbf{V}$ instead of $\tilde{\mathbf{V}}$. The Stokes-Leibenzon model suggests a point sink/source (x_1^0, x_2^0) of constant strength within the system. The rate of area (or mass) change is given as

$$\int\limits_{\partial U_\varepsilon} \rho\mathbf{V}\cdot\mathbf{n}\,ds = \text{const},$$

where $U_\varepsilon = \{(x,y) : (x_1 - x_1^0)^2 + (x_2 - x_2^0)^2 < \varepsilon^2\}$ for ε sufficiently small. Equality (1.11) and Green's theorem imply

$$\iint\limits_{U_\varepsilon} (-\frac{h^2\rho}{12\mu})\Delta p \; dx_1 dx_2 = \text{const},$$

for any ε. So $\Delta p = Q\delta_{(x_1^0,x_2^0)}$ for some constant Q, where $\delta_{(x_1^0,x_2^0)}$ is Dirac's distribution, and the potential function p has a logarithmic singularity at (x_1^0, x_2^0).

On the fluid boundary the balance of forces in the three-dimensional view gives that

$$p = \text{exterior air pressure} + \text{surface tension.}$$

The air pressure can be taken to be constant while the surface tension is roughly proportional to the curvature of the boundary. If the gap h is sufficiently small, then the curvature in the x_1, x_2 plane is negligible compared to the curvature in the x_3 direction. Due to capillary forces the boundary profile in the x_3 direction will be somewhat similar to the graph in Figure 1.4 which is more or less the same everywhere. Hence, the surface tension effect on p is more or less constant (at least with respect to x_1, x_2). Finally, rescaling p we can take $p = 0$ on the boundary.

1.4.2 The Polubarinova-Galin equation

Now let us study the motion of the boundary. Galin [117] and Polubarinova-Kochina [262], [263] first proposed a complex variable method by introducing the Riemann mapping from an auxiliary parametric plane (ζ) onto the phase domain in the (z)-plane and derived an equation for this parametric mapping. So the resulting equation is known as the *Polubarinova-Galin equation* (see e.g. [179], [171]) (see a survey on the Polubarinova-Kochina[2] contribution and its influence in natural sciences and industry in [253]).

We denote by $\Omega(t)$ the bounded simply connected domain in the phase z-plane occupied by the fluid at instant t, and we consider suction/injection through a single well placed at the origin as a driving mechanism (Figure 1.5). We assume

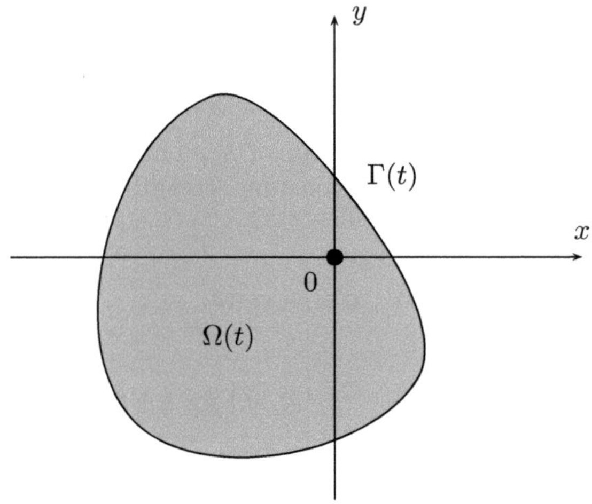

Figure 1.5: $\Omega(t)$ is a bounded simply connected phase domain with the boundary $\Gamma(t)$ and the sink/source at the origin

the sink/source to be of constant strength Q which is positive ($Q > 0$) in the case of suction and negative ($Q < 0$) in the case of injection. The dimensionless pressure p is scaled so that 0 corresponds to the atmospheric pressure. We put $\Gamma(t) \equiv \partial\Omega(t)$ and assume that it is given by the equation $\phi(x_1, x_2, t) \equiv \phi(z,t) = 0$, where $z = x_1 + ix_2$. The initial situation is represented at the instant $t = 0$ as $\Omega(0) = \Omega_0$, and the boundary $\partial\Omega_0 = \Gamma(0) \equiv \Gamma_0$ is defined by an implicit function

[2]Pelageia Yakovlevna Polubarinova-Kochina, 1899–1999. She attended the Pokrovskii Women's Gymnasium in St Petersburg, and after the 1917 Revolution, the University of Petrograd. After graduating, Pelageia Polubarinova worked at the Main Geophysical Laboratory under A. A. Friedmann. In 1925 Pelageia Polubarinova and Nikolai Yevgrafovich Kochin, 1901–1944, got married. Kochin graduated from the University of Petrograd in 1923. He was appointed to the Leningrad State University (former Petrograd) in 1924 and taught mathematics and mechanics there until 1934. The family moved to Moscow University in 1934. Kochin became the head of the mechanics section of the Mechanics Institute of the USSR Academy of Sciences from 1939 to 1944. The German armies reached the outskirts of Moscow in late 1941 and Pelageia and two daughters were evacuated to Kazan. Kochin, however, remained in Moscow carrying out military research. By 1943 the German army had suffered defeats by Soviet troops and Moscow was safe enough for Kochina to return. This she did, but Kochin became ill and died before the end of the war. At the time of his death he had been in the middle of lecture courses and Kochina took over the courses and completed delivering them.

$\phi(x_1, x_2, 0) = 0$. The potential function p is harmonic in $\Omega(t) \setminus \{0\}$ and

$$\Delta p = Q\delta_0(z), \quad z = x_1 + ix_2 \in \Omega(t), \tag{1.12}$$

where $\delta_0(z)$ is the Dirac distribution supported at the origin. The zero surface tension *dynamic* boundary condition is given by

$$p(z,t) = 0 \text{ as } z \in \Gamma(t). \tag{1.13}$$

The resulting motion of the free boundary $\Gamma(t)$ is given by the fluid velocity $\mathbf{V}$ on $\Gamma(t)$. This means that the boundary is formed by the same set of particles all the time. The normal velocity in the outward direction is

$$v_n = \mathbf{V}\Big|_{\Gamma(t)} \cdot \mathbf{n}\,(t),$$

where $\mathbf{n}\,(t)$ is the unit outer normal vector to $\Gamma(t)$. Rewriting this law of motion in terms of the potential function, and using (1.11) after suitable rescaling, we get the *kinematic* boundary condition

$$\frac{\partial p}{\partial n} = -v_n, \tag{1.14}$$

where $\frac{\partial p}{\partial n} = \mathbf{n} \cdot \nabla p$ denotes the outward normal derivative of p on $\Gamma(t)$.

Let us consider the complex potential $W(z,t)$, Re $W = p$. For each fixed t it is a multi-valued analytic function defined in $\Omega(t)$ whose real part solves the Dirichlet problem (1.12), (1.13). Making use of the Cauchy-Riemann conditions we deduce that

$$\frac{\partial W}{\partial z} = \frac{\partial p}{\partial x_1} - i\frac{\partial p}{\partial x_2}.$$

Since Green's function solves (1.12), (1.13), we have the representation

$$W(z,t) = \frac{Q}{2\pi}\log z + w_0(z,t), \tag{1.15}$$

where $w_0(z,t)$ is an analytic regular function in $\Omega(t)$.

To derive the equation for the free boundary $\Gamma(t)$ we consider an auxiliary parametric complex ζ-plane, $\zeta = \xi + i\eta$. The Riemann Mapping Theorem yields a unique conformal univalent map $f(\zeta,t)$ from the unit disk $U = \{\zeta : |z| < 1\}$ onto the phase domain $f : U \to \Omega(t)$, $f(0,t) = 0$, $f'(0,t) > 0$. The function $f(\zeta,0) = f_0(\zeta)$ parameterizes the initial boundary $\Gamma_0 = \{f_0(e^{i\theta}), \theta \in [0,2\pi)\}$ and the moving boundary is parameterized by $\Gamma(t) = \{f(e^{i\theta},t), \theta \in [0,2\pi)\}$. The normal velocity v_n of $\Gamma(t)$ in the outward direction is given by (1.14). From now on and throughout the monograph we use the notation $\dot{f} = \partial f/\partial t$, $f' = \partial f/\partial \zeta$. The normal outward vector is given by the formula

$$\mathbf{n} = \zeta\frac{f'}{|f'|}, \quad \zeta \in \partial U.$$

Therefore, the normal velocity is obtained as

$$v_n = \mathbf{V} \cdot \mathbf{n} = -\text{Re}\left(\frac{\partial W}{\partial z}\zeta\frac{f'}{|f'|}\right).$$

Because of the conformal invariance of Green's function we have the superposition

$$(W \circ f)(\zeta, t) = \frac{Q}{2\pi}\log\zeta,$$

and by taking the derivative we get

$$\frac{\partial W}{\partial z}f'(\zeta, t) = \frac{Q}{2\pi\zeta}.$$

On the other hand, in general for a moving boundary, we have

$$v_n = \text{Re}\,[\dot{f}\;\overline{\zeta f'/|f'|}],$$

and finally deduce that

$$\text{Re}\left[\dot{f}(\zeta, t)\overline{\zeta f'(\zeta, t)}\right] = -\frac{Q}{2\pi}, \quad \zeta = e^{i\theta}. \tag{1.16}$$

Galin [117] and Polubarinova-Kochina [262], [263] first derived the equation (1.16), so (1.16) is known as the *Polubarinova-Galin equation* (see e.g. [179], [171], [253]).

From (1.16) one can derive a Löwner-Kufarev type equation by the Schwarz-Poisson formula:

$$\dot{f}(\zeta, t) = -\zeta f'(\zeta, t)\frac{Q}{4\pi^2}\int_0^{2\pi}\frac{1}{|f'(e^{i\theta}, t)|^2}\frac{e^{i\theta}+\zeta}{e^{i\theta}-\zeta}d\theta, \tag{1.17}$$

where $\zeta \in U$. The equation (1.17) is equivalent to the kinematic condition (1.16) on the free boundary. Namely, one can take a limit in (1.17) as ζ tends to a point on the unit circle, and implementing the Sokhotskiĭ-Plemelj formulae [242], the equation (1.17) reduces to (1.16).

We call (1.17) a *Löwner–Kufarev type equation* because of the analogy with the linear partial differential equation that describes monotone deformations of simply connected univalent domains (see e.g. [13], [92], [270]). In the classical Löwner-Kufarev equation the integral in the right-hand side of (1.17) is to be replaced by an arbitrary time dependent analytic function with positive real part. This equation produces subordination Löwner chains whose properties have been deeply studied. Unlike the classical Löwner-Kufarev equation, the equation (1.17) is not even quasilinear and produces a special type of chains.

1.4.3 Local existence and ill/well-posedness

Under some assumptions on smoothness of $\partial\Omega(0)$ it is known that in the case of an expanding fluid ($Q < 0$) there exists a unique solution to the problem (1.12)–(1.14), or (1.16), in terms of analytic functions $f(\zeta, t)$ (strong or classical solution), locally forward in time. The first proof appeared in 1948 [339] by **Yurii P. Vinogradov** and **Pavel Parfen'evich Kufarev**(1909–1968). This proof was rather difficult, and later, Gustafsson [143] gave a simple proof in the case when a polynomial or a rational univalent function f_0 parameterizes the initial phase domain. In 1993 Reissig and von Wolfersdorf [280] made clear that this model could be interpreted as a particular case of an abstract Cauchy problem and that the strong solvability (locally in time) could be proved using a nonlinear abstract Cauchy-Kovalevskaya Theorem (see [249]). More precisely, they proved that if the initial function $f_0(z)$ is analytic and univalent in the disk $U_r = \{\zeta : |\zeta| < r\}$ for some $r > 1$, then there exists $t_0 > 0$, such that the solution $f(\zeta, t)$ to the Polubarinova-Galin equation exists and is unique in some time interval $t \in [0, t_0)$. In the multidimensional case a proof of local existence and uniqueness can be found in, e.g., [330].

Various aspects of planar Hele-Shaw viscous flows with zero surface tension have been investigated by a number of scientists. We note that the problem (1.12)–(1.14) is formally time reversible by changing $Q \to -Q$, $p \to -p$, $t \to -t$. However, the cases of suction and injection differ considerably. One of the main features of the problem (1.12)–(1.14) is that starting with an analytic boundary Γ_0 we obtain a one-parameter (t) chain of the solutions $p(z, t)$ (and equivalently $f(\zeta, t)$) that exists during an interval $t \in [0, t_0)$, developing possible cusps or double points (the boundary meets itself) at the boundary $\Gamma(t)$ in a blow-up time t_0. In the suction case the fluid can be completely removed from a finite region without blow-up when Ω_0 is a disk centered on the origin (see [171]). Let us note here that cusps or double points can be developed even in the problem with injection.

The zero surface tension Hele-Shaw model (1.12)–(1.14) with suction is Hadamard ill-posed. The blow-up time t_0 corresponds in the simplest cases (e.g., polynomial solutions) to the moment of cusp formation. The situation is quite subtle. Polynomial solutions that develop cusps of order $(4n - 1)/2$ at t_0 always blow up and the solution does not exist beyond t_0. The solutions that develop cusps of order $(4n+1)/2$ can sometimes continue to exist beyond t_0 (see [177] and [297] for complete classification). Moreover, if the initial function is a polynomial of degree $n \geq 2$, then cusp formation is guaranteed before the moving boundary reaches the sink [171]. Nonpolynomial solutions can produce other scenarios of evolution of the free boundary where, for instance, the blow-up time occurs at the moment when the free interface reaches the sink or the solution breaks down because $\Gamma(t)$ develops a corner or simply becomes nonanalytic in virtually any way.

An attempt to classify the solutions to the zero surface tension model for the Hele-Shaw flows in bounded and unbounded regions with suction was launched by Hohlov, Howison [171] and Richardson [284]. They also described cusp formation.

Another typical scenario is fingering that was first described in the classical work by Saffman and Taylor [293]. Recently it has became clear that in the model with injection fingering does not occur [146].

1.4.4 Regularizations

There are several proposals for regularization of the ill-posed problem. One of them is the "kinetic undercooling regularization" [172], where the condition (1.13) is replaced by

$$\beta\frac{\partial p}{\partial n}+p=0, \quad \text{on} \quad \Gamma(t), \quad \beta>0.$$

It was shown in [172], [282] that there exists a unique solution locally in time (even a strong solution) in both the suction and injection cases in a simply connected bounded domain $\Omega(t)$ with an analytic boundary. We remark that at the conference about Hele-Shaw flows, held in Oxford in 1998, V. M. Entov suggested using a nonlinear version of this condition motivated by applications. Reissig and Pleshchinskii discussed this model in [258] where local existence and uniqueness were obtained as well as some numerical results were presented.

Another proposal is to introduce surface tension as a regularization mechanism. The model with nonzero surface tension is obtained by modifying the boundary condition for the pressure p to be the product of the mean curvature κ of the boundary and surface tension $\gamma>0$. Let us rewrite the problem (1.12)–(1.14) with this new condition:

$$\Delta p = Q\delta_0(z), \quad \text{in} \quad z\in\Omega(t), \tag{1.18}$$

$$p = \gamma\kappa(z), \quad \text{on} \quad z\in\Gamma(t), \tag{1.19}$$

$$v_n = -\frac{\partial p}{\partial n}, \quad \text{on} \quad z\in\Gamma(t). \tag{1.20}$$

A similar problem appears in metallurgy in the description of the motion of phase boundaries by capillarity and diffusion [241]. The condition (1.19) is found in [234] (it is known as the Gibbs-Thomson law or the Laplace-Young condition). **Pierre-Simon Laplace** (1749–1827) and **Thomas Young** (1773-1829) obtained this law independently in 1805. Later **Josiah Willard Gibbs** (1839–1903) and **William Thomson** (Lord Kelvin) (1824–1907) in the 1870s derived an analogous relation. It takes into account how surface tension modifies the pressure through the boundary interface.

The problem of existence of a solution in the non zero surface tension case is more difficult. Duchon and Robert [91] proved the local existence in time for weak solution for all γ. Recently, Prokert [273] obtained even global existence in time and exponential decay (in the case of flow driven by surface tension) of the solution near equilibrium for bounded domains. The results are obtained in Sobolev spaces H^s with sufficiently big s. We refer the reader to the works by Escher and Simonett [106], [107] who proved the local existence, uniqueness and regularity of strong solutions to one- and two-phase Hele-Shaw problems with surface tension when the initial domain has a smooth boundary. The case of the initial domain bounded by a non-smooth boundary was considered in [15], [108]. The global existence in the case of the phase domain close to a disk was proved in [109]. If the domain occupied by the fluid is unbounded and its boundary extends to infinity, then the corresponding result about short-time existence and uniqueness for positive surface tension was obtained by Kimura [195] (he also shows that the problem is ill-posed in the case of suction). More results on existence for general parabolic problems can be found in [110]. Most of the authors work with weak formulation of the problem (see this formulation in Chapter 3 and in [81], [102], [144]). It is worth remarking that the weak solution to the problem with injection exists all the time and coincides with the strong one if the latter exists.

1.4.5 Numerical treatment

There exists a vast amount of works dedicated to the numerical study of Hele-Shaw problem. One may distinguish several directions of the development of numerical methods which are principally divided by well/ill-posedness and by presence/absence of the surface tension effect. A complete review of numerical methods exceeds the scope of this book but we give some remarks on them.

Numerical computation of receding evolution is known to be difficult due to the ill-posedness of the zero-surface tension problem [77], [236]. Reasonably accurate computation of solutions of the ill-posed problem is possible if one works with the conformal mapping from the flow domain into the unit circle rather than the usual free boundary problem formulation. Cusp formation can be described analytically in many cases and some of these cases are used as tests for the numerical scheme, which uses a finite element/boundary integral method, see, e.g., [10]. In addition, the possibility of singularity formation demands extremely high resolution. The cusp singularities are good candidates for potential topological singularities in the presence of small surface tension, but even perturbations at the low error level can lead to a rapid growth of the solution's high-frequency components [35], [71]. The boundary integral method with fourth-order time integration is typically employed, see [54]–[56]. It uses the small-scale decomposition technique of Hou *et al* [174] to remove the high-order stability constraint induced by surface tension. A good review of numerical methods may be found in, e.g., [175]. Some other relevant references are [22], [281], [343].

There are several works where the Helmholtz-Kirchhoff method is used and the free boundary problem is reformulated as an integro-differential equation. Numerical simulations are also performed and qualitative properties are demonstrated. For example, the Hele-Shaw problem is considered with the initial domain a non analytic perturbation of a disk [79], [80]. The authors develop the quasi-contour method in order to give a numerical simulation.

1.5 Complex moments

Let us consider the problem with injection ($Q < 0$) and let the strong solution to the Polubarinova-Galin equation (1.16) exist for $t \in [0, t_0)$. Since the free boundary moves in the normal direction and the the normal velocity on the boundary never vanishes, we have $\Omega(t) \subset \Omega(s)$ for $0 < t < s < t_0$. Richardson [283] introduced the *complex moments*

$$M_n(t) = \iint\limits_{\Omega(t)} z^n d\sigma_z = \iint\limits_{U} f^n(\zeta, t)|f'(\zeta, t)|^2 d\sigma_\zeta,$$

where $d\sigma_z$ and $d\sigma_\zeta$ denote area elements in the z- and ζ- planes respectively. He proved that

$$\begin{aligned} M_0(t) &= M_0(0) - Qt, \\ M_n(t) &= M_n(0), \quad \text{for } n \geq 1. \end{aligned}$$

More generally, let us consider the area integral

$$M_\Phi(t) = \iint\limits_{\Omega(t)} \Phi(z) d\sigma_z,$$

for any function Φ analytic in a neighbourhood of $\overline{\Omega(t)}$. The Reynolds Transport Theorem together with Green's formula imply

$$\begin{aligned} \tfrac{d}{dt} M_\Phi(t) &= \int\limits_{\Gamma(t)} \Phi(z)(\mathbf{V} \cdot \mathbf{n}) ds \\ &= -\int\limits_{\Gamma(t)} \Phi(z) \tfrac{\partial p}{\partial n} ds \\ &= -\int\limits_{\Gamma(t)} p \tfrac{\partial \Phi}{\partial n} ds - \iint\limits_{\Omega(t)} \Phi(z) \Delta p d\sigma_z = -Q\Phi(0). \end{aligned}$$

Integrating we obtain

$$\iint\limits_{\Omega(t)} \Phi(z) d\sigma_z = \iint\limits_{\Omega(0)} \Phi(z) d\sigma_z - Qt\Phi(0), \tag{1.21}$$

for all $t \in [0, t_0)$. It is easy to see that one can run the above arguments backward (see, e.g., [144]) to show that a smooth family $\Omega(t)$ of simply connected domains is a strong solution to the Hele-Shaw problem if and only if the equality (1.21) holds for every analytic and integrable function $\Phi(z)$ in $z \in \Omega(t_0)$.

1.6 Further remarks on the Polubarinova-Galin equation

Writing $\zeta = e^{i\theta}$ on ∂U we have $\frac{\partial f}{\partial \theta} = i\zeta\frac{\partial f}{\partial \zeta}$. Therefore the Polubarinova-Galin equation (1.16) can be written as

$$\operatorname{Im}\left(\frac{\partial f}{\partial t}\overline{\frac{\partial f}{\partial \theta}}\right) = \frac{Q}{2\pi}.$$

Decomposing f into its real and imaginary parts, $f = u+iv$, the equation becomes

$$\frac{\partial(u,v)}{\partial(\theta,t)} = \frac{Q}{2\pi}, \tag{1.22}$$

where

$$\frac{\partial(u,v)}{\partial(\theta,t)} = \frac{\partial u}{\partial \theta}\frac{\partial v}{\partial t} - \frac{\partial v}{\partial \theta}\frac{\partial u}{\partial t}$$

is the Jacobi determinant of the map $(\theta,t) \mapsto (u,v)$, or, from another point of view, the Poisson bracket of u and v as functions of (θ,t).

Equation (1.22) can be regarded as a differential equation for the two real-valued functions u and v defined on the circle. As such it expresses that the map $(\theta,t) \mapsto (u,v)$ shall be area preserving up to a constant factor. The two functions u and v in (1.22) are, however, not independent of each other, but are linked via the condition that, as a function of $e^{i\theta}$, $u+iv$ has an analytic continuation to all of U. In other words, v is to be the Hilbert transform of u.

Remarkably enough it is possible to write down the "general solution" of (1.22). To this end, following [60] (Anhang zum ersten Kapitel) we introduce new independent variables α and β and regard all of θ, t, u and v as functions of these. Then

$$\frac{\partial(u,v)}{\partial(\alpha,\beta)} = \frac{\partial(u,v)}{\partial(\theta,t)} \cdot \frac{\partial(\theta,t)}{\partial(\alpha,\beta)}$$

and (1.22) becomes

$$\frac{\partial(u,v)}{\partial(\alpha,\beta)} = \frac{Q}{2\pi}\frac{\partial(\theta,t)}{\partial(\alpha,\beta)}. \tag{1.23}$$

Now, if $Q<0$ the general solution of (1.23) is

$$\begin{cases} \theta = \alpha + \frac{\partial \omega}{\partial \beta}, & u = k\cdot(\beta + \frac{\partial \omega}{\partial \alpha}), \\ t = \beta - \frac{\partial \omega}{\partial \alpha}, & v = k\cdot(\alpha - \frac{\partial \omega}{\partial \beta}), \end{cases}$$

where $\omega = \omega(\alpha,\beta)$ is an arbitrary function satisfying

$$1 - (\frac{\partial^2 \omega}{\partial \alpha \partial \beta})^2 + \frac{\partial^2 \omega}{\partial \alpha^2}\cdot\frac{\partial^2 \omega}{\partial \beta^2} \neq 0, \tag{1.24}$$

and where $k^2 = -Q/2\pi$. The expression in the left member of (1.24) is simply $\frac{\partial(\theta,t)}{\partial(\alpha,\beta)}$. (If $Q > 0$ the general solution is obtained by modifying some signs above.)

The Poisson bracket point of view and its relation to integrable systems has recently been developed in a number of papers in which the Hele-Shaw problem (often named the Laplacian growth model) is embedded into a larger hierarchy of domain variations for which all the complex moments M_n (see Section 1.5) are treated as independent variables (generalized time variables). For us all of them are frozen except M_0, which is essentially ordinary time. See [3], [206], [231], [344].

1.7 The Schwarz function

This function appeared explicitly in a paper by Grave [137] in 1895, and was later employed by **Gustav Herglotz** in 1914 [169]. In the works by **Hermann Amandus Schwarz** (1843–1921) it does not seem to appear explicitly, whereas this designation (due to Philip Davis) is now immutably connected with his name. The definition of the Schwarz function is based on the Schwarz reflection principle. Let Γ be a non-singular analytic Jordan curve in $\mathbb{C}$, that is Γ possesses a real-analytic bijective parametrization with a non-vanishing derivative. Then there is a neighbourhood Ω of Γ and a uniquely determined analytic function $S(z)$, $z \in \Omega$, such that $S(z) = \bar{z}$ for $z \in \Gamma$. This function is called the *Schwarz function.* Thorough treatments of the Schwarz function are found in [76], [308].

A connection with the Hele-Shaw problem is as follows. Let $\phi(x_1, x_2, t) = 0$ be an implicit representation of the free boundary $\Gamma(t)$ which is supposed to be smooth analytic. Substituting $x_1 = (z+\bar{z})/2$ and $x_2 = (z-\bar{z})/2i$ into this equation and solving it for $\bar{z}$ we obtain

$$\bar{z} = S(z,t), \tag{1.25}$$

where the function $S(z,t)$ is defined and analytic in a neighbourhood of $\Gamma(t)$. This function satisfies the consistency condition $S(\overline{S(z,t)},t) \equiv \bar{z}$. Differentiating (1.25) with respect to an arc length parameter s on $\partial\Omega(t)$ for fixed t gives the expression

$$\frac{dz}{ds} = \frac{1}{\sqrt{S'(z,t)}},$$

for the unit tangent vector on $\partial\Omega(t)$.

The map $z \to \overline{S(z,t)}$ has the interpretation of being the anticonformal reflection in $\Gamma(t)$. Therefore, if $\Gamma(t)$ moves with the normal velocity v_n, then the point $\overline{S(z,t)}$ moves, for fixed z, with the double speed $|\dot{S}| = 2v_n$. Taking also the direction into account this gives

$$v_n = \frac{i\dot{S}(z,t)}{2\sqrt{S'(z,t)}}.$$

In the Hele-Shaw case the velocity vector $\frac{1}{2}\dot{\bar{S}}$ is equal to $-\overline{dW/dz}$, where $W(z,t)$ is the complex potential, hence the Hele-Shaw equation becomes

$$\frac{dW}{dz} = -\frac{1}{2}\dot{S}(z,t).$$

In general, one way to construct the Schwarz function is to consider the Cauchy integral

$$g(z) = \frac{1}{2\pi i}\int_{\partial\Omega}\frac{\bar{\zeta}d\zeta}{\zeta - z}.$$

It defines one analytic function, $g_e(z)$, in the exterior of Ω and one, $g_i(z)$, in the interior. On $\partial\Omega$ the jump condition

$$g_i(z) - g_e(z) = \bar{z}, \quad z \in \partial\Omega \tag{1.26}$$

holds for the boundary values. When $\partial\Omega$ is analytic both g_i and g_e extend analytically across the boundary so that $g_i(z) - g_e(z)$ is analytic in a full neighbourhood of $\partial\Omega$. Then the Schwarz function is defined as

$$S(z) = g_i(z) - g_e(z), \tag{1.27}$$

(see, e.g., [283]). Note also that, for $z \in \mathbb{C} \setminus \Omega$,

$$g_e(z) = \frac{1}{\pi}\iint_{\Omega}\frac{d\sigma_\zeta}{\zeta - z} \tag{1.28}$$

is the *Cauchy transform* of Ω. Similarly, $g_i(z)$ is (for $z \in \Omega$) a renormalized version of the Cauchy transform of $\mathbb{C} \setminus \Omega$.

Chapter 2

Explicit Strong Solutions

In this chapter we shall construct several explicit solutions to the Hele-Shaw problem, more precisely, to the Polubarinova-Galin equation, starting with the classical ones of Polubarinova-Kochina [262], [263], Galin [117] and Saffman, Taylor [293], [294]. Some properties of polynomial and rational solutions will be stated. In particular, we prove the existence theorem. Then we shall consider angular Hele-Shaw flows and give some new families of explicit solutions in terms of hypergeometric functions that contain, as particular cases, those constructed earlier by Ben Amar *et al.*[32], [33], [34], Arnéodo *et al.* [17], Kadanoff [188], etc.

2.1 Classical solutions

It is possible to construct many explicit solutions to the Hele-Shaw problem using the nonlinear Polubarinova-Galin equation (1.16). The main idea is to use a special form of the parametric univalent function $f(\zeta, t)$. The simplest solution is the expansion/shrinking of the disk centered on the sink/source. This is the only case when the fluid can be completely removed (see [117], [171]). The solution has the obvious form

$$f(\zeta, t) = \sqrt{\frac{|\Omega_0| - tQ}{\pi}}\zeta.$$

Here Ω_0 is a disk centered on the origin and $t \in [0, \infty)$ in the case of injection ($Q < 0$) and $t \in [0, |\Omega_0|/Q]$ in the case of suction $Q > 0$. In the case of injection it is possible to start with $\Omega_0 = \emptyset$.

2.1.1 Polubarinova and Galin's cardioid

The first non-trivial solution for the problem with suction ($Q > 0$) was constructed by Polubarinova-Kochina [262], [263] and Galin [117]. They chose a quadratic mapping

$$f(\zeta, t) = a_1(t)\zeta + a_2(t)\zeta^2,$$

$\zeta \in U$, with real coefficients $a_1(t)$ and $a_2(t)$. This mapping being substituted into equation (1.16) gives the following system for the coefficients

$$a_1^2(t)a_2(t) = a_1^2(0)a_2(0),$$

$$a_1^2(t) + 2a_2^2(t) = a_1^2(0) + 2a_2^2(0) - \frac{Qt}{\pi}.$$

For any initial condition such that $|a_2/a_1| < 1/2$ the solution $f(\zeta, t)$ is a univalent map locally in time $t \in [0, t_0)$. The blow-up time t_0 occurs exactly when the equality $|a_2/a_1| = 1/2$ is reached that corresponds to the vanishing boundary derivative of f and cusp formation at the boundary. This evolution is shown in Figure 2.1. As is observed, cusp formation occurs before the moving boundary

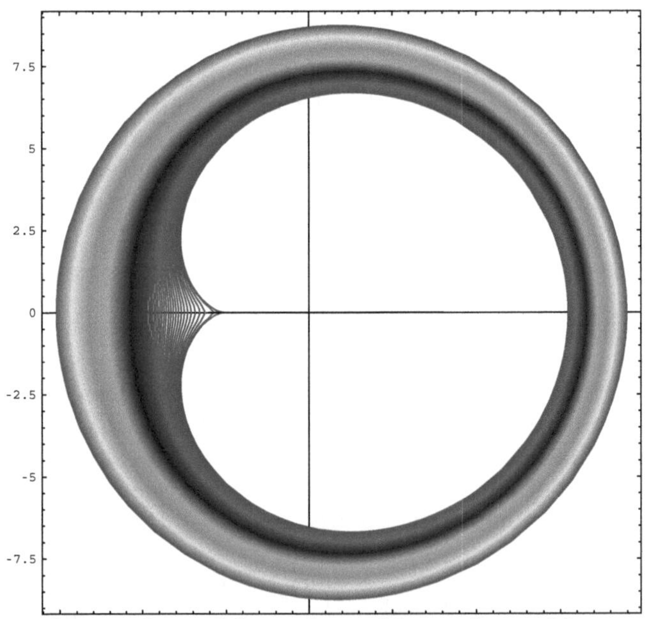

Figure 2.1: Polubarinova and Galin's cardioid

reaches the sink. This phenomenon is general for all polynomial solutions. It seems that Galin knew that, but did not prove it correctly. A correct proof appeared in [171]. Considering a general polynomial form of f,

$$f(\zeta, t) = \sum_{k=1}^{n} a_k(t)\zeta^k, \quad a_n(0) \neq 0, \; a_1(t) > 0,$$

one substitutes it in equation (1.16). By rotating $e^{i\alpha}f(e^{-i\alpha}\zeta,t)$ we make the coefficient $a_n(0)$ real. This leads to a system of n equations for the coefficients $a_k(t)$. The first one is

$$\frac{d}{dt}\sum_{k=1}^{n} k|a_k|^2 = -\frac{Q}{\pi}.$$

The last one is

$$\mathrm{Re}\left(n\bar{a}_n\frac{d\,a_1}{dt} + a_1\frac{d\,\bar{a}_n}{dt}\right) = 0.$$

Since f is univalent for all $t \in [0,t_0)$ and $a_1(t) > 0$, this equation is equivalent to

$$\frac{d}{dt}\mathrm{Re}\,(\bar{a}_n a_1^n) = 0,$$

where t_0 is the blow-up time. The initial conditions imply

$$\sum_{k=1}^{n} k|a_k(t)|^2 = \sum_{k=1}^{n} k|a_k(0)|^2 - \frac{Qt}{\pi}, \tag{2.1}$$

$$\mathrm{Re}\,(\bar{a}_n(t)a_1^n(t)) = a_n(0)a_1^n(0). \tag{2.2}$$

If the boundary reaches the sink at the moment t_0, then the kernel of the family $\Omega(t)$ degenerates: Ker $\{\Omega(t)\} = \{0\}$ for $t \to t_0$. The Carathéodory kernel theorem implies that $\lim_{t\to t_0} f(\zeta,t) \equiv 0$ which contradicts (2.1)–(2.2) ($a_n(0)a_1^n(0) \neq 0$).

Some sufficient conditions for the initial data $(a_1(0),\dots,a_n(0))$ for a polynomial strong solution to exist for all time were given in [215]. Several explicit solutions similar to Polubarinova-Galin's cardioid were obtained by Vinogradov and Kufarev in 1947 [340] but their work was wrongly forgotten.

2.1.2 Rational solutions of the Polubarinova-Galin equation

After this first non-trivial Polubarinova-Galin solution many other explicit solutions were constructed. Among them we distinguish a solution by Saffman and Taylor that will be discussed in the next section. It deals with a flow in a narrow channel. In this section we shall give examples of solutions by means of rational univalent functions. One finds them, e.g., in a paper by Hohlov and Howison [171]. The first explicit rational solutions were obtained by Kufarev in 1948–1950 [213], [214]. Unlike the previous case rational solutions can produce such evolution that the free boundary reaches the sink under suction before the total fluid is removed.

Let $Q > 0$ and consider the map

$$f(\zeta,t) = a(t)\frac{\zeta(1-b(t)\zeta)}{1-c(t)\zeta}, \tag{2.3}$$

where

$$a(t) = -\frac{2\alpha^4 - \alpha^2 - \frac{Qt}{\pi}}{2\alpha^3}, \quad b(t) = \frac{\alpha^3 - \frac{\alpha Qt}{\pi}}{2\alpha^4 - \alpha^2 - \frac{Qt}{\pi}}, \quad c(t) = \frac{1}{\alpha},$$

and $\alpha = \alpha(t)$ is the root of the algebraic equation

$$2\alpha^6 - \left(5 - \frac{2Qt}{\pi}\right)\alpha^4 + \left(\frac{Qt}{\pi}\right)^2 = 0,$$

satisfying the condition $\lim_{t\to\pi/Q}\alpha(t) = -1$. The initial domain is given by the mapping

$$f(\zeta, 0) = \frac{\zeta(4 - \sqrt{\frac{5}{2}}\zeta)}{2\zeta - \sqrt{10}}.$$

The solution $f(\zeta, t)$ exists and is univalent during the time interval $[0, \pi/Q)$. At this moment the moving boundary reaches the sink at the origin and the residual fluid occupies the disk $|z + 1| < 1$, see Figure 2.2.

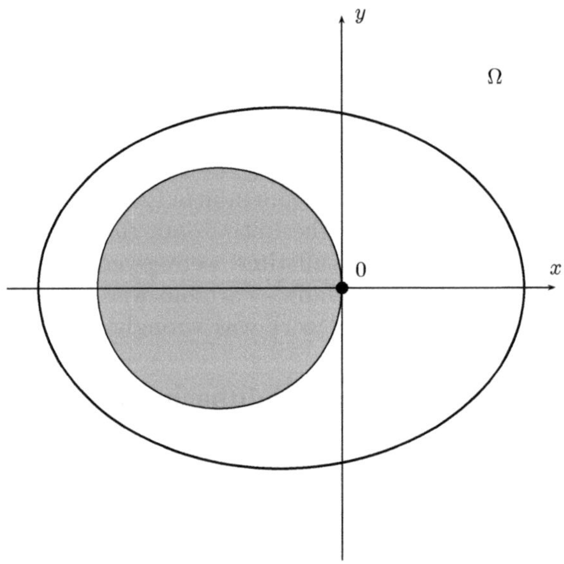

Figure 2.2: Rational solution

The next example is a rational map

$$f(\zeta, t) = a(t)\frac{\zeta(1 - b(t)\zeta^2)}{1 - c(t)\zeta^2},$$

with the parameters a, b, c chosen so that the final domain in a blow-up time consists of two equal disks touching at the sink. Due to complicated details we give only a sketch in Figure 2.3.

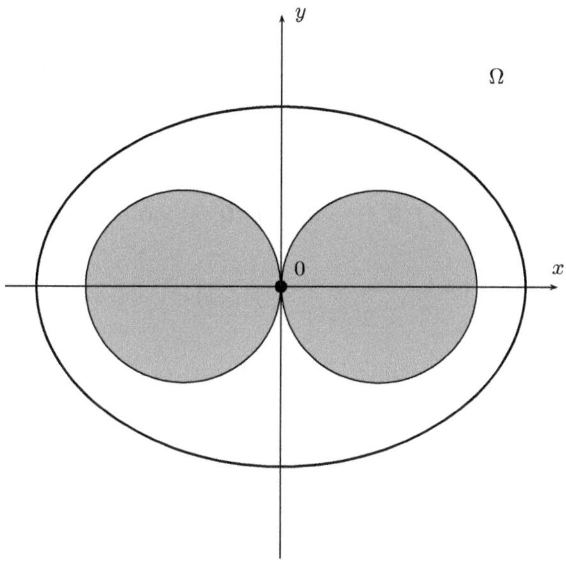

Figure 2.3: Symmetric rational solution

Let us now discuss rational solutions in general. When speaking about a strong, or classical, solution of a differential equation one generally means that all functions and boundaries appearing should be smooth enough and that the equations involved should hold in a pointwise sense. For the Hele-Shaw problem it is convenient to introduce the notion of a smooth family of domains [330]. We call a family of domains $\{\Omega(t)\}$ smooth if the boundaries $\partial\Omega(t)$ are smooth (C^∞) for each t, and the normal velocity v_n continuously depends on t at any point of $\partial\Omega(t)$.

Then a *strong solution* of the Hele-Shaw problem is defined to be a smooth family $\{\Omega(t) : 0 \leq t < t_0\}$ such that (1.12)–(1.14) hold in a pointwise sense (the function p will be uniquely determined by $\Omega(t)$ and will be smooth up to $\partial\Omega(t)$). If the domains $\Omega(t)$ are simply connected it is equivalent to require the Polubarinova-Galin equation (1.16) to hold.

In the above definition the interval $[0, t_0)$ may be replaced by any open, closed or half-open interval.

Given a domain $\Omega(0)$ with smooth boundary it is known that in the well-posed case $Q < 0$ there exists a strong solution of (1.12)–(1.14) on some interval $[0, t_0)$. In the ill-posed case $Q > 0$ such a statement is true only if $\partial\Omega(0)$ is analytic (see, e.g., [330]).

Since we do not know any reasonably short proof of these general existence results we shall not include any such proof here, but just refer to the literature:

[108], [280], [339]. Instead we shall discuss some general properties of solutions in the simply connected case, and also provide an elementary proof of existence of solutions when the initial domain is the conformal image of U under a rational function. We shall first make some general observations.

Assume that $f(\zeta,t)$ is analytic and univalent in a neighbourhood of $\overline{U}$ for each t and is normalized by

$$f(0,t)=0, \quad f'(0,t)>0. \tag{2.4}$$

It is useful to set

$$g(\zeta,t)=\frac{\dot f(\zeta,t)}{\zeta}.$$

In view of (2.4), g is holomorphic in U with $g(0,t)>0$. The Polubarinova-Galin equation (1.16) becomes

$$\operatorname{Re}[\overline{g}\cdot f']=-\frac{Q}{2\pi}. \tag{2.5}$$

On dividing by $|f'|^2$ we get

$$\operatorname{Re}[\frac{g(\zeta,t)}{f'(\zeta,t)}]=-\frac{Q}{2\pi|f'(\zeta,t)|^2} \quad (\zeta=e^{i\theta}). \tag{2.6}$$

Here the left member is a harmonic function in U and (2.6) gives its boundary values on ∂U. This Dirichlet problem can be solved explicitly in terms of a Poisson integral, and taking also the imaginary part into account we get g solved in terms of f' as

$$g=\mathcal{G}(f'),$$

where $\mathcal{G}$ is a nonlinear operator defined by

$$\mathcal{G}(f')(\zeta)=\frac{-Qf'(\zeta)}{4\pi^2}\int_0^{2\pi}|f'(e^{i\theta})|^{-2}\frac{e^{i\theta}+\zeta}{e^{i\theta}-\zeta}d\theta. \tag{2.7}$$

(We suppress t from notation whenever convenient.)

Now we observe the following properties of $\mathcal{G}(f')$:

1. If f' is holomorphic in U_R for some $R>1$, then also $\mathcal{G}(f')$ is holomorphic in U_R.
2. If f' is a rational function with poles of order k_j at some points z_j (outside $\overline{U}$), then the same is true for $\mathcal{G}(f')$. One of the points may be the point at infinity.

It is important for these conclusions that f' is holomorphic in a neighbourhood of $\overline{U}$ and has no zeros there (whereas the univalency of f is not needed in itself).

To prove 1) and 2) we first write (2.7) as

$$\mathcal{G}(f')(\zeta)=-\frac{Q}{4\pi^2 i}f'(\zeta)\int_{\partial U}\frac{1}{f'(z)\overline{f'(\frac{1}{\bar z})}}\frac{z+\zeta}{z-\zeta}\frac{dz}{z}. \tag{2.8}$$

With $\zeta \in U$ the integrand above is holomorphic in some neighbourhood of ∂U. It follows that the path of integration can be replaced by a contour slightly outside ∂U. This shows that $g = \mathcal{G}(f')$ is analytic in some neighbourhood of $\overline{U}$ (to start with).

Next we go back to (2.5) and write it as

$$\operatorname{Re}\left(\overline{g} \cdot f' + \frac{Q}{2\pi}\right) = 0$$

holding on ∂U. Spelling out the real part and using that $\zeta = 1/\overline{\zeta}$ on ∂U gives that

$$\overline{g(1/\overline{\zeta})} f'(\zeta) + g(\zeta) \overline{f'(1/\overline{\zeta})} + \frac{Q}{\pi} = 0 \tag{2.9}$$

on ∂U. But here the left member is a holomorphic function in some neighbourhood of ∂U, hence (2.9) continues to hold identically in any such neighbourhood. Since now both g and f' are holomorphic in a neighbourhood of $\overline{U}$ and f' has no zeros there, it follows from (2.9) that the singularities of g outside $\overline{U}$, i.e., the singularities of $\overline{g(\frac{1}{\overline{\zeta}})}$ for $\zeta \in U$, are no worse than the singularities of $f'(\frac{1}{\overline{\zeta}})$ for $\zeta \in U$. This proves 1) and 2).

From the above remarks we easily deduce the following theorem.

Theorem 2.1.1. *Assume $f(\zeta, 0)$ is a rational function which is holomorphic and univalent in some neighbourhood of $\overline{U}$ and is normalized by (2.4). Then in some time interval around $t = 0$ there exists a rational solution $f(\zeta, t)$ of (1.16). Each $f(\zeta, t)$ is analytic and univalent in a neighbourhood of $\overline{U}$ and normalized by (2.4). The pole structure of $f(\zeta, t)$ is the same as that of $f(\zeta, 0)$, but all poles except the one at infinity may move around. Poles can not collide or disappear, with the sole exception that the pole at infinity may disappear for one value of t.*

Remark 2.1.2. *We shall see later (see Theorem 3.4.1) that, in the well-posed case $Q < 0$, the radius of analyticity $R(t)$ of $f(\zeta, t)$, i.e., the largest number R such that $f(\zeta, t)$ is holomorphic in U_R, is a strictly increasing function of t. If the solution exists for all $0 \leq t < \infty$ we shall even have that $R(t) \to \infty$ as $t \to \infty$. Thus, the poles of $f(\zeta, t)$ will not cause any breakdown of the solution. If the solution breaks down in finite time it will be because univalency will be lost, either due to zeros of $f'(\zeta, t)$ reaching ∂U or of $f(\zeta, t)$) taking the same value twice on $\partial U \equiv S^1$.*

This remark applies in general to strong solutions of (2.6), not only when $f(\zeta, 0)$ is rational.

Proof. In order to avoid too many summation signs, let us assume that $f(\zeta, 0)$ has only two poles, one finite pole and one pole at infinity:

$$f(\zeta, 0) = \sum_{k=1}^{m} \frac{b_k}{(\zeta - a)^k} + \sum_{j=0}^{n} c_j \zeta^j,$$

where $b_m \neq 0$, $m, n \geq 1$. The general case is obtained by replacing a by a_l, letting m depend on l and summing over l. Then we make the "Ansatz", for $f(\zeta, t)$,

$$f(\zeta,t) = \sum_{k=1}^{m} \frac{b_k(t)}{(\zeta - a(t))^k} + \sum_{j=0}^{n} c_j(t)\zeta^j. \tag{2.10}$$

Here it is necessary to postulate $n \geq 1$, even if $c_1 = 0$, because the Hele-Shaw injection/suction will in any case create a pole at infinity. This gives

$$f'(\zeta,t) = -\sum_{k=1}^{m} \frac{kb_k(t)}{(\zeta - a(t))^{k+1}} + \sum_{j=1}^{n} jc_j(t)\zeta^{j-1},$$

$$\dot{f}(\zeta,t) = \sum_{k=1}^{m} \frac{\dot{b}_k(t)}{(\zeta - a(t))^k} + \sum_{k=1}^{m} \frac{k\dot{a}(t)b_k(t)}{(\zeta - a(t))^{k+1}} + \sum_{j=0}^{n} \dot{c}_j(t)\zeta^j$$

$$= \frac{m\dot{a}(t)b_m(t)}{(\zeta - a(t))^{m+1}} + \sum_{k=1}^{m} \frac{\dot{b}_k(t) + (k-1)\dot{a}(t)b_{k-1}(t)}{(\zeta - a(t))^k} + \sum_{j=0}^{n} \dot{c}_j(t)\zeta^j.$$

By the properties 1), 2) of $\mathcal{G}$, $\mathcal{G}(f')$ will be of the form

$$\mathcal{G}(f')(\zeta,t) = \sum_{k=1}^{m+1} \frac{B_k(t)}{(\zeta - a(t))^k} + \sum_{j=1}^{n} C_j(t)\zeta^{j-1}$$

for suitable coefficients $B_k(t)$ and $C_j(t)$. It is not hard to see, for example from the formula (2.8), that these finitely many coefficients depend smoothly on the coefficients of f' (see [143] for details).

Now the Polubarinova-Galin equation in terms of present notation is

$$\dot{f}(\zeta,t) = \zeta\mathcal{G}(f')(\zeta).$$

Inserting here the above expressions for $\dot{f}$ and f' (or rather $\mathcal{G}(f')$) and identifying coefficients gives a system of differential equations for $a(t)$, $b_k(t)$, $c_j(t)$, and one sees immediately from the last expression for $\dot{f}$ that this system can be solved for the time derivatives $\dot{a}(t)$, $\dot{b}_k(t)$, $\dot{c}_j(t)$ as long as $b_m(t) \neq 0$. Thus, the Polubarinova-Galin equation reduces to a finite dimensional system of ordinary differential equations of standard form, which by Picard's theorem has a unique solution, at least for a short two-sided interval around $t = 0$. This means that the "Ansatz" (2.10) was successful so that the rational solution (2.10) survives in the same form for a little while. This proves the theorem, except for the statement about collision, which will be discussed in Section 3.7 (Theorem 3.6.3). □ □

2.1.3 Saffman-Taylor fingers

The most famous solutions to the original Hele-Shaw problem are the travelling-wave fingers of Saffman and Taylor (1958) [293], [294]. When a low viscosity fluid (for example, water) is injected into a more viscous one, such as glycerin, an instability occurs. In fact, Hele-Shaw (1898) [166] proposed the model of air injection into a narrow channel. An important reason for studying this problem is that it is closely related to many technologically relevant ones, such as a flow in porous media. One of the features of the channel model is that we should change the Dirichlet problem (1.12), (1.13) to a mixed boundary problem[1] for the potential function p.

Let us consider an infinite channel with parallel sides

$$\operatorname{Re} z \in (-\infty, \infty), \quad \operatorname{Im} z \in (-\pi, \pi),$$

in which an inviscid fluid is injected from the left (or the viscous fluid is extracted from the right) at a constant rate $Q > 0$, see Figure 2.4. The function $p(z,t)$ is

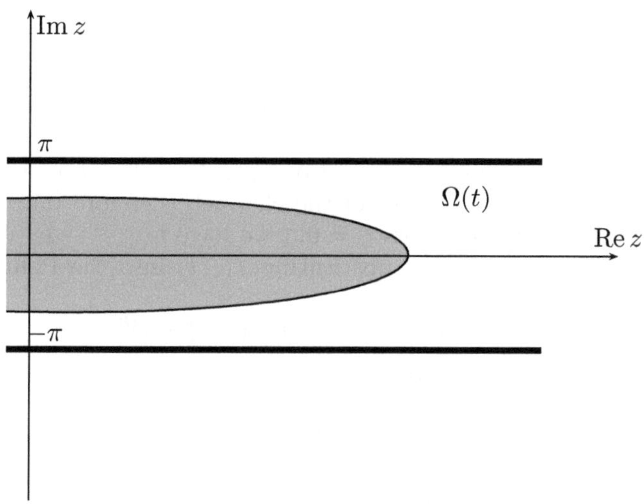

Figure 2.4: The Saffman-Taylor finger

harmonic in the region $\Omega(t)$ occupied by the viscous fluid and vanishes on the free

[1]This type of boundary conditions, known also as Robin's boundary conditions, named so after the French mathematical physicist **Gustave Robin** (1855–1897) by Bergman and Schiffer [36], appeared in connection with the third type of boundary conditions (after Dirichlet's and Neumann's). Robin completed a doctoral thesis in 1886 under Emile Picard and it is most probable that this attribution does not correspond to Robin's own works (see [141]) though his name in this context is widely used nowadays.

boundary $\Gamma(t)$. It satisfies the condition of non-penetration at the walls $\operatorname{Im} z = \pm\pi$. Therefore, we have the mixed boundary value problem

$$\begin{aligned} \Delta p &= 0 \quad \text{in } \Omega(t), \\ p\big|_{\Gamma(t)} &= 0, \\ \frac{\partial p}{\partial n}\Big|_{\operatorname{Im} z=\pm\pi} &= 0, \\ \frac{\partial p}{\partial n}\Big|_{\Gamma(t)} &= -v_n, \end{aligned}$$

with the normalization at infinity $p \sim -\frac{Q}{2\pi}\operatorname{Re} z$, as $\operatorname{Re} z \to +\infty$. We choose an auxiliary parametric domain $D = U\setminus(-1,0]$ and construct the conformal univalent mapping $z = f(\zeta,t)$ from D onto $\Omega(t)$ assuming that the slit along the negative axis is mapped onto the walls. For the flow outside the bubble we require $\arg f(\zeta,t) \sim -\arg\zeta$. The pressure in terms of this auxiliary variable ζ is written as just

$$(p\circ f)(\zeta) = \frac{Q}{2\pi}\log|\zeta|.$$

Applying the standard technique, as was done in Section 1.4.2, we come to the Polubarinova-Galin equation for the free boundary $\Gamma(t)$:

$$\operatorname{Re}[\dot f(\zeta,t)\overline{\zeta f'(\zeta,t)}] = -\frac{Q}{2\pi}, \quad \zeta = e^{i\theta}, \quad \theta\in(-\pi,\pi).$$

We are looking for travelling-wave solutions $f(\zeta,t) = At + h(\zeta)$ with $A > 0$. The slit in D is mapped onto the walls of the channel, therefore taking into account possible singularities at the points $\zeta = 0,1$ we have $h(\zeta) \sim -\log\zeta$ as $\zeta \to 0$ and $h(\zeta) \sim \log(1+\zeta)$ as $\zeta \to -1$. Substituting $f(\zeta,t)$ into the Polubarinova-Galin equation we have

$$\operatorname{Re}(A\zeta h'(\zeta)) = -Q/2\pi, \quad \zeta = e^{i\theta}.$$

Differentiating with respect to θ leads to

$$\operatorname{Im}(\zeta(\zeta h'(\zeta))') = 0.$$

The singularities of h suggest the form of the function

$$\zeta(\zeta h'(\zeta))' = \frac{c_1\zeta}{(1+\zeta)^2},$$

where c_1 is some real constant. The solution to this equation, neglecting a horizontal shift, is

$$h(\zeta) = (c_2 - c_1)\log\zeta + c_1\log(1+\zeta).$$

To determine the constants we use the Polubarinova-Galin equation again and obtain

$$A\operatorname{Re}(c_2 - \frac{c_1}{2}) = -\frac{Q}{2\pi}.$$

One can choose the ratio $\lambda \in (0,1)$ of the width of the finger at $\operatorname{Re}\zeta \to -\infty$ as a parameter and derive $c_1 = 2(1-\lambda)$, $c_2 = 1-2\lambda$, $A = \frac{Q}{2\pi\lambda}$. Finally,

$$f(\zeta,t) = \frac{Q}{2\pi\lambda}t - \log\zeta + 2(1-\lambda)\log(1+\zeta).$$

This function gives a travelling-wave solution, moving with the speed $\frac{Q}{2\pi\lambda}$, for any value $\lambda \in (0,1)$. The way in which Saffman and Taylor derived the solution shows that it is the only possible form for a steady, travelling wave. In terms of analytic functions this corresponds to the uniqueness of the solution to the mixed boundary value problem with given singularities.

Curiously enough, Saffman-Taylor's work seems to have been underestimated when it appeared. For example, a MR review says that "...the authors' analysis does not seem to be completely rigorous, mathematically. Many details are lacking. Besides, the authors do not seem to be aware of the fact that there exists a vast amount of literature concerning viscous fluid flow into porous (homogeneous and non-homogeneous) media in Russian and Romanian". Nowadays, the Saffman-Taylor fingers are widely known in many fields of mechanics, chemistry and industry.

Saffman and Taylor found experimentally that an unstable planar interface evolves through finger competition to a steady translating finger with $\lambda = 1/2$. Recently, Tanveer and Xie [325], [326] proved that even a small surface tension effect implies non-existence of a strong solution when the relative finger width λ is smaller than 1/2. They also solved [325] the selection problem for $\lambda > 1/2$.

2.2 Corner flows

Having handled these first steps many authors have been constructing non-trivial solutions. We should say that mostly these explicit solutions are either polynomials and rational functions, or else, logarithmic solutions linked to Saffman-Taylor fingers. Another type of explicit solutions was proposed by, e.g., Howison, King [181], Cummings [67], who reduced the problem to solving the Poisson equation, eliminating time by applying the Baiocchi transformation. The solutions were given making use of the Riemann P-function and hypergeometric functions.

Corner flows of an inviscid incompressible fluid were studied intensively, e.g., in [191], [192], [235], [256], [332] (see also the references therein). In particular, we mention here papers [17], [32], [33], [34], [188], [286], [287], [328]. A solution constructed by Kadanoff [188] is directly linked with ours.

In this section we shall construct explicit solutions in an infinite corner of arbitrary angle such that the viscous fluid is glued to one of the walls, the interface extends to infinity along it and has fluid-wall angle $\pi/2$ at a moving contact point at the other wall. These solutions will present a logarithmic deformation of the trivial (circular) solution. In the case of a right angle we get Kadanoff's solution [188]. Then we present an analogous solution in the corner with a source at its vertex. Finally, we construct self-similar solutions in a wedge analogous to Saffman-Taylor fingers.

2.2.1 Mathematical model

In this subsection we deal with a general case of corner flows. We suppose that the viscous fluid occupies a simply connected domain $\Omega(t)$ in the phase z-plane. The boundary $\Gamma(t)$ consists of two walls $\Gamma_1(t)$ and $\Gamma_2(t)$ of the corner and a free interface $\Gamma_3(t)$ between them at a moment t. The inviscid fluid (or air) fills the complement to $\Omega(t)$. The simplifying assumption of constant pressure at the interface between the fluids means that we omit the effect of surface tension. The velocity must be bounded close to the contact wall-fluid point that yields the contact angle between the walls of the corner and the moving interface to be $\pi/2$ (see Figure 2.5). A limiting case corresponds to one finite contact point and the other tending to infinity. By a shift we can place the point of the intersection of the wall extensions at the origin. To simplify matters, we set the corner of angle α between the walls so that the positive real axis x contains one of the walls and fix this angle as $\alpha \in (0, \pi]$.

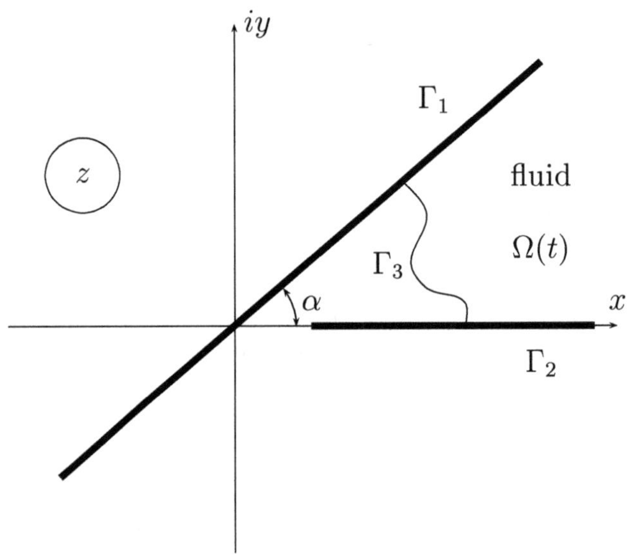

Figure 2.5: $\Omega(t)$ is the phase domain within an infinite corner and the homogeneous sink/source at ∞

Let us mention here that the model can be studied in the presence of surface tension and the macroscopic contact angle between the walls and the free interface can then be different from $\pi/2$. Let us denote it by β. The contact angle β at a moving contact line obeys interesting properties that were studied by Ablett [2] (see also [61], [99]) in a particular case of water in contact with a paraffin surface. It turns out that the steady angle β depends on the velocity of the contact line. The angle β increases with the velocity increased for the liquid advancing over the surface up to a certain value β_0 and, then, remains the same for a greater velocity. Reciprocally, β decreases with the velocity increased for the liquid receding over the surface up to a certain value β_1 (different from β_0) and, then, remains the same for a greater velocity.

In our zero surface tension model we have Robin's boundary value problem for fluid pressure $p(z,t) \equiv p(x,y,t)$,

$$\Delta p = 0 \quad \text{in the flow region } \Omega(t), \tag{2.11}$$

and the fluid velocity $\mathbf{V}$ averaged across the gap is $\mathbf{V} = -\nabla p$. The free boundary conditions

$$p\Big|_{\Gamma_3} = 0, \quad \frac{\partial p}{\partial t}\Big|_{\Gamma_3} = |\nabla p|^2 \tag{2.12}$$

are imposed on the free boundary $\Gamma_3 \equiv \Gamma_3(t)$. This implies that the normal velocity v_n of the free boundary Γ_3 outwards from $\Omega(t)$ is given by

$$\frac{\partial p}{\partial n}\Big|_{\Gamma_3} = -v_n. \tag{2.13}$$

On the walls $\Gamma_1 \equiv \Gamma_1(t)$ and $\Gamma_2 \equiv \Gamma_2(t)$ the boundary conditions are given as

$$\frac{\partial p}{\partial n}\Big|_{\Gamma_1 \cup \Gamma_2} = 0. \tag{2.14}$$

We suppose that the motion is driven by a homogeneous source/sink at infinity. Since the angle between the walls at infinity is also α, the pressure behaves about infinity as

$$p \sim \frac{-Q}{\alpha} \log |z|, \quad \text{as } |z| \to \infty,$$

where Q corresponds to the constant strength of the source ($Q < 0$) or sink ($Q > 0$). Finally, we assume that $\Gamma_3(0)$ is a given analytic curve.

We introduce Robin's complex analytic potential $W(z,t) = p(z,t) + i\psi(z,t)$, where $-\psi$ is the stream function. Let us consider an auxiliary parametric complex ζ-plane, $\zeta = \xi + i\eta$. We set

$$\begin{aligned}
D &= \{\zeta : |\zeta| > 1,\ 0 < \arg \zeta < \alpha\}, \\
D_3 &= \{z : z = e^{i\theta},\ \theta \in (0,\alpha)\}, \\
D_1 &= \{z : z = re^{i\alpha},\ r > 1\}, \\
D_2 &= \{z : z = r,\ r > 1\}, \\
\partial D &= D_1 \cup D_2 \cup D_3,
\end{aligned}$$

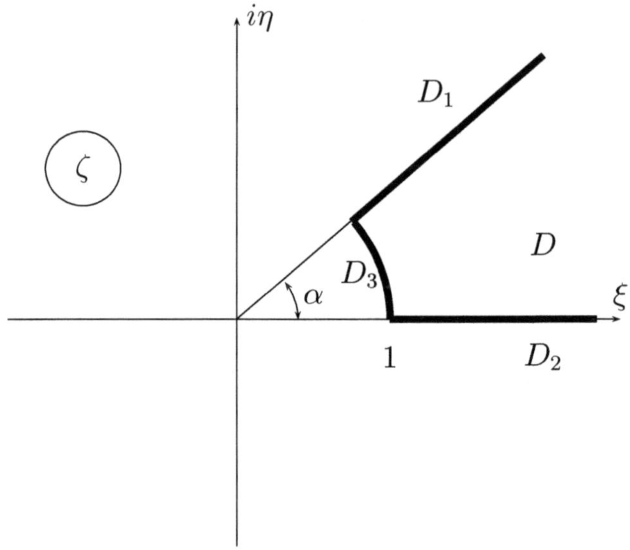

Figure 2.6: The parametric domain D

and construct a conformal univalent time-dependent map $z = f(\zeta, t)$, $f : D \to \Omega(t)$, so that being continued onto ∂D, $f(\infty, t) \equiv \infty$, and the circular arc D_3 of ∂D is mapped onto Γ_3 (see Figure 2.6). This map has an expansion

$$f(\zeta, t) = \zeta \sum_{n=0}^{\infty} a_n(t) \zeta^{-\frac{\pi n}{\alpha}}$$

near infinity and $a_0(t) > 0$. The function f parameterizes the boundary of the domain $\Omega(t)$ by $\Gamma_j = \{z : z = f(\zeta, t), \, \zeta \in D_j\}$, $j = 1, 2, 3$.

The normal unit vector in the outward direction is given by

$$\mathbf{n} = -\zeta \frac{f'}{|f'|} \text{ on } \Gamma_3, \quad \mathbf{n} = -i \text{ on } \Gamma_2, \text{ and } \quad \mathbf{n} = ie^{i\alpha} \text{ on } \Gamma_1.$$

Therefore, the normal velocity is obtained as

$$v_n = \mathbf{V} \cdot \mathbf{n} = -\frac{\partial p}{\partial n} = \begin{cases} -\operatorname{Re}\left(\dfrac{\partial W}{\partial z}\dfrac{\zeta f'}{|f'|}\right) & \text{for } \zeta \in D_3, \\ 0 & \text{for } \zeta \in D_1, \\ 0 & \text{for } \zeta \in D_2. \end{cases} \tag{2.15}$$

The superposition $W \circ f$ is the solution to the mixed boundary problem (2.11), (2.12), (2.14) in D, therefore, it is Robin's function given by $(W \circ f)(\zeta) = -\frac{Q}{\alpha} \log \zeta$.

On the other hand,

$$v_n = \begin{cases} \operatorname{Re}(\dot{f}\overline{\zeta f'}/|f'|) & \text{for } \zeta \in D_3, \\ \operatorname{Im}(\dot{f}e^{-i\alpha}) & \text{for } \zeta \in D_1, \\ -\operatorname{Im}(\dot{f}) & \text{for } \zeta \in D_2. \end{cases} \tag{2.16}$$

The first lines of (2.15), (2.16) give that

$$\operatorname{Re}(\dot{f}\,\overline{\zeta f'}) = \frac{Q}{\alpha}, \quad \text{for } \zeta \in D_3. \tag{2.17}$$

The remaining lines of (2.15), (2.16) imply

$$\operatorname{Im}(\dot{f}e^{-i\alpha}) = 0 \quad \text{for } \zeta \in D_1, \quad \operatorname{Im}(\dot{f}) = 0 \quad \text{for } \zeta \in D_2. \tag{2.18}$$

One of the typical properties of the problem (2.11)–(2.14) is that starting with an analytic boundary component $\Gamma_3(0)$, the one-parameter evolutionary chain of solutions develops possible cusps at a finite blow-up time t_0. Another typical scenario is fingering. The strong solution exists locally in time in the case of an analytic boundary Γ_3. We only refer the reader to some relevant works [107], [176], [179], [273], [283], [293].

2.2.2 Logarithmic perturbations of the trivial solution

We consider the case of a sink at infinity ($Q > 0$). The simplest explicit solution in this case is

$$f(\zeta, t) = \sqrt{\frac{2Qt}{\alpha}}\zeta,$$

that produces a circular dynamics of the free boundary. Our aim is to perturb this trivial solution by a function independent of t, say we are looking for a solution in the form

$$f(\zeta, t) = \sqrt{\frac{2Qt}{\alpha}}\zeta + \zeta g(\zeta),$$

where $g(\zeta)$ is regular in D with the expansion

$$g(\zeta) = \sum_{n=0}^{\infty} \frac{a_n}{\zeta^{\frac{\pi n}{\alpha}}}$$

near infinity. The branch is chosen so that g, being continued symmetrically into the reflection of D, is real at real points. Equation (2.17) implies that the function g satisfies

$$\operatorname{Re}(g(\zeta) + \zeta g'(\zeta)) = 0, \quad \zeta \in D_3.$$

Taking into account the expansion of g we are looking for a solution satisfying the equation

$$g(\zeta) + \zeta g'(\zeta) = \frac{\zeta^{\frac{\pi}{\alpha}} - 1}{\zeta^{\frac{\pi}{\alpha}} + 1}, \quad \zeta \in D, \tag{2.19}$$

although other forms may be possible. The general solution to (2.19) can be given in terms of the Gauss hypergeometric function $\mathbf{F} \equiv {}_2\mathbf{F}_1$ as

$$\zeta g(\zeta) = \zeta - 2\zeta \mathbf{F}\left(\frac{\alpha}{\pi}, 1, 1+\frac{\alpha}{\pi}; -\zeta^{\frac{\pi}{\alpha}}\right) + C.$$

We note that f' vanishes only when $\zeta^{\frac{\pi}{\alpha}} = (2/(1+\sqrt{2Qt/\alpha})) - 1$, therefore, the function f is locally univalent, the cusp problem appears only at the initial time $t = 0$ and the solution exists during infinite time. The resulting function is homeomorphic on the boundary ∂D, hence it is univalent in D. This presents an example (apart from the trivial one) of the long-time existence of the solution in the problem with suction (ill-posed problem). To complete our solution we need to determine the constant C. First of all we choose the branch of the function ${}_2\mathbf{F}_1$, so that the points of the ray $\zeta > 1$ have real images. This implies that $\operatorname{Im} C = 0$.

We continue verifying the asymptotic properties of the function $f(e^{i\theta}, t)$ as $\theta \to \alpha - 0$. The slope is

$$\lim_{\theta\to\alpha-0} \arg[ie^{i\theta} f'(e^{i\theta}, t)] = \alpha + \frac{\pi}{2} + \lim_{\theta\to\alpha-0} \arg\left(\sqrt{\frac{2Qt}{\alpha}} + \frac{e^{i\frac{\pi\theta}{\alpha}} - 1}{e^{i\frac{\pi\theta}{\alpha}} + 1}\right) = \alpha + \pi.$$

To calculate the shift we choose C such that

$$\lim_{\theta\to\alpha-0} \operatorname{Im}\,[e^{-i\alpha} f'(e^{i\theta}, t)] = 0.$$

Using the properties of hypergeometric functions we have

$$\lim_{\gamma\to 0+0} \operatorname{Im} \mathbf{F}\left(\frac{\alpha}{\pi}, 1, 1+\frac{\alpha}{\pi}; e^{i\gamma}\right) = \frac{\alpha}{2}.$$

Therefore, $C = \alpha$. We present numerical simulation in Figure 2.7.

The special case with angle $\alpha = \pi/2$ was considered by Kadanoff [188]. The hypergeometric function reduces to an arctangent and we obtain

$$f(\zeta, t) = (\sqrt{4Qt/\pi} + 1)\zeta + i \log\frac{1+i\zeta}{1-i\zeta} + \frac{\pi}{2}, \quad Q > 0.$$

This function maps the domain $\{|\zeta| > 1,\ 0 < \arg\zeta < \pi/2\}$ onto the infinite domain bounded by the imaginary axis (Γ_1), the ray $\Gamma_2 = \{r : r \geq \sqrt{4Qt/\pi} + 1\}$ of the real axis and an analytic curve Γ_3 which is the image of the circular arc, see Figure 2.8.

By the analogy with an infinite sink we are able to give solutions for a finite source (see Figure 2.9). The phase domain is a simply connected finite domain at the vertex of the corner which is a source. We locate the corner so that one of the walls lies on the real axis and the other forms the corner of angle α at the origin. We set $G = \{\zeta : |\zeta| < 1, 0 < \arg\zeta < \alpha\}$, $G_3 = \{z : z = e^{i\theta}, \theta \in (0, \alpha)\}$,

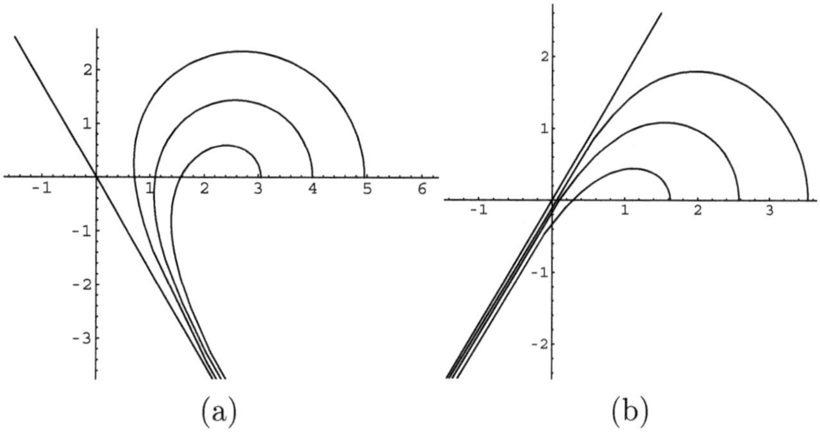

Figure 2.7: Evolution in the corner of angle: (a) $\alpha = 2\pi/3$; (b) $\alpha = \pi/3$

$G_1 = \{z : z = re^{i\alpha}, r < 1\}$, $G_2 = \{z : z = r, r < 1\}$, $\partial G = G_1 \cup G_2 \cup G_3$, and construct a conformal univalent time-dependent map $z = f(\zeta, t)$, $f : G \to \Omega(t)$. This map has an expansion

$$f(\zeta, t) = \zeta \sum_{n=0}^{\infty} a_n(t) \zeta^{\frac{\pi n}{\alpha}}$$

near the origin and $a_0(t) > 0$. The equations for this function at the boundary of G are

$$\mathrm{Re}\,(\dot{f}\,\overline{\zeta f'}) = -\frac{Q}{\alpha}, \quad \text{for } \zeta \in G_3,$$

where $Q < 0$, and

$$\mathrm{Im}\,(\dot{f} e^{-i\alpha}) = 0 \quad \text{for } \zeta \in G_1, \quad \mathrm{Im}\,(\dot{f}) = 0 \quad \text{for } \zeta \in G_2.$$

We give a solution analogous to the infinite case by

$$f(\zeta, t) = \sqrt{\frac{2|Q|t}{\alpha}}\zeta - \zeta + 2\zeta \mathbf{F}\left(\frac{\alpha}{\pi}, 1, 1 + \frac{\alpha}{\pi}; -\zeta^{\frac{\pi}{\alpha}}\right).$$

The numerical simulation is presented in Figure 2.10.

Remark 2.2.1. *By the proposed method we can perturb several known self-similar solutions even for more general flows. The idea is as follows. Let $f_0(\zeta, t) = H(t)F(\zeta)$ be a known solution to the problem, the basic equation of which is the Polubarinova-Galin equation $\mathrm{Re}\,(\dot{f}\,\overline{\zeta f'})$ =const (positive or negative) and where ζ belongs to a circular component of the parametric domain. We are looking for a new solution of the form $f(\zeta, t) = f_0(\zeta, t) + g(\zeta)$, where $g(\zeta)$ is an analytic function with an*

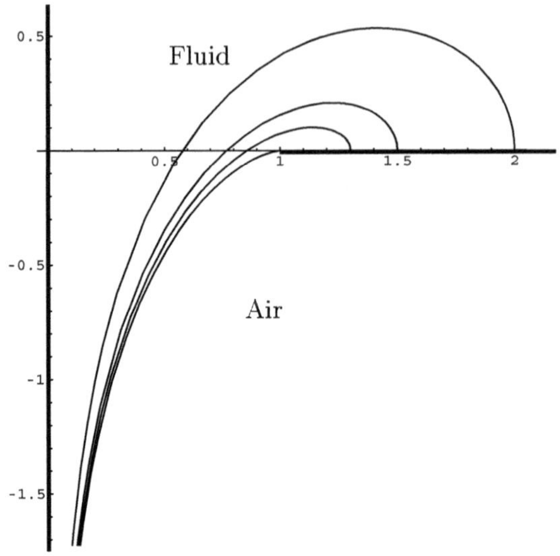

Figure 2.8: Kadanoff's solution

appropriate expansion. Then, on the circular component this function satisfies the equation

$$\operatorname{Re}\frac{\zeta g'(\zeta)}{F(\zeta)} = 0.$$

So one must solve the equation $\zeta g'(\zeta) = F(\zeta)P(\zeta)$*, where* $P(\zeta)$ *is a function with vanishing real part at the points of the circular component.*

2.2.3 Self-similar bubbles

In this subsection we discuss deformation of two-dimensional bubbles in a corner flow in which there is a replacement of two immiscible fluids, one of which is viscous and the other is effectively inviscid. We shall give self-similar (homothetic) drop-shaped solutions in a corner that include Ben Amar's solution [32] as well as those constructed in [17], [328] as particular cases. Y. Tu [331] also analysed viscous fingering in corners applying the hodograph method for the complex velocity potential. In the symmetric case this leads to Ben Amar's solution [32] given in terms of hypergeometric functions, whereas in the non-symmetric case no explicit solution was given.

The bubbles are assumed to originate at the vertex as in Figure 2.11 and the bubble-wall contact angles are $\beta \in (0, \alpha/2)$. We let the positive real axis contain one of the walls and fix the angle between the walls as $\alpha \in (0, 2\pi)$.

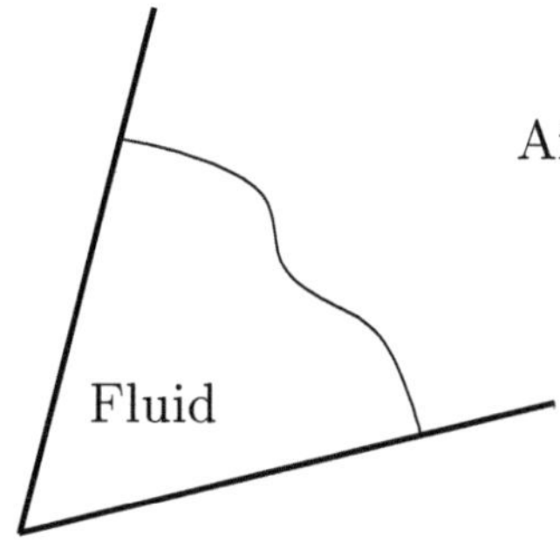

Figure 2.9: Finite source

Mathematically, this model is described by Robin's boundary value problem (2.11)–(2.14), where the potential function $p(z,t)$ behaves near infinity as

$$p \sim \frac{-Q}{\alpha} \log |z|, \quad \text{as } |z| \to \infty,$$

and where Q is the constant strength of the source ($Q < 0$) or sink ($Q > 0$).

Let us consider an auxiliary parametric complex ζ-plane, $\zeta = \xi + i\eta$. We set $D = \{\zeta : |\zeta| > 1, 0 < \arg\zeta < \pi\}$, $D_3 = \{z : z = e^{i\theta}, \theta \in (0,\pi)\}$, $D_1 = \{z : z = -r, r > 1\}$, $D_2 = \{z : z = r, r > 1\}$, $\partial D = D_1 \cup D_2 \cup D_3$. Construct a conformal univalent time-dependent map $z = f(\zeta,t)$, $f : D \to \Omega(t)$, such that being continued onto ∂D, $f(\infty,t) \equiv \infty$, and the circular arc D_3 of ∂D is mapped onto Γ_3 (see Figure 2.12). This map has an expansion $f(\zeta,t) = \zeta^{\alpha/\pi} \sum_{k=0}^{\infty} a_k(t)\zeta^{-k}$ near infinity, and $a_0(t) > 0$. The function f parameterizes the boundary of the domain $\Omega(t)$ by $\Gamma_j = \{z : z = f(\zeta,t), \zeta \in D_j\}$, $j = 1,2,3$.

The free boundary condition is expressed in terms of the function f as in the preceding subsection by

$$\operatorname{Re}(\dot{f}\,\overline{\zeta f'}) = \frac{Q}{\pi}, \quad \text{for } \zeta \in D_3, \tag{2.20}$$

and the wall conditions imply that

$$\operatorname{Im}(\dot{f}e^{i\alpha}) = 0 \quad \text{for } \zeta \in D_1; \quad \operatorname{Im}(\dot{f}) = 0 \quad \text{for } \zeta \in D_2. \tag{2.21}$$

We are going to construct an analogue of the Saffman-Taylor fingers for the corner flows (*self-dilating drops* whose interface contains the vertex). Analytic solutions were discovered first in the case $\alpha = \pi/2$ in [328] and then for general values of angles in [32], [33], [34]. We give a generalization that, in fact, presents possible self-similar solutions, and in particular, we obtain exact solutions for non-symmetric drops.

To simplify matters we scale the angles α, β by $\alpha \to \alpha\pi$, $\beta \to \beta\pi/2$. Let us analyse the auxiliary mapping $f(\zeta,t)$. In the case of self-dilating solutions the

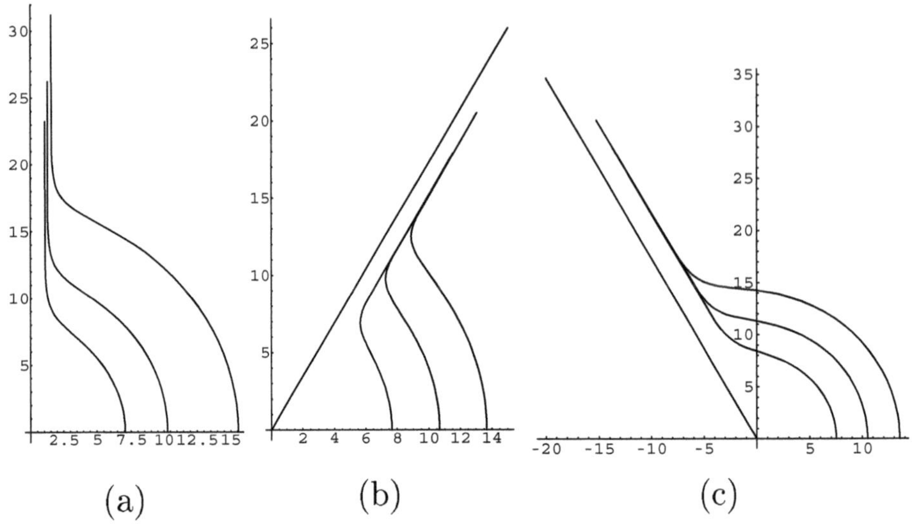

Figure 2.10: Long-pin dynamics of the advancing fluid in the corner of angle: (a) $\alpha = \pi/2$; (b) $\alpha = \pi/3$; (c) $\alpha = 2\pi/3$

phase domain $\Omega(t)$ is a dilation of an initial domain $\Omega(0)$. Then the solution $f(\zeta, t)$ to the equations (2.20)–(2.21) is represented as $f(\zeta, t) = G(t)F(\zeta)$. Since Q does not depend on t, the equation (2.20) implies that $G(t) = C\sqrt{t}$, where C is a constant. Reducing the mapping f to a regular function we represent it as

$$f(\zeta, t) = \sqrt{t}\zeta^\alpha g(\zeta),$$

where $g(\zeta)$ is an analytic function which is regular at infinity.

The boundary Γ_3 starts and ends at the origin under the same bubble-wall contact angles $\beta \in (0, \alpha)$, and forms a self-similar drop-shaped bubble. Therefore, the function $g(\zeta)$ can be represented as

$$g(\zeta) = \left(1 - \frac{1}{\zeta^2}\right)^\beta h(\zeta),$$

where $h(\zeta)$ is a regular function in the closure of D. We differentiate equation (2.20) with respect to θ, taking into account $\zeta = e^{i\theta}$, $\theta \in (0, \pi)$ and obtain

$$\operatorname{Im}\left[(2\alpha + 1)\frac{\zeta g'(\zeta)}{g(\zeta)} + \frac{\zeta^2 g''(\zeta)}{g(\zeta)}\right] = 0, \quad \zeta = e^{i\theta}.$$

In terms of the function h we have $\operatorname{Im} G(\zeta) = 0$, where

$$G(\zeta) \equiv \frac{2\beta(2\alpha+1)}{\zeta^2 - 1} + \frac{4\beta(\beta-1)}{(\zeta^2-1)^2} - \frac{6\beta}{\zeta^2-1} + \left((2\alpha+1) + \frac{4\beta}{\zeta^2-1}\right)\frac{\zeta h'(\zeta)}{h(\zeta)} + \frac{\zeta^2 h''(\zeta)}{h(\zeta)}.$$

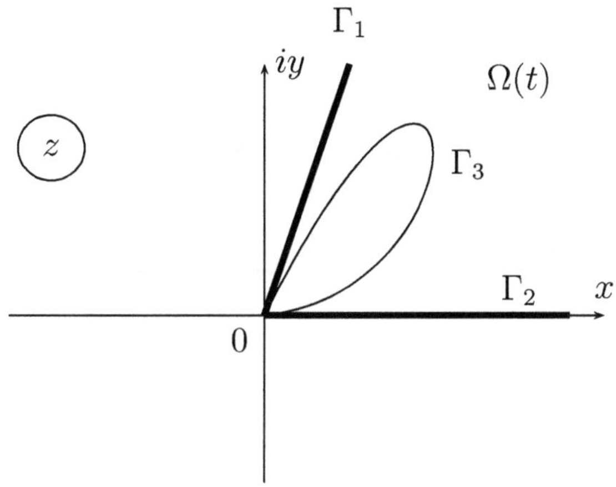

Figure 2.11: $\Omega(t)$ is the phase domain within an infinite corner and the homogeneous sink/source at ∞

Equations (2.21) imply that the equation $\operatorname{Im} G(\zeta) = 0$ is satisfied on the whole boundary $D_1 \cup D_2 \cup D_3$. The function $h(\zeta)$ is regular at ± 1, therefore

$$G(\zeta) \sim \frac{1}{(\zeta^2 - 1)^2} \quad \text{as } \zeta \to \pm 1.$$

Taking into account the regularity of $h(\zeta)$ near infinity we propose that the function G has the form

$$G(\zeta) = \frac{4\beta(\beta - 1)\zeta^2}{(\zeta^2 - 1)^2},$$

although other forms may be possible. Our intention is to obtain a complex differential equation for which we can construct explicit solutions. So we have it in the form

$$\frac{4\beta(\alpha - \beta)}{\zeta^2 - 1} + \left((2\alpha + 1) + \frac{4\beta}{\zeta^2 - 1}\right)\frac{\zeta h'(\zeta)}{h(\zeta)} + \frac{\zeta^2 h''(\zeta)}{h(\zeta)} = 0.$$

Changing variables $w = 1/\zeta^2$, $Y(w) \equiv h(1/\sqrt{w})$ we come to the hypergeometric equation

$$(1 - w)wY'' + (1 - \alpha - (1 + 2\beta - \alpha)w)\, Y' - \beta(\beta - \alpha)Y = 0. \tag{2.22}$$

The general solution of (2.22) can be given in terms of the Gauss hypergeometric function ${}_2\mathbf{F}_1$. We thus have two linearly independent solutions

$$h_1(\zeta) = \mathbf{F}\left(\beta - \alpha, \beta, 1 - \alpha; \frac{1}{\zeta^2}\right), \quad h_2(\zeta) = \frac{1}{\zeta^{2\alpha}}\mathbf{F}\left(\beta, \beta + \alpha, 1 + \alpha; \frac{1}{\zeta^2}\right).$$

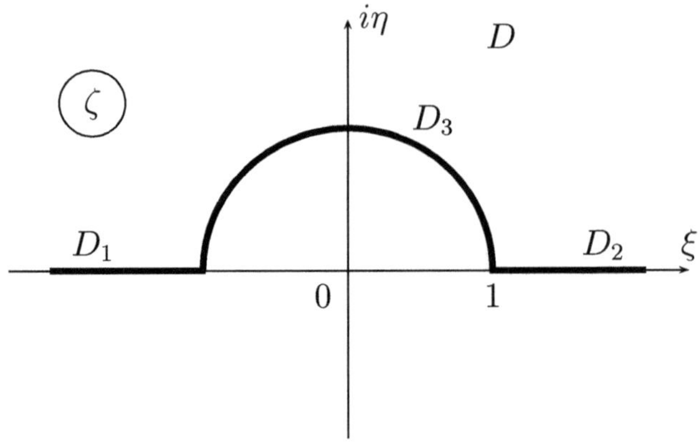

Figure 2.12: The parametric domain D

Finally, we find $f(\zeta,t)$ in the form

$$f(\zeta,t)=\sqrt{t}\zeta^{\alpha}\left(1-\frac{1}{\zeta^2}\right)^{\beta}(C_1h_1(\zeta)+C_2h_2(\zeta)), \tag{2.23}$$

for real constants C_1, C_2 and we choose the branch so that $f(r)>0$ and $h(r)>0$ for $r>1$. Since the primitive

$$\int \operatorname{Im}\left(|f|^2G(e^{i\theta})\Big|_{h=C_1h_1+C_2h_2}\right)d\theta=\operatorname{Re}\dot{f}(e^{i\theta},t)\overline{e^{i\theta}f'(e^{i\theta},t)}$$

is constant, we can choose C_1, C_2 such that it is exactly $Q/\pi>0$ and $f(\zeta,t)$ satisfies the equation (2.20) in the arc $\{e^{i\theta},\,\theta\in(0,\pi)\}$. By construction we have that the function f maps the rays $(-\infty,-1]$ and $[1,\infty)$ onto the walls Γ_1 and Γ_2 respectively. In order to check the univalence of f we note that given a positive Q and f of the form (2.23), we choose the constants C_1, C_2 as mentioned above. The function f is starlike with respect to the origin because $Q>0$ and, hence, univalent. If the constant C_2 vanishes, then the equality $f(-\bar{\zeta},t)=e^{i\alpha\pi}\overline{f(\zeta,t)}$ is easily verified. This means that the solution is symmetric with respect to the bisectrix of the phase angle, namely the ray $z=re^{i\alpha/2}$, $r>0$.

In Figures 2.13, 2.14 we present asymmetric drops in angles $\pi/3$ and $2\pi/3$ (a,c), as well as the symmetric case (b).

In the case $\alpha=1/2$ the hypergeometric functions reduce to a simpler form:

$$h_1(\zeta)=\frac{1}{2}\left(\left(1+\frac{1}{\zeta}\right)^{1-2\beta}+\left(1-\frac{1}{\zeta}\right)^{1-2\beta}\right),$$

$$h_2(\zeta)=\frac{1}{2(1-2\beta)}\left(\left(1+\frac{1}{\zeta}\right)^{1-2\beta}-\left(1-\frac{1}{\zeta}\right)^{1-2\beta}\right),$$

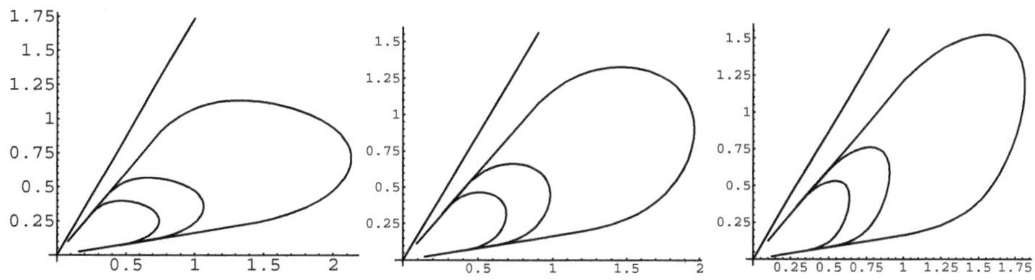

(a) (b) (c)

Figure 2.13: Finger dynamics in the wedge angle $\pi/3$ and the bubble-wall angles $\pi/20$: (a) $C_1 = 1$, $C_2 = 0.9$; (b) $C_1 = 1$, $C_2 = 0$; (c) $C_1 = 1$, $C_2 = -1$

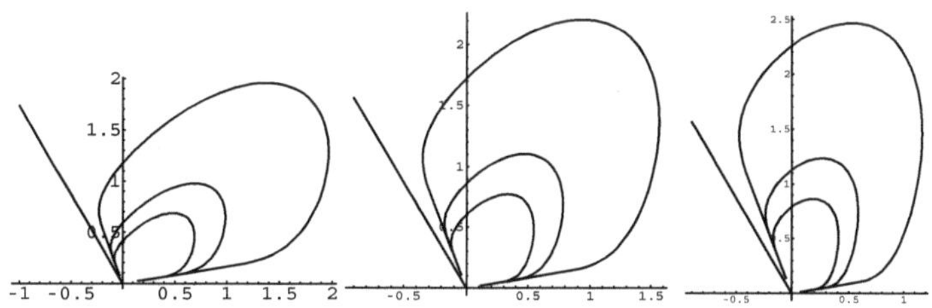

(a) (b) (c)

Figure 2.14: Finger dynamics in the wedge angle $2\pi/3$ and the bubble-wall angle $\pi/20$: (a) $C_1 = 1$, $C_2 = 0.9$; (b) $C_1 = 1$, $C_2 = 0$; (c) $C_1 = 1$, $C_2 = -1$

and we have

$$f(\zeta,t) = \sqrt{\frac{t}{\zeta}}\left(A(\zeta+1)^{1-\beta}(\zeta-1)^{\beta} + B(\zeta-1)^{1-\beta}(\zeta+1)^{\beta}\right), \tag{2.24}$$

where $\beta \in (0,1/2)$, $Q = 4AB(1-2\beta)\sin(\frac{\pi}{2}(1-2\beta))$, $A, B > 0$. We remark here that the map $f(\zeta,t)$ becomes non-univalent for other choices of A, B, β.

The function $f(\zeta,t)$ obviously satisfies the equations (2.20), (2.21). It maps D onto $\Omega(t)$ that is a complement of a bubble for any time t. The boundary Γ_3 starts and ends at the origin under the same bubble-wall angle $\pi\beta/2$, and forms a self-similar drop-shaped bubble. If $A = B$, then the bubble is symmetric with respect to the bisectrix of the corner (Figures 2.13(b), 2.14(b) and 2.15) and the solution is known [17], [328]. If $A \neq B$, then we have non-symmetric dynamics (Figures 2.13(a,c), 2.14(a,c), 2.16, 2.17). It is interesting that although the bubble-wall angles are the same, we have a two-parameter (A/B, β) continuum of possible developments of fingers. For angles greater than π the procedure is the same.

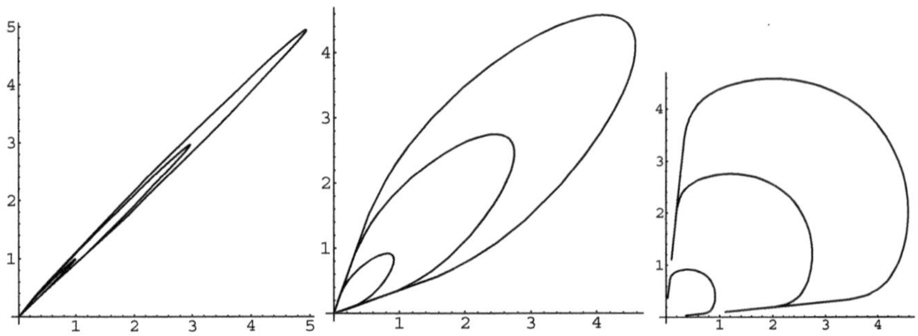

Figure 2.15: Finger dynamics: (a) $A = 1$, $B = 1$, $\beta = 0.16$; (b) $A = 1$, $B = 1$, $\beta = 0.1$; (c) $A = 1$, $B = 1$, $\beta = 0.05$

A corner of angle π implies other linearly independent solutions of the equation (2.22)

$$\begin{aligned} h_1(\zeta) &= \frac{1}{\zeta^2}\mathbf{F}\left(\beta, \beta+1, 2; \frac{1}{\zeta^2}\right), \\ h_2(\zeta) &= \frac{-2\log\zeta}{\zeta^2}\mathbf{F}\left(\beta, \beta+1, 2; \frac{1}{\zeta^2}\right) \\ &+ \sum_{k=1}^{\infty} \frac{\prod_{j=0}^{k-2}(\beta+j)^2(\beta-1)(\beta+k-1)}{\zeta^{2k+2}(k!)^2(k+1)} \left(2\left(\sum_{j=1}^{k-1}\frac{1}{\beta+j} - \sum_{j=2}^{k}\frac{1}{j}\right)\right. \\ &+ \left.\frac{1}{\beta} + \frac{1}{\beta+k} - 1 - \frac{1}{k+1}\right) - \frac{1}{\beta(\beta+1)}, \end{aligned}$$

that can be treated similarly.

Most of the results presented in this section are found in [229], [230].

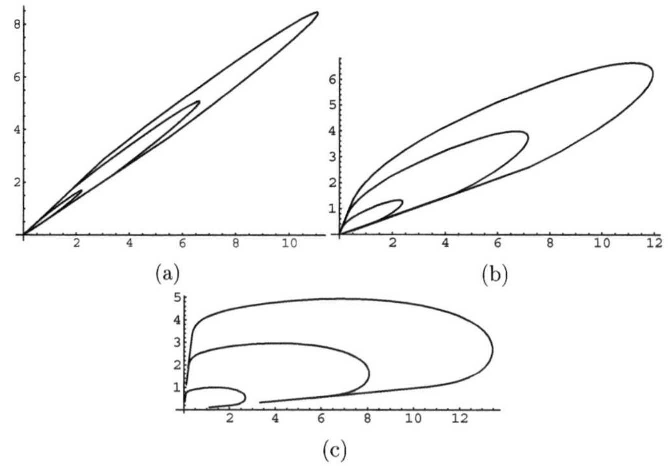

Figure 2.16: Finger dynamics: (a) $A = 1$, $B = 3$, $\beta = 0.16$; (b) $A = 1$, $B = 3$, $\beta = 0.1$; (c) $A = 1$, $B = 3$, $\beta = 0.05$

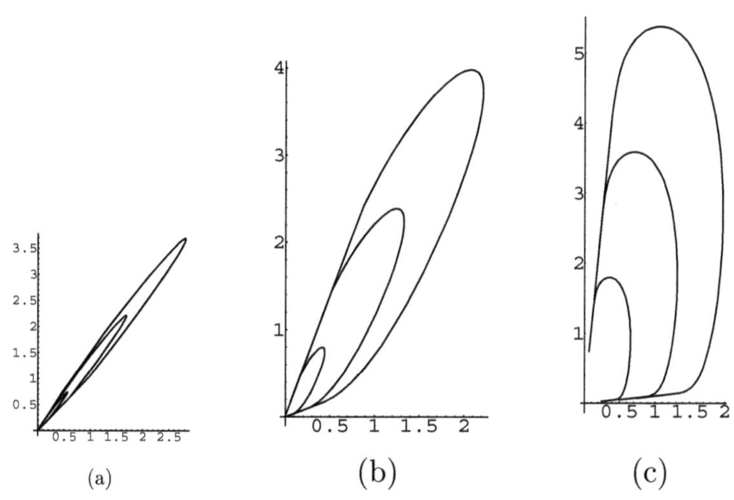

Figure 2.17: Finger dynamics: (a) $A = 1$, $B = 1/3$, $\beta = 0.16$; (b) $A = 1$, $B = 3$, $\beta = 0.1$; (c) $A = 1$, $B = 1$, $\beta = 0.05$

Chapter 3

Weak Solutions and Balayage

In the previous chapter we discussed strong solutions, which for their definition required smooth analytic boundaries. This section is devoted to *weak solutions* and their relations to potential theory.

3.1 Definition of weak solution

For the well-posed version ($Q < 0$) of the Hele-Shaw problem (without surface tension) there is a good notion of weak solution. It is based on the *Baiocchi transform*, replacing the pressure p in (1.12) by

$$u(z,t) = \int_0^t p(z,\tau)\,d\tau. \tag{3.1}$$

This type of transformation, with time t replaced by the vertical coordinate y, was used by C. Baiocchi in [20] to obtain a variational inequality formulation of the so-called dam problem. For the Hele-Shaw problem, weak or variational inequality formulations (in somewhat different disguises) were obtained around 1980 by Elliott, Janovský [102], Sakai [295], [298], Gustafsson [144]. See also [103].

To arrive at the concept of a weak solution, let $\Omega(t)$ be a strong solution of (1.12)–(1.14) with $Q < 0$, i.e., $\Omega(t)$ is a smooth family of domains such that (1.12)–(1.14) hold. For simplicity we take $Q = -1$, so that $p(\cdot,t)$ simply is Green's function of $\Omega(t)$ with the singularity

$$p(z,t) = -\frac{1}{2\pi}\log|z| + \text{harmonic}.$$

Using the Reynolds Transport Theorem and Green's formula we have

$$\frac{d}{dt}\iint\limits_{\Omega(t)} \varphi d\sigma_z = \int\limits_{\partial\Omega(t)} v_n \varphi ds = -\int\limits_{\partial\Omega(t)} \frac{\partial p}{\partial n}\varphi ds$$

$$= -\int_{\partial\Omega(t)} p\frac{\partial\varphi}{\partial n}ds - \iint_{\Omega(t)} \varphi\Delta p d\sigma_z + \iint_{\Omega(t)} p\Delta\varphi d\sigma_z \geq \varphi(0),$$

for any test function $\varphi \in C^\infty(\mathbb{C})$ which is subharmonic in $\Omega(t)$. Here we used that $-\Delta p = \delta_0$ and that $p \geq 0$.

We have already remarked that $\{\Omega(t)\}$ is a monotone increasing family. Integrating the above inequality from s to t, where $s < t$, gives

$$\iint_{\Omega(t)} \varphi d\sigma_z - \iint_{\Omega(s)} \varphi d\sigma_z \geq (t-s)\varphi(0),$$

for all φ which are subharmonic in $\Omega(t)$. In particular,

$$\iint_{\Omega(t)} \varphi d\sigma_z - \iint_{\Omega(0)} \varphi d\sigma_z \geq t\varphi(0), \tag{3.2}$$

which already is the weak formulation given by Sakai [295]. By approximation any integrable subharmonic function φ in $\Omega(t)$ is allowed in (3.2). Sakai shows that given $\Omega(0)$ and $t > 0$ there is a unique, up to null-sets, domain $\Omega(t)$ satisfying (3.2) for these φ. If φ is harmonic in $\Omega(t)$, then both $\pm\varphi$ are subharmonic so we get (3.2) with equality. Therefore, (3.2) contains (1.21) as a special case (with $Q = -1$).

To go further, we keep $t > 0$ fixed and define

$$u(z,t) = \iint_{\Omega(t)} \log|\zeta - z| d\sigma_\zeta - \iint_{\Omega(0)} \log|\zeta - z| d\sigma_\zeta - t\log|z| \tag{3.3}$$

for any $z \in \mathbb{C}$. Notice that this is the difference between the left and the right member in (3.2) with φ chosen to be

$$\varphi(\zeta) = \log|\zeta - z|.$$

Since this φ is integrable and subharmonic in $\Omega(t)$, (3.2) gives that

$$u \geq 0 \quad \text{everywhere.} \tag{3.4}$$

For z outside $\Omega(t)$ also $\varphi(\zeta) = -\log|\zeta - z|$ is subharmonic in $\Omega(t)$, so we obtain $u \leq 0$ outside $\Omega(t)$, hence

$$u(z,t) = 0 \quad \text{for } z \notin \Omega(t). \tag{3.5}$$

Finally, by definition (3.3), u satisfies

$$\chi_{\Omega(t)} = \chi_{\Omega(0)} + t\delta_0 + \Delta u \tag{3.6}$$

in the sense of distributions.

Equations (3.4)–(3.6) comprise the requirements we shall have on a weak solution. The function u is a kind of potential (indeed, it is the logarithmic potential of the measure $\chi_{\Omega(0)} - \chi_{\Omega(t)} + t\delta$) and it is uniquely determined by $\Omega(t)$, even at every point if the natural representative, given by (3.3), is chosen. Away from the origin Δu is a bounded function by (3.6), therefore u is continuously differentiable in the space variables. Since u attains its minimum ($u = 0$) on $\mathbb{C} \setminus \Omega(t)$, in particular on $\partial\Omega(t)$, it follows that also $\nabla u = 0$ there.

As an open set just satisfying (3.4)–(3.6), $\Omega(t)$ is not always uniquely determined. Indeed, equation (3.6) allows for arbitrary changes as to null-sets of $\Omega(t)$, whereas (3.5) is more sensitive. A point $z \in \Omega(t)$ may be removed from $\Omega(t)$ only if $u(z,t) = 0$, while points may be added as long as the area does not increase and $\Omega(t)$ remains an open set. As a matter of normalization, in order to make $\Omega(t)$ uniquely determined as a set, we shall usually take it to be *saturated* with respect to area measure. This means that we add to $\Omega(t)$ all discs $U_r(z)$ such that $U_r(z) \setminus \Omega(t)$ has area measure zero. This gives a domain which contains $\Omega(t)$, has the same area as $\Omega(t)$, and which cannot be further enlarged keeping these properties.

In summary we state the following definition.

Definition 3.1.1. *A weak solution of the Hele-Shaw problem* (1.12)–(1.14) *with* $Q = -1$ *is a family* $\{\Omega(t) : 0 \le t < t_0\}$ *of bounded open sets containing the origin such that there exists, for each* t, *a function* $u = u(z,t)$ *so that* (3.4)–(3.6) *hold.*

When u exists satisfying (3.4)–(3.6) it is given by (3.3), because being $= 0$ in a neighbourhood of infinity it is the logarithmic potential of $(-\nabla u)$.

It is clear from the above derivation that a strong solution (if it exists) is always a weak solution. In the case a weak solution is derived from a strong solution we may differentiate (3.6) with respect to t to obtain that $\delta_0 + \Delta \frac{\partial u}{\partial t} = 0$ in $\Omega(t)$. Using that $u = |\nabla u| = 0$ on $\partial\Omega(t)$, showing that u grows only quadratically near $\partial\Omega(t)$, it also follows that $\frac{\partial u}{\partial t} = 0$ on $\partial\Omega(t)$. Thus, $\frac{\partial u}{\partial t} = p$ (in (1.12)), so the function u is indeed the Baiocchi transform of p. Note that $u(z,0) = 0$.

There are other types of solutions to Hele-Shaw problems. For example, Crandall and Lions [62] introduced the notion of *viscosity solutions* which was successfully used to study nonlinear elliptic and parabolic equations. Caffarelli and Vázquez [50] proved the existence and uniqueness of the viscosity solution for the porous medium equation, and later, Kim [194] adapted this notion for the Hele-Shaw problem.

3.2 Existence and uniqueness of weak solutions

For weak solutions we have the following remarkably good existence theorem.

Theorem 3.2.1. *Given any bounded open set Ω_0 there exists a unique weak solution $\{\Omega(t) : 0 \le t < \infty\}$ with $\Omega(0) = \Omega_0$ (uniqueness in the strict sense only if the $\Omega(t)$ are required to be saturated).*

Proof. Let $t > 0$ be fixed. In order to construct $\Omega(t)$ we shall relax (3.4)–(3.6) further, to

$$u \ge 0, \tag{3.7}$$

$$\chi_{\Omega(0)} + t\delta_0 + \Delta u \le 1, \tag{3.8}$$

$$\iint_{\mathbb{C}} u(1 - \chi_{\Omega(0)} - t\delta_0 - \Delta u) d\sigma_z = 0. \tag{3.9}$$

Here $\Omega(t)$ has disappeared from the formulation, but u remains and $\Omega(t)$ can finally be recovered. The system (3.7)–(3.9) is sometimes called a *linear complementarity problem* because it states that two linear inequalities must hold and that at each point there shall be equality in at least one of them.

There are several ways of constructing the solution of (3.7)–(3.9). The most direct way of obtaining u is simply to say that u shall be the smallest among all functions satisfying (3.7), (3.8) alone. This function will satisfy (3.9) as well.

To see that such a smallest function exists we choose a function ψ satisfying

$$\Delta\psi = \chi_{\Omega(0)} + t\delta_0 - 1 \quad \text{in } \mathbb{C},$$

for example,

$$\psi(z) = -\frac{1}{2\pi} \iint_{\Omega(0)} \log|\zeta - z| d\sigma_\zeta - \frac{t}{2\pi} \log|z| - \frac{1}{4}|z|^2, \tag{3.10}$$

and set

$$v = u + \psi. \tag{3.11}$$

Then v is to be the smallest among all functions satisfying

$$\begin{cases} v & \ge & \psi, \\ -\Delta v & \ge & 0. \end{cases}$$

We can think of ψ as an obstacle function, and the problem becomes that of finding the smallest superharmonic function v passing the obstacle. It is well-known from general potential theory (see, e.g., [16], [278]) that such a v exists. It is the lower semicontinuous regularization of the pointwise infimum of all superharmonic functions $\ge \psi$. Superharmonic functions are usually normalized to be lower semicontinuous.

Thus v, and hence u, as above exists. Now we have to show that u satisfies (3.9). For this we continue to work with v. Suppose we have the strict inequality $v(z) > \psi(z)$ at some point z. Then since v is lower semicontinuous and ψ is continuous (outside the origin) there is an $\varepsilon > 0$ and a disk $U_r(z) = \{w : |w-z| < r\}$, $(r > 0)$, such that $v \geq \psi+\varepsilon$ in all $U_r(z)$. Therefore, if v is not already harmonic in $U_r(z)$, then it can be made smaller by making a Poisson modification of it (i.e., by replacing v in $U_r(z)$ by the harmonic function with the same boundary values on $U_r(z)$). If the radius $r > 0$ is sufficiently small, then the modified v will be $\geq \psi$ in $U_r(z)$.

From this we realize that $\Delta v = 0$ in the open set $\{v > \psi\}$, or that

$$\Delta u = \Delta\psi = \chi_{\Omega(0)} + t\delta_0 - 1,$$

in $\{u > 0\}$. This proves (3.9).

Thus we have produced a solution u of (3.7)–(3.9). One easily sees that $u = 0$ outside some large disk U_R. In fact, by comparing with an expanding disk solution one sees that if $\Omega(0) \subset U_{R_0}$, then any $R > 0$ with $\pi R_0^2 + t < \pi R^2$ will do. Thus R depends on t.

To prove that u is unique and to obtain some further properties of it, let us indicate two other ways of constructing u, or v. By minimizing Dirichlet integrals (measuring energies) of u and v, keeping one of the inequalities (3.7) or (3.8) as a side condition, these two problems are:

(i) Minimize $\int_{U_R} |\nabla u|^2 d\sigma_z$ among all u vanishing on ∂U_R and satisfying

$$\Delta u + \chi_{\Omega(0)} + t\delta_0 \leq 1;$$

(ii) Minimize $\int_{U_R} |\nabla v|^2 d\sigma_z$ among all functions v which agree with ψ outside U_R and satisfy $v \geq \psi$ everywhere.

In order to have finite integrals above one should, for the first integral, smooth out δ_0 a little (e.g., replace δ_0 by $\tilde{\delta} = \frac{1}{|U_\varepsilon|}\delta_{U_\varepsilon}$ for some small $\varepsilon > 0$). The proper settings then are that one works in the Sobolev space $H^1(U_R)$ (or $H_0^1(U_R)$ in the case of u). Both problems (i) and (ii) then have unique solutions which can be characterized by their variational formulations (variational inequalities).

The variational formulation for (ii) is

$$\iint_{U_R} \nabla v \, \nabla(w - v) d\sigma_z \geq 0,$$

to hold for all $w \geq \psi$ having the same boundary values as ψ on ∂U_R, or using Stokes' theorem

$$\iint_{U_R} \Delta v \, (w - v) d\sigma_z \leq 0,$$

for all w as above. Since any $w \geq v$ is allowed here, we get $\Delta v \leq 0$. Choosing then $w = \psi$ gives

$$\iint_{U_R} \Delta v\,(v - \psi) d\sigma_z \geq 0,$$

hence actually,

$$\iint_{U_R} \Delta v\,(v - \psi) d\sigma_z = 0.$$

Thus, in the two inequalities $\Delta v \leq 0$ and $v \geq \psi$ there is nowhere strict inequalities in both. The problem (i) is treated similarly. In both cases we find that u and v satisfy linear complementary problems, which expressed in terms of u are (3.7)–(3.9). Conversely, one can go backward in the above reasoning, so the complementary problem (3.7)–(3.9) is equivalent to the two variational inequalities and the two minimum problems. In particular, it follows that the solution u of (3.7)–(3.9) is unique and that it has finite Dirichlet integral $\iint |\nabla u|^2 d\sigma_z$ (leaving out a neighbourhood of the origin).

Next invoking general regularity theory for the obstacle problem [115], [196], it follows that u actually is in the higher order Sobolev space $H^{2,p}(U_R)$ for any $p < \infty$. Now define $\Omega(t)$ to be the largest open set in which

$$\Delta u + \chi_{\Omega(0)} + t\delta_0 = 1.$$

In other words, $\Omega(t)$ is the complement of the closed support of the distribution $1 - \Delta u - \chi_{\Omega(0)} - t\delta_0$. By (3.9), $u = 0$ outside $\Omega(t)$. It is known [196] that this implies that $\Delta u = 0$ almost everywhere outside $\Omega(t)$ (when $u \in H^{2,p}$). Therefore,

$$\Delta u + \chi_{\Omega(0)} + t\delta_0 = \chi_{\Omega(0)} + t\delta_0 < 1$$

almost everywhere outside $\Omega(t)$, hence actually $\Delta u + \chi_{\Omega(0)} + t\delta_0 = 0$ there (because $\chi_{\Omega(0)} + t\delta_0 \geq 1$, where it is not zero). Finally, we conclude that

$$\Delta u + \chi_{\Omega(0)} + t\delta_0 = \chi_{\Omega(t)},$$

which means that we have established all properties of a weak solution. Note also that $\Omega(t)$ was defined to be saturated. □

3.3 General properties of weak solutions

It is clear from the way the concept of a weak solution was defined that a strong solution always is a weak solution. This guarantees the uniqueness of a strong solution by the uniqueness of the weak one. However, a weak solution need not be a strong one, e.g., because a weak solution may change topology, a possibility which is not allowed even in the concept of a strong solution.

A remarkable property of the weak solution is that the time variable t only occurs as a parameter in it. No derivative with respect to t occurs and one may jump to compute $\Omega(t)$ for only $t > 0$ directly, without computing it for any smaller values of t.

The following proposition shows that a weak solution has the monotonicity and semigroup properties one expects.

Proposition 3.3.1. *Let $\{\Omega(t)\}$ and $\{\tilde{\Omega}(t)\}$ be weak solutions with the initial domains, or just open sets, $\{\Omega(0)\}$ and $\{\tilde{\Omega}(0)\}$ respectively, and let $u = u_t$ and $\tilde{u} = \tilde{u}_t$ be the corresponding potentials. Then,*

(a) *if $\Omega(0)$ is connected, then so is $\Omega(t)$ for any $t > 0$;*

(b) *if $0 < s < t$, then $u_s \leq u_t$ and $\Omega(s) \subset \Omega(t)$;*

(d) *if $\Omega(0) \subset \tilde{\Omega}(0)$, then $\Omega(t) \subset \tilde{\Omega}(t)$ for all $t > 0$;*

(e) *if $\tilde{\Omega}(0) = \Omega(s)$ for some $s > 0$, then $\tilde{\Omega}(t) = \Omega(t+s)$ for all $t > 0$.*

(e) *Inequality (3.2) holds for all integrable subharmonic functions φ in $\Omega(t)$. In particular, (taking $\varphi = \pm 1$) we have $|\Omega(t)| = |\Omega(0)| + t$.*

Proof. (a) Let D be a connected component of $\Omega(t)$. Then $u = 0$ on ∂D. If D does not meet $\Omega(0)$, then $\Delta u = 1$ in D, which in view of the maximum principle contradicts $u \geq 0$. Thus every component of $\Omega(t)$ intersects $\Omega(0)$.

(b) u_s is the smallest function satisfying $u_s \geq 0$, $\chi_{\Omega(0)} + s\delta_0 + \Delta u_s \leq 1$, and similarly for u_t. Since $\chi_{\Omega(0)} + s\delta_0 \leq \chi_{\Omega(0)} + t\delta_0$, it follows immediately that $u_s \leq u_t$. Outside $\Omega(t)$ we have $u_t = 0$, hence also $u_s = 0$, and so

$$\begin{aligned}
\chi_{\Omega(s)} &= \chi_{\Omega(0)} + s\delta_0 + \Delta u_s \\
&= \chi_{\Omega(0)} + s\delta_0 + \Delta u_t \\
&\leq \chi_{\Omega(0)} + t\delta_0 + \Delta u_t \\
&= \chi_{\Omega(t)} = 0
\end{aligned}$$

there. Thus $\Omega(s) \subset \Omega(t)$.

(c) This is proved the same way as (b).

(d) Fix $s > 0$, $t > 0$, and set $w = u_{s+t} - u_s$. Then

$$\Delta w = \chi_{\Omega(s+t)} - \chi_{\Omega(s)} - t\delta_0.$$

By (b) $w \geq 0$ and $w = 0$ outside $\Omega(s+t)$. Thus $\{\Omega(s+t),\ 0 \leq t < \infty\}$ is the weak solution with the initial domain $\Omega(s)$, which is exactly what is stated in (d).

(e) By (b) $\Omega(0) \subset \Omega(t)$ so (3.2) makes sense. The definition of a weak solution amounts exact to (3.2) holding for all φ of the form $\varphi(\zeta) = \log|\zeta - z|$ for $z \in \mathbb{C}$, and $\varphi(\zeta) = -\log|\zeta - z|$ for $z \notin \Omega(t)$ and it is known [295] that the positive linear combinations of these are dense in the set of integrable subharmonic functions in $\Omega(t)$. Now (e) follows.

□

3.4 Regularity of the boundary

Let $\{\Omega(t) : 0 \le t < \infty\}$ be a weak solution. Then $\Omega(s) \subset \Omega(t)$ for all $0 \le s < t$, but it is not always true that $\overline{\Omega(0)} \subset \Omega(t)$. If we, for example, choose the initial domain $\Omega(0)$ such that $\partial\Omega(0)$ is of positive area measure, then it will take $\partial\Omega(t)$ a finite time to move through $\partial\Omega(0)$. Even if $\Omega(0)$ has piecewise smooth boundary containing a corner at z_0 with the interior angle smaller than $\pi/2$, it is known [197], [198], [302], that $\partial\Omega(t)$ stays at the corner for some positive time.

On the other hand, if $\Omega(0)$ is C^1-smooth (not only piecewise), then $\overline{\Omega(0)} \subset \Omega(t)$. The following regularity theorem is mainly due to Sakai [296], [297], and shows that the situation outside $\overline{\Omega(0)}$ is rather pleasant. The last statement is taken from [151].

Theorem 3.4.1. *Assume $\overline{\Omega(0)} \subset \Omega(t)$. Then $\partial\Omega(t)$ consists of finitely many analytic curves which may have finitely many singularities in the form of inward cusps or double points, but no other singularities. In case the $\Omega(t)$ are simply connected, the Riemann map $f(\zeta, t)$ parameterizing the phase domain $\Omega(t)$ extends analytically to a disk $U_{R(t)}$ where the radius of analyticity $R(t) > 1$ is non-decreasing as a function of t.*

Remark 3.4.2. *It is important that $\Omega(t)$ is saturated, otherwise the formulation becomes more complicated. The most difficult part of the proof is actually to show that $\Omega(t)$ is finitely connected. We shall not take this difficulty here, but just prove the theorem in the case $\Omega(t)$ is finitely connected, say simply connected.*

Proof. So assume $\Omega(t)$ is simply connected and let $f : U \to \Omega(t)$ be the Riemann map, $z = f(\zeta, t)$. Using the potential u in (3.4)–(3.6) we can define a one-sided Schwarz function, defined in $\overline{\Omega(t)} \setminus \Omega(0)$, by

$$S(z,t) = \bar{z} - 4\frac{\partial u}{\partial z}.$$

We see immediately from (3.6) that $S(z,t)$ is analytic in $\Omega(t) \setminus \Omega(0)$. Since u is continuously differentiable away from the origin, $u \ge 0$ attains its minimum on $\mathbb{C} \setminus \Omega(t)$, and $|\nabla u| = 0$ there, $S(z,t)$ is continuous up to $\partial\Omega(t)$ with $S(z,t) = \bar{z}$ on $\partial\Omega(t)$.

The conjugate of $S(z)$ can be interpreted as the anticonformal reflection in $\partial\Omega(t)$ and we use it to extend f in the following way. We extend the function f by

$$f(1/\bar{\zeta}, t) = \overline{S(f(\zeta, t))},$$

for those $\zeta \in U$ for which $f(\zeta, t) \in \Omega(t) \setminus \overline{\Omega(0)}$. This defines f analytically in U and in an annulus $1 < |\zeta| < R(t)$. Here we take $R(t) > 1$ largest possible, which means that $R(t) = 1/r(t)$ where $0 < r(t) < 1$ is the smallest radius such that $f^{-1}(\Omega(0), t) \subset U_{r(t)}$. Across ∂U we have a certain form of continuity because of the continuity of $S(z,t)$. Indeed, as $|\zeta| \to 1$ with $\zeta \in U$ we have

$$|f(\zeta,t) - f(1/\bar{\zeta},t)| = |f(\zeta,t) - \overline{S(f(\zeta,t),t)}| \to 0, \tag{3.12}$$

where $z = \overline{S(z,t)}$ on $\partial\Omega(t)$, and therefore, given $\varepsilon > 0$, we have $|z - \overline{S(z,t)}| < \varepsilon$ for $z \in \Omega(t)$ in some neighbourhood of $\partial\Omega(t)$. By now the function $f(\zeta,t)$ is defined in U as well as in the annulus $1 < |\zeta| < R(t)$, hence, almost everywhere in the disk $|\zeta| < R(t)$. Let us prove that the distributional derivative $\partial f(\zeta,t)/\partial\bar{\zeta}$ vanishes in $|\zeta| < R(t)$ using (3.12). Obviously, we must verify this across the circle $|\zeta| = 1$. Given a test function φ with compact support in $|\zeta| < R(t)$ we have

$$\begin{aligned}
\langle \frac{\partial f}{\partial\bar{\zeta}}, \varphi\rangle &= -\iint\limits_{\mathbb{C}} f(\zeta,t)\frac{\partial\varphi}{\partial\bar{\zeta}} d\sigma_\zeta \\
&= -\frac{1}{2i}\iint\limits_{U} f(\zeta,t)\frac{\partial\varphi}{\partial\bar{\zeta}} d\bar{\zeta}d\zeta - \frac{1}{2i}\iint\limits_{|\zeta|>1} f(\zeta,t)\frac{\partial\varphi}{\partial\bar{\zeta}} d\bar{\zeta}d\zeta \\
&= -\frac{1}{2i}\lim_{\varepsilon\downarrow 0}\left(\int\limits_{|\zeta|=1-\varepsilon} f(\zeta,t)\varphi(\zeta)\,d\zeta - \int\limits_{|\zeta|=1+\varepsilon} f(\zeta,t)\varphi(\zeta)\,d\zeta\right) \\
&= -\frac{1}{2i}\lim_{\varepsilon\downarrow 0}\int\limits_{|\zeta|=1-\varepsilon} (f(\zeta,t) - f(1/\bar{\zeta},t))\varphi(\zeta)\,d\zeta = 0.
\end{aligned}$$

In the above curve integrals we take the counterclockwise direction on the circles. Thus, the function $f(\zeta,t)$ is analytic in the disk $|\zeta| < 1/r(t)$.

For any pair of numbers s,t such that $0 < s < t \le T$, we have that the function $h(\zeta,s,t) \equiv f^{-1}(f(\zeta,s),t)$ maps the unit disk into itself and $h(0,s,t) \equiv 0$. A simple application of the Schwarz Lemma to the function h shows that

$$f^{-1}(\Omega(0),t) \subset U_{r(s)}.$$

Therefore, $r(t) \le r(s)$, hence $R(s) \le R(t)$.

We have $f(\partial U,t) = \partial\Omega(t)$ as sets, f is univalent in U but need not be univalent on $\bar{U}$. Therefore, $\partial\Omega(t)$ is analytic with possible singularities as stated. □

Remark 3.4.3. *More generally, the arguments of the proof work for any isolated component of $\partial\Omega(t)$. A different approach to the regularity of $\partial\Omega(t)$ was given in* [147]*. There it was shown that the analytic continuation of a certain exponential transform directly gives a real analytic defining function for the boundary.*

3.5 Balayage point of view

At this point it may be apparent that in the treatment of weak solutions the expression $\chi_{\Omega(0)} + t\delta_0$ always appears as one quantity. The weak solution itself is the family $\{\Omega(t)\}$, or better $\{\chi_{\Omega(t)}\}$. Moreover, time t only plays the role of a parameter, and for any fixed $t > 0$ the whole construction really amounted to the construction of a map

$$\chi_{\Omega(0)} + t\delta_0 \mapsto \chi_{\Omega(t)}.$$

This map finally came out to be just the addition of the term Δu, where u solves the complementary problem (3.7)–(3.9).

For a further systematic treatment of weak solutions it is really advantageous to take this operator theoretic point of view. Everything looks more natural if we replace the number 1 in the right member of (3.8) by a more general function, say $\rho \geq 0$. It will have the interpretation of a density. We shall use letters μ or similar for what used to be $\chi_{\Omega(0)} + t\delta_0$ (also a density, or a measure denoted as a density).

Assume that $\mu \geq 0$ is a measure with compact support and that

$$0 < c_1 \leq \rho \leq c_2 < \infty.$$

Then we define

$$\mathrm{Bal}\,(\mu, \rho) = \mu + \Delta u,$$

where u is the smallest function satisfying

$$u \geq 0 \quad \text{in } \mathbb{C}, \tag{3.13}$$

$$\mu + \Delta u \leq \rho \quad \text{in } \mathbb{C}. \tag{3.14}$$

Such a function exists as before. We also define a corresponding saturated set Ω (which will be bounded),

$$\Omega = \{\text{the largest open set in which } \mu + \Delta u = \rho\}. \tag{3.15}$$

Then the complementary condition

$$\{u > 0\} \subset \Omega \tag{3.16}$$

holds.

The interpretation of Bal is that it performs a kind of balayage[1] – *partial balayage*. Indeed, let $\nu = \mathrm{Bal}\,(\mu, \rho)$, and let

$$U^{\mu}(z) = -\frac{1}{2\pi} \iint \log |z - \zeta| \, d\mu(\zeta)$$

[1]The idea of balayage (sweeping in English translation) goes back [120] to Johann Carl Friedrich Gauss, 1777–1855, see [84] for historical accounts in general. The method of successive sweeping operations was invented by Poincaré [259], [260] for solving the Dirichlet problem.

be the logarithmic potential of μ, and similarly for other measures. Then since $\nu = \mu + \Delta u$ and $u = 0$ outside Ω (in particular, in a neighbourhood of infinity) it follows that

$$u = U^\mu - U^\nu.$$

The vanishing of u outside Ω, therefore, means that μ and ν are graviequivalent in a certain sense, which explains the word "balayage" (sweeping of measures without changing exterior potentials). For details we refer to [146], [154], see also [86].

In view also of the minimization problem (i) in the proof of Theorem 3.2.1, what the operator $\mu \mapsto \nu = \mathrm{Bal}\,(\mu, \rho)$ does, is that it replaces any measure μ by a measure ν satisfying $\nu \leq \rho$ (everything denoted as densities) using as little work

$$\iint |\nabla u|^2 d\sigma_z = \text{energy of } \nu - \mu$$

as possible. The result of the whole thing is a measure ν and an open set Ω, such that

$$U^\nu = U^\mu \quad \text{outside } \Omega,$$

$$\nu = \rho \quad \text{in } \Omega$$

(by (3.16) and (3.15) respectively). Thus ν has the desired potential U^μ outside Ω and the desired density ρ in Ω.

In terms of $\mathrm{Bal}\,(\mu, \rho)$, a weak solution of the Hele-Shaw problem now is just a family of open sets $\{\Omega(t) : 0 \leq t < \infty\}$ satisfying

$$\mathrm{Bal}\,(\chi_{\Omega(0)} + t\delta_0, 1) = \chi_{\Omega(t)}.$$

By (d) of Proposition 3.3.1 we also have, for arbitrary $s < t$,

$$\mathrm{Bal}\,(\chi_{\Omega(s)} + (t-s)\delta_0, 1) = \chi_{\Omega(t)}.$$

This is then an instance of a general property of Bal, namely that

$$\mathrm{Bal}\,(\mu_1 + \mu_2, \rho) = \mathrm{Bal}\,(\mathrm{Bal}\,(\mu_1, \rho) + \mu_2, \rho).$$

Take $\rho = 1$, $\mu_1 = \chi_{\Omega(0)} + t\delta_0$, $\mu_2 = (t-s)\delta_0$ to get the previous statement. A more general statement is also true:

$$\mathrm{Bal}\,(\mu_1 + \mu_2, \rho_1) = \mathrm{Bal}\,(\mathrm{Bal}\,(\mu_1, \rho_2) + \mu_2, \rho_1), \tag{3.17}$$

whenever $\rho_1 \leq \rho_2 + \mu_2$.

Similarly, parts (b) and (c) in Proposition 3.3.1 are special cases of the implication

$$\mu_1 \leq \mu_2 \Rightarrow \mathrm{Bal}\,(\mu_1, \rho) \leq \mathrm{Bal}\,(\mu_2, \rho). \tag{3.18}$$

Given μ, taking here $\mu_1 = \min\,(\rho, \mu)$, $\mu_2 = \mu$, gives the lower bound in the estimate

$$\min\,(\rho, \mu) \leq \mathrm{Bal}\,(\mu, \rho) \leq \rho, \tag{3.19}$$

because $\mathrm{Bal}(\min(\rho,\mu),\rho) = \min(\rho,\mu)$. The upper bound is just by definition.

The inequality (3.19) can be viewed as a regularity statement for the functions u and v in (i), (ii) (in the proof of Theorem 3.2.1). With, for example, $\rho = 1$, $\mu = \chi_{\Omega(0)} + t\delta_0$, we get, for u, that $0 \leq \Delta u \leq 1$ away from the origin, which gives the previously used regularity $u \in H^{2,p}$ for all $p < \infty$.

The general structure of $\mathrm{Bal}(\mu,\rho)$ is

$$\mathrm{Bal}(\mu,\rho) = \rho\chi_\Omega + \mu\chi_{\mathbb{C}\setminus\Omega}. \tag{3.20}$$

Indeed, (3.20) is true in Ω by definition (3.15) of Ω, and outside Ω we have $u = 0$, hence $\Delta u = 0$ there, at least under some regularity assumptions (e.g., $\mu \in L^\infty$); the general case can be handled by an approximation argument [312]. Thus $\mathrm{Bal}(\mu,\rho) = \mu$ outside Ω.

In the special case of Hele-Shaw dynamics ($\rho = 1$, $\mu = \chi_{\Omega(0)} + t\delta_0$) we have $\mu \geq \rho$ everywhere where μ does not vanish. This guarantees

$$\mathrm{Bal}(\mu,\rho) = \rho\chi_\Omega,$$

as is immediate from (3.20) together with the definition (3.15) of Ω. For any kind of injection Hele-Shaw problem, if the sources are located within the initial domain $\Omega(0)$, and if the accumulated sources up to time $t > 0$ are represented by the measure $\mu(t) \geq 0$, then the weak solution $\Omega(t)$ is given by

$$\mathrm{Bal}(\chi_{\Omega(0)} + \mu, 1) = \chi_{\Omega(t)}.$$

If there are sources outside $\Omega(0)$ and these are sufficiently weak (meaning that $\mu(t) < 1$ outside $\Omega(0)$ and for some time $t > 0$), then there will also be the second term in the right member of (3.20),

$$\mathrm{Bal}(\chi_{\Omega(0)} + \mu, 1) = \chi_{\Omega(t)} + \mu\chi_{\mathbb{C}\setminus\Omega(t)},$$

corresponding to some kind of "mushy region".

As a useful application of (3.17) we have the following. Given $t > 0$ choose $r > 0$ so small that $\pi r^2 < t$ and let

$$\tilde{\delta} = \frac{1}{|U_r|}\chi_{U_r} = \mathrm{Bal}(\delta_0, \frac{1}{|U_r|})$$

be the Dirac measure swept out to a uniform density on U_r. Then also

$$t\tilde{\delta} = \mathrm{Bal}(t\delta_0, \frac{t}{|U_r|}),$$

for any $t > 0$. Since $\frac{t}{|U_r|} > 1$, (3.17) with $\rho_1 = 1$, $\rho_2 = \frac{t}{|U_r|}$ shows that

$$\begin{aligned}\mathrm{Bal}(t\delta_0 + \chi_{\Omega_0}, 1) &= \mathrm{Bal}(\mathrm{Bal}(t\delta_0, \frac{t}{|U_r|}) + \chi_{\Omega_0}, 1)\\ &= \mathrm{Bal}(t\tilde{\delta} + \chi_{\Omega_0}, 1),\end{aligned}$$

i.e., the Hele-Shaw evolutions with δ_0 and $\tilde{\delta}$ are exactly the same.

3.6 Existence and non-branching backward of weak solutions

In this section we discuss the existence and uniqueness of weak and strong solutions backward in time. For the strong case Tian [329], [330] proved the local backward existence, uniqueness for an analytic smooth initial boundary, and the fact that if the initial boundary is not analytic (but still smooth e.g.), then the backward strong solution will not exist. As to existence of a backward weak solution $\{\Omega(t)\}$ for some interval $-\varepsilon < t < 0$, satisfying $\overline{\Omega(t)} \subset \Omega(0)$ we necessarily need the analyticity of the initial boundary.

Theorem 3.6.1. *Assume that $\Omega(0)$ has a smooth analytic boundary and contains the origin. Then there exists, for some $\varepsilon > 0$, a weak solution $\{\Omega(t)\}$, $-\varepsilon < t < 0$, with $\overline{\Omega(t)} \subset \Omega(0)$, such that*

$$\operatorname{Bal}(\chi_{\Omega(s)} + (t-s)\delta_0, 1) = \chi_{\Omega(t)}$$

holds for any $s < t$, and in particular,

$$\operatorname{Bal}(\chi_{\Omega(t)} - t\delta_0, 1) = \chi_{\Omega(0)}.$$

Proof. We first recall that δ_0 may be replaced by a smoothed out version of it, say

$$\tilde{\delta} = \frac{1}{|U_r|}\chi_{U_r},$$

where $r > 0$ is so small that $U_r \subset \overline{U_r} \subset \Omega(0)$.

Next we construct a domain D satisfying

$$U_r \subset D \subset \bar{D} \subset \Omega(0),$$

and a measure μ on $\bar{D}$ which satisfies $d\mu = (1+\beta)d\sigma_z$ on $\bar{D}$ for some $\beta > 0$, $\mu = 0$ outside $\bar{D}$, and for which $\operatorname{Bal}(\mu, 1) = \chi_{\Omega(0)}$. This is done as follows (we just outline the construction, more details can be found in [145]). Using the Cauchy-Kovalevskaya theorem we solve the Cauchy problem

$$\begin{cases} \Delta u &= 1, \text{ in } \Omega(0), \text{ near } \partial\Omega(0), \\ u &= 0, \text{ on } \partial\Omega(0), \\ \nabla u &= 0, \text{ on } \partial\Omega(0), \end{cases}$$

in some neighbourhood of $\partial\Omega(0)$ in $\Omega(0)$. This requires the analyticity of $\partial\Omega(0)$. The solution can analytically be gotten directly from the Schwarz function $S(z)$ of $\partial\Omega(0)$ as

$$u(z) = \frac{1}{2}\operatorname{Re}\int (S(z) - \bar{z})dz,$$

where the integration is performed from any point on $\partial\Omega(0)$ (and to the point z in $u(z)$).

The function $u(z)$ will grow quadratically with the distance from $\partial\Omega(0)$

$$u(z) \sim \frac{1}{2}\text{dist}^2(z, \partial\Omega(0))$$

and we take ∂D to be a level set $u = \alpha$ for u, with $\alpha > 0$ so small that the normal derivative in the direction out from D satisfies

$$\frac{\partial u}{\partial n} \leq c_1 < 0.$$

Then in D we take u to solve the Dirichlet problem

$$\begin{cases} \Delta u &= -\beta, \quad \text{in } D, \\ u &= \alpha, \quad \text{on } \partial D, \end{cases}$$

with $\beta > 0$ sufficiently small, so that the outward normal derivative of $u|_D$ satisfies

$$c_1 < c_2 \leq \frac{\partial u}{\partial n} < 0.$$

Extending u by zero outside $\Omega(0)$ we have

$$\Delta u = \chi_{\Omega(0)} - \mu,$$

where μ is a positive measure on $\bar{D}$, which has density $1 + \beta$ in D and also has a contribution on ∂D corresponding to the jump of the normal derivative of u.

Since $u \geq 0$, and $u = 0$ outside $\Omega(0)$, we have $\text{Bal}\,(\mu, 1) = \chi_{\Omega(0)}$. Now take $\varepsilon = \pi r^2 \beta = |U_r|\beta$. Then for $-\varepsilon \leq t < 0$ we still have $\mu + t\tilde{\delta} \geq 1$ in D. Therefore,

$$\text{Bal}\,(\mu + t\tilde{\delta}, 1) = \chi_{\Omega(t)}, \tag{3.21}$$

for some domains $\Omega(t)$, $-\varepsilon \leq t < 0$, and this will be the weak solution.

One may notice, using (3.17), that

$$\begin{aligned} \text{Bal}\,(\mu + t\tilde{\delta}, 1) &= \text{Bal}\,(\mu - \varepsilon\tilde{\delta} + (t + \varepsilon)\tilde{\delta}, 1) \\ &= \text{Bal}\,(\text{Bal}\,(\mu - \varepsilon\tilde{\delta}, 1) + (t + \varepsilon)\tilde{\delta}, 1) \\ &= \text{Bal}\,(\chi_{\Omega(-\varepsilon)} + (t + \varepsilon)\tilde{\delta}, 1), \end{aligned}$$

so the family $\chi_{\Omega(t)}$ really is an ordinary weak solution started with $\Omega(-\varepsilon)$. By construction of μ, taking $t = 0$ in (3.21) gives the given initial domain $\Omega(0)$, and for $t > 0$ one gets the usual forward solution. Thus (3.21) defines a weak solution on all $-\varepsilon \leq t < \infty$. □

Remark 3.6.2. *By the formula* (3.21) *we have, for t in a small interval around $t = 0$, the solution $\{\Omega(t)\}$ represented by smooth perturbations of a measure μ, sitting compactly in $\Omega(0)$. Moreover, $\partial\Omega(0)$ is smooth real analytic. It is known* [47], [304], *that the solution $\{\Omega(t)\}$ in such a case also will vary smoothly in t. Therefore, the solution obtained will in fact be a strong solution for the Hele-Shaw problem.*

A weak solution can branch at any time in the backward direction. This occurs when a simply connected domain $G(t_0) = \Omega(t_0)$ for some $0 < t_0 < \infty$ appears as a result of a strong simply connected dynamics $\Omega(t)$ and at the same time as a result of a weak dynamics $G(t)$, where $G(t)$ for $t < t_0$ is multiply connected with some holes to be filled in as $t \to t_0^-$, see Figure 3.1. Our next result says that branching can take place only when such changes of topology occur, see [152].

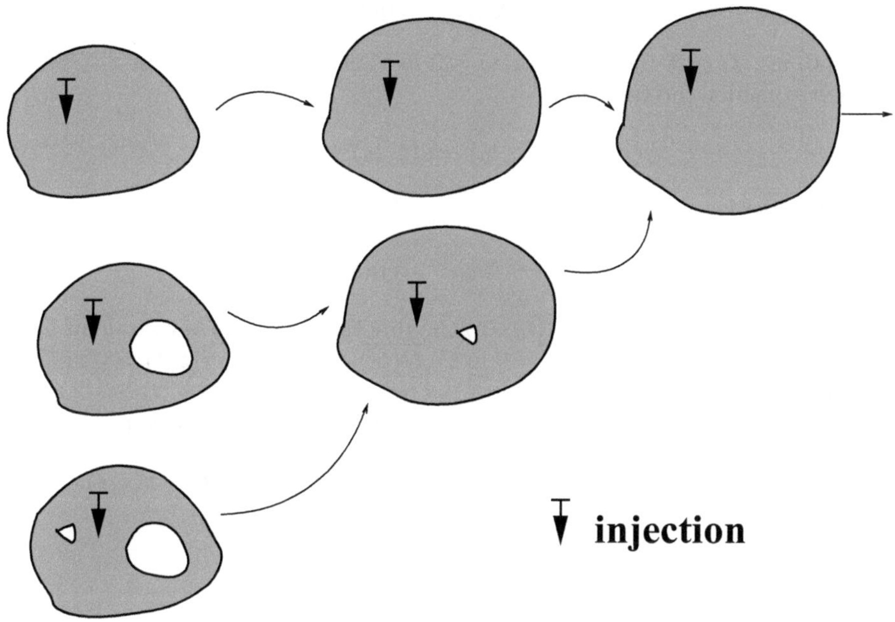

Figure 3.1: Branching weak solutions

Theorem 3.6.3. *Let $G(0)$ and $H(0)$ be two initial domains in $\mathbb{C}$, and $G(t)$ and $H(t)$ be the corresponding weak solutions, $0 \le t < \infty$. We assume that*

- $\overline{G(0)} \subset G(t)$ *and* $\overline{H(0)} \subset H(t)$ *for all* $t > 0$;
- $\mathbb{C} \setminus \overline{G(t)}$ *and* $\mathbb{C} \setminus \overline{H(t)}$ *are connected for any* $t > 0$;
- *there exists* $0 < t_0 < \infty$ *such that* $G(t_0) = H(t_0)$;
- *there exists* $\varepsilon > 0$ *such that* $\overline{H(0)} \subset G(t_0 - \varepsilon)$.

Then, $G(t) = H(t)$ *for all* $t \in (t_0 - \varepsilon, t_0]$.

Remark 3.6.4. *If the initial domain $G(0)$ is bounded by a smooth analytic curve, then the strong solution exists locally in time and coincides with the weak one $G(t)$.*

Since in the strong case the normal velocity at the boundary does not vanish, the first assumption of the theorem is satisfied whenever the boundaries of the initial domains are smooth analytic.

Proof. Let us consider $t \in (t_0 - \varepsilon, t_0)$ and construct the functions $u(z,t)$ and $v(z,t)$ that correspond to the domains $G(t)$ and $H(t)$ respectively. Then,

$$\begin{aligned} \Delta u &= \chi_{G(t)} - \chi_{G(0)} + t\delta_0, \\ \Delta v &= \chi_{H(t)} - \chi_{H(0)} + t\delta_0, \end{aligned}$$

$u(z,t) \geq 0$, $v(z,t) \geq 0$ in $\mathbb{C}$ and $u(z,t) = 0$ in $\mathbb{C} \setminus G(t)$, $v(z,t) = 0$ in $z \in \mathbb{C} \setminus H(t)$.

Next consider the function

$$\gamma(z,t) = v(z,t_0) - u(z,t_0) + u(z,t).$$

One easily calculates

$$\Delta\gamma(z,t) = \chi_{G(t)} - \chi_{H(0)} + t\delta_0.$$

Under the assumption $\overline{H(0)} \subset G(t_0 - \varepsilon)$ the function $\gamma(z,t)$ is harmonic in $\mathbb{C} \setminus \overline{G(t)}$ for any $t \in (t_0 - \varepsilon, t_0)$, and $\gamma(z,t) = 0$ in $\mathbb{C} \setminus \overline{G(t_0)}$. Therefore, $\gamma(z,t) = 0$ in $\mathbb{C} \setminus \overline{G(t)}$ by harmonic continuation (using that $\mathbb{C} \setminus \overline{G(t)}$ is connected).

We have $v(z,t) \geq 0$ in $\mathbb{C}$ and $v(z,t) = 0$ in $\mathbb{C} \setminus \overline{H(t)}$. Let us set

$$w(z,t) = v(z,t_0) - v(z,t) - u(z,t_0) + u(z,t) = \gamma(z,t) - v(z,t).$$

This function is non-positive in $\mathbb{C} \setminus \overline{G(t)}$, and

$$\Delta w = \chi_{G(t)} - \chi_{H(t)}. \tag{3.22}$$

Therefore, $\Delta w \geq 0$ in $G(t)$. Hence, $w \leq 0$ in $\mathbb{C}$. Moreover, the function w is subharmonic in the connected set $\mathbb{C} \setminus \overline{H(t)}$. Therefore, $w < 0$ in $\mathbb{C} \setminus \overline{H(t)}$, or else, $w \equiv 0$ in $\mathbb{C} \setminus \overline{H(t)}$ by the maximum principle. Since $w(z) = 0$ for z of a sufficiently big norm, only the second option is valid. In particular, $\Delta w = 0$ in $\mathbb{C} \setminus \overline{H(t)}$, which by the equation (3.22) implies that

$$G(t) \subset \overline{H(t)}. \tag{3.23}$$

By Proposition 3.3.1 (e) we have $|H(t)| = |G(t)|$. Since $G(t)$ and $H(t)$ are the saturated sets which satisfy (3.4)–(3.6) for $G(0)$ and $H(0)$ respectively, and $|\partial H(t)| = |\partial G(t)| = 0$, it follows from (3.23) that $G(t) = H(t)$ for all $t \in (t_0 - \varepsilon, t_0)$. This ends the proof. □

Problem 3.6.5. *What happens when a strong or weak solution in the ill-posed direction (suction) breaks down? What will the typical boundary look like, and is it possible to continue the solution in any reasonable way beyond the break down time? Compare* [170], [177].

One common point of view is that a strong solution in the ill-posed direction will typically break down through the development of an (inward) cusp on the boundary since this is what happens with most explicit solutions. On the other hand one may start with any domain whatsoever and construct a weak solution in the well-posed direction with this domain as an initial domain. The result will, after a sufficiently long time (which can be estimated), be an almost circular domain with analytic boundary. Running time backwards it follows that starting with an almost circular domain, the ill-posed Hele-Shaw evolution from this may break down in virtually any way (cf. [81]*). For example, the boundary of the terminal region may have Hausdorff dimension greater than 1. Conjectures concerning the typical Hausdorff dimension have been proposed in the context of DLA and some estimates have been given, see, e.g.,* [52], [290].

3.7 Hele-Shaw flow and quadrature domains

Closely related to partial balayage, and hence to Hele-Shaw flow, is the notion of a quadrature domain. Ideas related to quadrature domain theory have already been used implicitly in the previous sections. Here we shall spell out some basic definitions and thus make the connections more explicit.

If $\mu \geq 0$ is a measure with compact support, then a bounded domain $\Omega \subset \mathbb{C}$ containing $\operatorname{supp}\mu$ is called a quadrature domain for subharmonic functions for μ if the inequality

$$\iint_{\Omega} \varphi \, d\sigma_z \geq \iint \varphi \, d\mu \tag{3.24}$$

holds for all integrable subharmonic functions φ in Ω. See [295]. Thus equation (3.2), says that $\Omega(t)$ is a quadrature domain for subharmonic functions for the measure $\mu(t) = \chi_{\Omega(0)} + t\delta_0$ and this is an equivalent way of expressing that the family of domains $\Omega(t)$ is a weak Hele-Shaw solution. In general, a domain Ω (if assumed saturated) is a quadrature domain for subharmonic functions if and only if $\operatorname{Bal}(\mu, 1) = \chi_\Omega$. If $\operatorname{Bal}(\mu, 1)$ is not of this form, i.e., if there is also a remainder term $\mu\chi_{\Omega^c}$, then there exists no quadrature domain for subharmonic functions for μ.

In case φ is harmonic the inequality (3.24) becomes an equality, because both φ and $-\varphi$ are then subharmonic. Replacing the inequality sign (3.24) by equality we may also consider analytic (hence complex-valued) test functions. A particularly rich theory then arises for measures of the form $\mu = \sum_{k=1}^{n} c_k \delta_{z_k}$, i.e., for measures with support in a finite number of points. Allowing, more generally, μ to be an arbitrary distribution with support in a finite number of points one arrives at the following classical concept of a quadrature domain: a bounded domain Ω is called a *(classical) quadrature domain* if there exist finitely many points $z_1, \dots, z_m \in \Omega$ and coefficients $c_{kj} \in \mathbb{C}$ $(0 \leq j \leq n_{k-1}, 1 \leq k \leq m$, say), such that the *quadrature identity*

$$\iint_{\Omega} \Phi \, d\sigma_z = \sum_{k=1}^{m} \sum_{j=0}^{n_k - 1} c_{kj} \Phi^{(j)}(z_k) \tag{3.25}$$

holds for every integrable analytic function Φ in Ω. The integer $n = \sum_{k=1}^{m} n_k$ is then called the order of the quadrature identity (assuming $c_{k,n_k-1} \neq 0$).

Notions of quadrature domains and identities as above were introduced in the 1970s by Davis [76] and Aharonov and Shapiro [4]. For general developments after that, see e.g., [308], [153]. In [333] quadrature domains as above are named *algebraic domains.*

The relationship between classical quadrature domains and Hele-Shaw flow is immediate from the generalized moment property (1.21): if $\Omega(0)$ is a quadrature domain as in (3.25), then all domains $\Omega(t)$ in a Hele-Shaw evolution with injection or suction at the origin are quadrature domains as well. To be precise, if (3.25) holds for $\Omega(0)$ and the suction rate is Q, then the $\Omega(t)$ satisfy the quadrature identity

$$\iint_{\Omega(t)} \Phi \, d\sigma_z = \sum_{k=1}^{m} \sum_{j=0}^{n_k - 1} c_{kj} \Phi^{(j)}(z_k) - Qt\Phi(0) \tag{3.26}$$

for any t.

Theorem 3.7.1. *Let $\Omega \subset \mathbb{C}$ be a bounded domain. Then the following are equivalent.*

(i) *Ω is a (classical) quadrature domain.*

(ii) *The exterior part $g_e(z)$ of the Cauchy transform* (1.28) *of Ω is a rational function.*

(iii) *There exists a meromorphic function $S(z)$ in Ω, continuous up to $\partial\Omega$, such that*

$$S(z) = \bar{z} \quad \text{on } \partial\Omega. \tag{3.27}$$

If Ω is simply connected a further equivalent property is:

(iv) *any Riemann mapping function $f : U \to \Omega$ is a rational function.*

Clearly the function $S(z)$ will be the Schwarz function of $\partial\Omega$. More precisely, it can be shown [4], [142] that the boundary of a quadrature domain always is an algebraic curve. This curve may have certain singular points, namely such points which in the simply connected case are images $f(\zeta_0)$ of points $\zeta_0 \in \partial U$ such that f is not univalent in a full neighbourhood of ζ_0 ($f'(\zeta_0) = 0$, or $f(\zeta_1) = f(\zeta_0)$ for another $\zeta_1 \in \partial U$). Away from these singular points $S(z)$ is analytic in a full neighbourhood of $\partial\Omega$, hence is a true Schwarz function. At singular points $S(z)$ is only a "one-sided Schwarz function".

Proof. (i) implies (ii): Just choose $\Phi(\zeta) = \frac{1}{\zeta - z}$ for $z \in \mathbb{C} \setminus \Omega$ in the quadrature identity.

(ii) implies (iii): Assuming a little regularity of $\partial\Omega$ we simply define $S(z)$ by the formula (1.27) Since $g_e(z)$ is rational (by assumption) and $g_i(z)$ (see (1.27)) is always holomorphic in all of Ω, $S(z)$ then is meromorphic in Ω and the statement follows.

(iii) implies (i): Using the residue theorem we have, when $S(z)$ is meromorphic in Ω and Φ is analytic,

$$\iint_\Omega \Phi\, d\sigma = \frac{1}{2i} \iint_\Omega \Phi(z)\, d\bar{z} dz = \frac{1}{2i} \int_{\partial\Omega} \Phi(z)\bar{z} dz$$

$$= \frac{1}{2i} \int_{\partial\Omega} \Phi(z) S(z) dz = \pi \sum_{z \in \Omega} \operatorname{Res} \Phi(z) S(z),$$

which is a quadrature identity of the form (3.25).

(iii) implies (iv): In the presence of $S(z)$ any conformal map $f : U \to \Omega$ can be extended to the Riemann sphere by

$$\overline{f(1/\bar{\zeta})} = S(f(\zeta)) \tag{3.28}$$

for $\zeta \in U$, i.e., for $\frac{1}{\bar{\zeta}} \in U^*$ (cf. the proof of Theorem 3.4.1). This makes f meromorphic in $\overline{\mathbb{C}}$, hence rational.

(iv) implies (iii): If f is rational we can define $S(z)$ for $z = f(\zeta) \in \Omega$ by (3.28) and it is easy to see that it becomes meromorphic in Ω with $S(z) = \bar{z}$ on $\partial\Omega$. □

It is clear from the above proof that when Ω is a quadrature domain, the relationship between the data in (3.25) and the poles of $g_e(z)$, $S(z)$ and $f(\zeta)$ is as follows:

$$g_e(z) = \frac{1}{\pi} \sum_{k=1}^{m} \sum_{j=0}^{n_k - 1} \frac{j! c_{kj}}{(z - z_k)^{j+1}},$$

$$S(z) = \frac{1}{\pi} \sum_{k=1}^{m} \sum_{j=0}^{n_k - 1} \frac{j! c_{kj}}{(z - z_k)^{j+1}} + \text{regular},$$

$$f(\zeta) = \sum_{k=1}^{m} \sum_{j=0}^{n_k - 1} \frac{b_{kj}}{(\zeta - 1/\bar{\zeta}_k)^{j+1}} + \text{regular}.$$

Here $\zeta_k \in U$ are the points which are mapped onto the quadrature nodes: $f(\zeta_k) = z_k$. The expressions for the coefficients b_{kj} in terms of c_{kj} and z_k are somewhat complicated because of the nonlinear nature of (3.28) as an equation for f.

Quadrature domain theory is in many cases helpful for understanding properties of Hele-Shaw evolutions, and also for construction of explicit solutions. For example, Theorem 3.7.1 gives a new proof of the fact that the Polubarinova-Galin equation preserves rational mapping functions (Theorem 2.1.1). Indeed, if $f(\zeta, 0)$ is a rational function, then $\Omega(0)$ is a quadrature domain, say satisfies (3.25). Hence all the $\Omega(t)$ are quadrature domains as in (3.26), and therefore $f(\zeta, t)$ is rational for every t. In addition, from the above relationships between the poles of $f(\zeta, t)$ and the quadrature data $\{z_k\}$, $\{c_{kj}\}$, which by (3.26) remain fixed during the Hele-Shaw evolution (except for the coefficient at $z = 0$), it becomes clear that the poles of $f(\zeta, t)$ cannot collide or disappear. The only possible exception here is the pole at infinity which, being linked to the source/suction point $z = 0$, may disappear for one value of t.

Let us next revisit the first example in Section 2.1.2 and try to explain why it is possible, in the suction case, to have a rational solution $f(\zeta, t)$ of the Polubarinova-Galin equation such that the free boundary reaches the sink, whereas this is not possible in the pure polynomial case.

In Figure 2.2 the residual fluid domain after suction of fluid at the origin is the disk $\Omega = \{|z + 1| < 1\}$, for which the quadrature identity

$$\iint_{\Omega} \Phi \, d\sigma_z = \pi \Phi(-1)$$

holds. From this disk the whole Hele-Shaw family may be recovered by injecting fluid at $z = 0$. This gives a family of quadrature domains $\Omega(t)$ with quadrature nodes at $z = -1$ (for the original disk) and $z = 0$ (due to injection there). Letting the time parameter t be the same as in Section 2.1.2 the quadrature identities are

$$\iint_{\Omega(t)} \Phi \, d\sigma_z = \pi \Phi(-1) + (\pi - Qt)\Phi(0),$$

$0 \leq t \leq \pi/Q$. The corresponding mapping functions $f(\zeta, t)$ will be the rational functions (2.3) with one pole at infinity, corresponding to the quadrature node $z = 0$, and one finite pole $\zeta = 1/c(t)$ having the property that the reflected point $1/\bar{\zeta} = c(t)$ is mapped onto the other quadrature node $z = -1$. All this is perfectly fine, and the same argument can be used to show that rational solutions of the Polubarinova-Galin equation may end up in virtually any simply connected quadrature domain Ω with $0 \in \partial\Omega$.

However, these rational solutions can never be of pure polynomial type, because for polynomial solutions the fluid domains will satisfy quadrature identities of the kind

$$\iint_{\Omega(t)} \Phi \, d\sigma_z = (c_0 - Qt)\Phi(0) + c_1 \Phi'(0) + \cdots + c_{n-1}\Phi^{(n-1)}(0),$$

and then $z = 0$ can never be on $\partial\Omega(t)$. The corresponding mapping functions will be of the form $f(\zeta, t) = a_1(t)\zeta + \cdots + a_n(t)\zeta^n$ with $a_1(t) > 0$. An important remark here is that the quadrature identity remains valid for the limiting domain in the Hele-Shaw evolution, even if the Polubarinova-Galin solution breaks down there. Notice also that the above coefficient of $\Phi(0)$ equals the area of $\Omega(t)$ (choose $\Phi = 1$), hence vanishes only if all fluid has been sucked, which occurs only in the shrinking disk case.

Returning to quadrature domains in general, let us be a little more explicit concerning the algebraic boundary of a (classical) quadrature domain. Using the so-called exponential transform [147] one can show [148] that if Ω is a quadrature domain such that (3.25) holds, then the equation for $\partial\Omega$ can be written in the form

$$|P_n(z)|^2 = \sum_{k=0}^{n-1} |P_k(z)|^2, \tag{3.29}$$

where each $P_k(z)$ is a polynomial of degree k (exactly). The two highest polynomials $P_n(z)$ and $P_{n-1}(z)$ make up the rational function $g_e(z)$:

$$g_e(z) = \sqrt{\frac{|\Omega|}{\pi}}\,\frac{P_{n-1}(z)}{P_n(z)},$$

hence $P_n(z) = \prod_{k=1}^{m}(z - z_k)^{n_k}$ (up to a constant).

It follows in particular that for a rational solution of the Polubarinova-Galin equation, the fluid fronts $\partial\Omega(t)$ are given, in the notations of (3.26), by equations of the form

$$|P_{n+1}(z)|^2 = \sum_{k=0}^{n} |P_k(z,t)|^2,$$

with $P_{n+1}(z) = z\prod_{k=1}^{m}(z - z_k)^{n_k}$ and $P_n(z,t)$ determined by

$$\sqrt{\frac{|\Omega(0)| - Qt}{\pi}}\,\frac{P_n(z,t)}{P_{n+1}(z)} = g_e(z,t) = \frac{1}{\pi}\sum_{k=1}^{m}\sum_{j=0}^{n_k-1}\frac{j!c_{kj}}{(z-z_k)^{j+1}} - \frac{Qt}{\pi z}.$$

To get hold of the remaining polynomials $P_k(z,t)$ $(0 \le k \le n-1)$ seems to be quite difficult in general. See [64] for some studies of such questions. An overview of applications of quadrature domains to problems in fluid dynamics is given in [65].

3.8 Multidimensional Hele-Shaw flow and other generalizations

The governing law (Darcy's law) for the evolution of the fluid boundary in a Hele-Shaw cell makes sense, in its mathematical form, in any number of dimensions. In three space dimensions (and higher) there are then other physical models. Some examples are porous medium flow ([263], [264]), electro-chemical deposition ([341]), dendritic growth, growth processes in biology [238], [255].

The classical formulation of the zero surface tension version in any number of dimensions is that a family of bounded domains $\Omega(t)$, smooth both in time and space, makes up a Hele-Shaw evolution with a source/sink at a point a, if the boundary $\partial\Omega(t)$ moves in the normal direction with speed proportional to the normal derivative of the Green function for the Laplacian, $G_{\Omega(t)}(x,a)$. A convenient way of formulating the moving boundary condition is to say that for any test function φ the identity

$$\frac{d}{dt}\int_{\Omega(t)} \varphi(x)dx = Q\int_{\partial\Omega(t)} \varphi(x)\frac{\partial G_{\Omega(t)}(x,a)}{\partial n}ds$$

is to hold, where Q is the constant of proportionality and ds is surface measure on $\partial\Omega(t)$.

Certainly, conformal mapping and other complex variable methods do not apply in the higher dimensional case, but the potential theory tools work practically unchanged. For example, the construction and basic properties of weak solutions in Sections 3.1–3.3, 3.5–3.6 work in the same way with minor modifications (see, e.g., [29], [81], [145], [146]). When it comes to geometry and regularity of the free boundary (Sections 3.4 and 4.6) the situation in higher dimension is basically the same as in two dimensions, but the theory is somewhat less complete. For example, there is no such precise statement as Theorem 3.4.1 in higher dimensions, but it is known, in the well-posed (injection) case, that $\partial\Omega(t) \setminus \overline{\Omega(0)}$ is a (smooth) real analytic manifold except possibly for a small exceptional set, and that outside the convex hull of $\overline{\Omega(0)}$ there are no exceptional points at all. These statements are based on the known regularity results for the obstacle problem [45], [46], [49], [48], [115] combined with geometric results (relevant parts of Theorem 4.6.2 remain true [146]).

Classical (strong) solutions to the higher dimensional Hele-Shaw problem, with and without surface tension, has been studied in a number of papers, for example [21], [107], [108], [330]. Compare also [58], [91]. What one can basically prove is the existence and uniqueness, locally in time in both time directions, of strong solutions under the assumption that the initial domain has a smooth real analytic boundary. Under certain assumptions, like strong starlikeness of the initial domain, it is possible even to prove the existence globally in time in the well-posed time direction (injection) [152]. Another such example is found in [237].

A generalization of the Hele-Shaw problem in another direction is to replace the Laplace operator by a more general elliptic operator in divergence form. This allows for considering irregular and anisotropic media in for example a porous medium interpretation of the model. A selection of papers going in this direction is [30], [124], [226], [232], [279].

Chapter 4

Geometric Properties

In this chapter we deal with geometric properties of general Hele-Shaw flows. Special classes of univalent functions that admit explicit geometric interpretations are considered to characterize the shape of the free interface under injection. In particular, we are concerned with the following question: which geometrical properties are preserved during the time evolution of the moving boundary? We also discuss the geometry of weak solutions.

4.1 Distance to the boundary

In this section, using some simple observations found in [171], we shall estimate the minimal distance from the source to the free boundary.

Let us consider the problem of injection ($Q < 0$) into a bounded domain $\Omega(t)$ parameterized by a univalent function $f(\zeta, t)$ that maps the unit disk U onto $\Omega(t)$, normalized as $f(\zeta, t) = a(t)\zeta + a_2(t)\zeta^2 + \cdots$, $a(t) > 0$.

Using the Löwner-Kufarev type equation (1.17) for the function f we obtain

$$\dot{a}(t) = -\frac{Q}{2\pi}a(t)\frac{1}{2\pi}\int_0^{2\pi}\frac{1}{|f'(e^{i\theta},t)|^2}d\theta.$$

This immediately gives the inequality

$$\dot{a}(t) \geq -\frac{Q}{2\pi}\frac{1}{a(t)},$$

or

$$a^2(t) \geq a^2(0) - \frac{Qt}{\pi}.$$

The 1/4 Koebe theorem (see, e.g., [127]) yields the inequality

$$\operatorname{dist}(\partial\Omega(t), 0) \geq \frac{1}{4}\sqrt{a^2(0) - \frac{Qt}{\pi}}.$$

A more general result will be given at the end of this chapter.

4.2 Special classes of univalent functions

Let us define some special classes of univalent functions which will parameterize our phase domains.

A domain $\Omega \subset \mathbb{C}$, $0 \in \Omega$ is said to be *starlike* (with respect to the origin) if each ray starting at the origin intersects Ω in a connected set. If a function $f(\zeta)$ maps U onto a domain which is starlike, $f(0) = 0$, then we say that $f(\zeta)$ is a *starlike function.* If a function $f(\zeta)$ maps U onto a domain which is convex, $f(0) = 0$, then we say that $f(\zeta)$ is a *convex function.* We denote the class of starlike functions by S^* and the class of convex functions by C. A necessary and sufficient condition for a function $f(\zeta)$, $\zeta \in U$, $f(0) = 0$ to be starlike is that the inequality

$$\operatorname{Re} \frac{\zeta f'(\zeta)}{f(\zeta)} > 0, \quad \zeta \in U \tag{4.1}$$

holds. Similarly, a necessary and sufficient condition for a function f to be convex is the inequality

$$\operatorname{Re} \left(1 + \frac{\zeta f''(\zeta)}{f'(\zeta)}\right) > 0, \quad \zeta \in U. \tag{4.2}$$

These standard characterizations can be found, e.g., in [13], [92], [132], [270].

A simple way to generalize the class S^* is to introduce a class of so-called starlike functions of order α, $0 < \alpha \leq 1$, obtained by replacing 0 in the right-hand side of (4.1) by the constant α. Let us denote it by S^*_α. It is known that any convex function is in $S^*_{1/2}$ (see [233]). Unfortunately, the classes S^*_α do not admit any clear geometric interpretation. A more reasonable generalization was given by Brannan, Kirwan [44] and Stankiewicz [314]. A function $f : U \to \mathbb{C}$, $f(0) = 0$ is said to be *strongly starlike of order* α in U, $0 < \alpha \leq 1$, if for all $\zeta \in U$,

$$\left| \arg \frac{\zeta f'(\zeta)}{f(\zeta)} \right| < \alpha \frac{\pi}{2}. \tag{4.3}$$

The set of all such functions is denoted by $S^*(\alpha)$. This class of functions has a better geometric description: every *level line* $L_r = \{f(re^{i\theta}),\ \theta \in [0, 2\pi)\}$, $f \in S^*(\alpha)$ is reachable from outside by the radial angle $\pi(1-\alpha)$ (see Figure 4.1).

The inequalities (4.1–4.3) also give sufficient conditions for an analytic function f to be univalent (see [19] for a collection of sufficient conditions of univalence). Kaplan [190] proved that if $f(\zeta)$ and $g(\zeta)$ are analytic in U, $g \in C$, and

$$\operatorname{Re} \frac{f'(\zeta)}{g'(\zeta)} > 0, \quad \zeta \in U,$$

then f is univalent in U. Kaplan gave the name *close-to-convex* to univalent functions f that satisfy the above condition. The close-to-convex functions have a nice

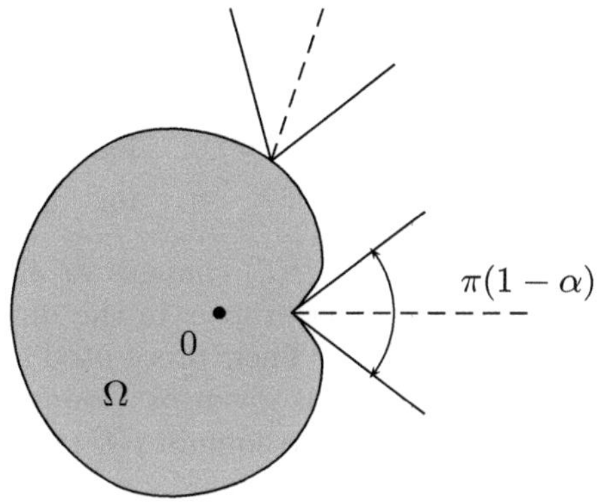

Figure 4.1: Strongly starlike functions of order α ($S^*(\alpha)$)

geometric characterization: every level line L_r of a close-to-convex function f has no "large hairpin" turns, that is there are no sections of the curve L_r in which the tangent vector turns backward through an angle greater than π.

We say that a simply connected hyperbolic domain Ω is *convex in the direction of the real axis* $\mathbb{R}$ if each line parallel to $\mathbb{R}$ either misses Ω, or the intersection with Ω is a connected set. The study of this class goes back to Fejér [111] and Robertson [288]. If a function $f(\zeta)$ maps U onto a domain which is convex in the direction of the real axis, $f(0) = 0$, then we say that $f(\zeta)$ is a *convex function in the direction of the real axis* and denote the class of such functions by $C_{\mathbb{R}}$ (see Figure 4.2). The criterion that characterizes these functions is as follows: the unit disk

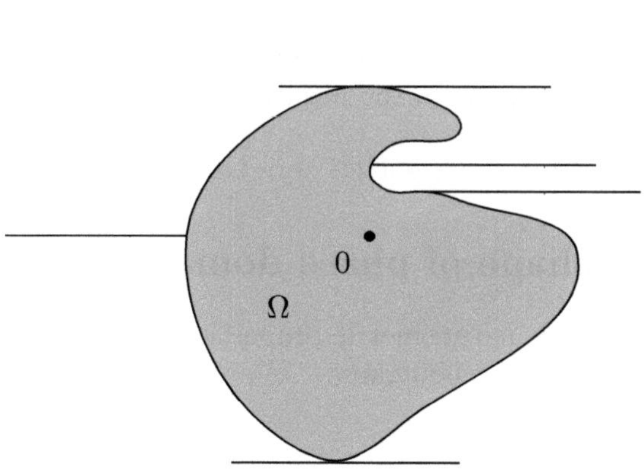

Figure 4.2: Convex functions in the direction of the real axis ($C_{\mathbb{R}}$)

can be divided into two disjoint arcs I and J, $I = \{\zeta = e^{i\theta},\ \theta \in [0, \varphi] \cup [\psi, 2\pi]\}$, $J = \{\zeta = e^{i\theta},\ \theta \in [\varphi, \psi]\}$, such that

$$\begin{aligned} \text{Re}\ \zeta f'(\zeta) &\geq 0, \quad \text{for} \quad \zeta \in I, \\ \text{Re}\ \zeta f'(\zeta) &\leq 0, \quad \text{for} \quad \zeta \in J. \end{aligned}$$

The harmonic function Re $\zeta f'(\zeta)$ changes its sign in U. Therefore, the level lines L_r for $0 < r < 1$ need not be convex in the direction of $\mathbb{R}$. Hengartner and Schober [168] proved this in 1973. Their proof used a slightly modified argument. An example of such a function was given by Goodman and Saff [131]. Fejér and Szegö in 1951 [112] proved that if the domain $f(U)$ is symmetric with respect to the imaginary axis, then the above conditions for a holomorphic function f, $f(0) = 0$, are sufficient for its univalence and the level lines are convex in the direction of $\mathbb{R}$. Prokhorov [274] proved in 1988 that, in general, for any $r \in (0, \sqrt{2} - 1)$ the level lines are still convex in the direction of $\mathbb{R}$. Independently this result (even a more general one) was obtained by Ruscheweyh and Salinas [291]. An example by Goodman and Saff [131] shows that the constant $\sqrt{2} - 1$ can not be improved.

Now we discuss conformal maps from the right half-plane. A simply-connected domain Ω with a boundary that contains more than two points in the extended complex plane $\overline{\mathbb{C}}$ is said to be *convex in the negative direction* of the real axis $\mathbb{R}^-$ if its complement can be covered by a family of non-intersecting parallel rays starting at the same direction of $\mathbb{R}^-$. A holomorphic univalent map $f(\zeta)$, $\zeta \in H^+$, $H^+ = \{\zeta : \text{Im}\,\zeta > 0\}$, is said to be convex in the negative direction if $f(H^+)$ is as above. We denote this class by $H_{\mathbb{R}}^-$. This is somehow an analogue of the class of starlike functions for the half-plane. The criterion for this property is

$$\text{Re}\ f'(\zeta) > 0, \quad \zeta = \xi + i\eta \in H^+.$$

We define a subclass $H_{\mathbb{R}}^-(\alpha)$ of $H_{\mathbb{R}}^-$ of functions whose level lines $L_a = \{f(a + i\eta),\ \eta \in (-\infty, \infty)\}$ are reachable by the angles $\pi(1 - \alpha)$ with their bisectors co-directed with $\mathbb{R}^-$. We call these functions *convex of order α in the negative direction* (see Figure 4.3). The necessary and sufficient condition for a holomorphic function f to be convex of order α in the negative direction is that

$$\left|\arg f'(\zeta)\right| < \alpha\frac{\pi}{2}, \quad 0 < \alpha \leq 1, \quad \zeta \in H^+. \tag{4.4}$$

4.3 Hereditary shape of phase domains

In this section we shall find some geometric properties which are preserved during the time evolution of the moving boundary.

4.3.1 Bounded dynamics

Simple examples show that virtually no geometric properties are preserved in the case of suction, $Q > 0$. So we henceworth assume that $Q < 0$.

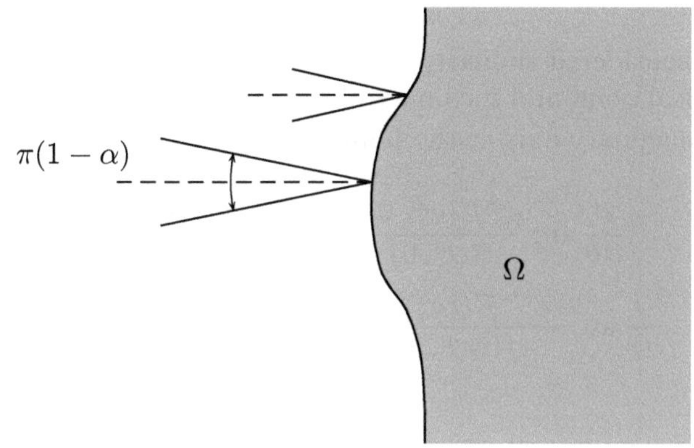

Figure 4.3: Convex functions of order α in the negative direction ($H^-_{\mathbb{R}}(\alpha)$)

Starlike dynamics

Let us start with starlike dynamics. We suppose that the initial function f_0 is analytic in the closure of U to guarantee the local in time existence of solutions (see Section 1.4.3). The following theorem was proved in [173] (see also [334]). Here we use a slightly modified argument.

Theorem 4.3.1. *Let $Q < 0$, $f_0 \in S^*$, and be analytic and univalent in a neighbourhood of $\bar{U}$. Then the family of domains $\Omega(t)$ (in the sequel, the family of univalent functions $f(\zeta, t)$) remain in S^* as long as the solution to the Polubarinova-Galin equation exists.*

Proof. If we consider f in the closure of U, then the inequality sign in (4.1) is to be replaced by ($\geq$) where equality can be attained only for $|\zeta| = 1$.

The proof is based on consideration of a critical map $f \in S^*$, such that the image of U under the map $\zeta f'(\zeta, t)/f(\zeta, t)$, $|\zeta| \leq 1$ touches the imaginary axis, say there exist $t' \geq 0$ and $\zeta_0 = e^{i\theta_0}$, such that

$$\arg \frac{\zeta_0 f'(\zeta_0, t')}{f(\zeta_0, t')} = \frac{\pi}{2} \quad (\text{or } -\tfrac{\pi}{2}), \tag{4.5}$$

and for any $\varepsilon > 0$ there are $t > t'$ and $\theta \in (\theta_0 - \varepsilon, \theta_0 + \varepsilon)$ such that

$$\arg \frac{e^{i\theta} f'(e^{i\theta}, t)}{f(e^{i\theta}, t)} \geq \frac{\pi}{2} \quad (\text{or } \leq -\tfrac{\pi}{2}). \tag{4.6}$$

For definiteness we consider the sign (+) in (4.5). Without loss of generality, let us assume $t' = 0$. Since $f'(e^{i\theta}, t) \neq 0$, our assumption about the sign in (4.5) yields that

$$\operatorname{Im} \frac{\zeta_0 f'(\zeta_0, 0)}{f(\zeta_0, 0)} > 0, \tag{4.7}$$

(the negative case is considered similarly).

Since ζ_0 is a critical point and the image of the unit disk U under the mapping $\frac{\zeta f'(\zeta,0)}{f(\zeta,0)}$ touches the imaginary axis at the point $\zeta_0 = e^{i\theta_0}$, we deduce that

$$\begin{aligned} \frac{\partial}{\partial \theta} \arg \frac{e^{i\theta} f'(e^{i\theta}, 0)}{f(e^{i\theta}, 0)}\bigg|_{\theta=\theta_0} &= 0, \\ \frac{\partial}{\partial r} \arg \frac{re^{i\theta_0} f'(re^{i\theta_0}, 0)}{f(re^{i\theta}, 0)}\bigg|_{r=1} &\geq 0. \end{aligned}$$

From this we calculate

$$\operatorname{Re} \left[1 + \frac{\zeta_0 f''(\zeta_0, 0)}{f'(\zeta_0, 0)} - \frac{\zeta_0 f'(\zeta_0, 0)}{f(\zeta_0, 0)}\right] = 0, \tag{4.8}$$

$$\operatorname{Im} \left[1 + \frac{\zeta_0 f''(\zeta_0, 0)}{f'(\zeta_0, 0)} - \frac{\zeta_0 f'(\zeta_0, 0)}{f(\zeta_0, 0)}\right] \geq 0. \tag{4.9}$$

By straightforward calculations one derives

$$\frac{\partial}{\partial t} \arg \frac{\zeta f'(\zeta, t)}{f(\zeta, t)} = \operatorname{Im} \frac{\partial}{\partial t} \log \frac{f'(\zeta, t)}{f(\zeta, t)} = \operatorname{Im} \left(\frac{\frac{\partial}{\partial t} f'(\zeta, t)}{f'(\zeta, t)} - \frac{\frac{\partial}{\partial t} f(\zeta, t)}{f(\zeta, t)} \right). \tag{4.10}$$

We now differentiate the equation (1.16) with respect to θ,

$$\operatorname{Im} \left(\overline{f'(\zeta, t)} \frac{\partial}{\partial t} f'(\zeta, t) - \overline{\zeta f'(\zeta, t)} \dot{f}(\zeta, t) - \overline{\zeta^2 f''(\zeta, t)} \dot{f}(\zeta, t) \right) = 0, \tag{4.11}$$

for $|\zeta| = 1$. This equality is equivalent to the following:

$$\begin{aligned} &|f'(\zeta, t)|^2 \operatorname{Im} \left(\frac{\frac{\partial}{\partial t} f'(\zeta, t)}{f'(\zeta, t)} - \frac{\frac{\partial}{\partial t} f(\zeta, t)}{f(\zeta, t)} \right) \\ &= \operatorname{Im} (\overline{\zeta f'(\zeta, t)} \dot{f}(\zeta, t)) \left(\overline{\frac{\zeta f''(\zeta, t)}{f'(\zeta, t)}} - \frac{\zeta f'(\zeta, t)}{f(\zeta, t)} + 1 \right). \end{aligned}$$

Substituting (1.16) and (4.8) in the latter expression we finally have

$$\begin{aligned} &\frac{\partial}{\partial t} \arg \frac{\zeta f'(\zeta, t)}{f(\zeta, t)}\bigg|_{\zeta=e^{i\theta_0}, t=0} \\ &= \frac{Q}{2\pi |f'(e^{i\theta_0}, 0)|^2} \operatorname{Im} \left(\frac{e^{i\theta_0} f'(e^{i\theta_0}, 0)}{f(e^{i\theta_0}, 0)} + \frac{e^{i\theta_0} f''(e^{i\theta_0}, 0)}{f'(e^{i\theta_0}, 0)} \right). \end{aligned}$$

The right-hand side of this equality is strictly negative because of (4.7), (4.9). Therefore,

$$\arg \frac{e^{i\theta} f'(e^{i\theta}, t)}{f(e^{i\theta}, t)} < \frac{\pi}{2}$$

for $t > 0$ (close to 0) in some neighbourhood of θ_0. This contradicts the assumption that $\Omega(t)$ fails to be starlike for some $t > 0$ and ends the proof for the class S^*. □

The property of preservation of starlikeness is especially interesting in view of Novikov's theorem [251], that says: if two bounded domains are starlike with respect to a common point and have the same exterior gravity potential, then they coincide.

We continue with strongly starlike functions of order α. We shall prove that starting with a bounded phase domain Ω_0 which is strongly starlike of order α and bounded by an analytic curve, we obtain a subordination chain of domains $\Omega(t)$ (and functions $f(\zeta, t)$) under injection at the origin which remain strongly starlike of order $\alpha(t)$ with a decreasing order $\alpha(t)$. The following monotonicity theorem is found in [151]. A similar result has recently and independently been obtained in [216]

Theorem 4.3.2. *Let $f_0 \in S^*(\alpha)$, $\alpha \in (0, 1]$, be analytic and univalent in a neighbourhood of $\bar{U}$. Then the strong solution $f(\zeta, t)$ to the Polubarinova-Galin equation* (1.16) *forms a subordination chain (see the definition in Section* 6.1*) of strongly starlike functions of order $\alpha(t)$ with a strictly decreasing $\alpha(t)$ during the time of existence, $\alpha(0) = \alpha$.*

Proof. Let t_0 be such that the strong solution $f(\zeta, t)$ exists during the time $t \in [0, t_0)$, $t_0 > 0$. Since all functions $f(\zeta, t)$ have analytic univalent extension into a neighbourhood of $\bar{U}$ during the time of the existence of the strong solution to (1.16), their derivatives $f'(\zeta, t)$ are continuous and do not vanish in $\bar{U}$. Moreover, $f(\zeta, t)$ are starlike in U (see Theorem 4.3.1). Therefore, there exists $\alpha(t)$, $0 < \alpha(t) \leq 1$, such that $f(\zeta, t) \in S^*(\alpha(t))$ and $f(\zeta, t) \notin S^*(\alpha(t) - \varepsilon)$ for any $\varepsilon > 0$.

Let us fix $t' \in [0, t_0)$ and consider the set A of all points ζ, $|\zeta| = 1$ for which $|\arg \frac{\zeta f'(\zeta, t')}{f(\zeta, t')}| = \alpha\pi/2$. First, we deal with the subset A^+ of A where

$$\arg \frac{\zeta f'(\zeta, t')}{f(\zeta, t')} = \frac{\alpha\pi}{2}. \tag{4.12}$$

The sets A^+ and $A^- = A \setminus A^+$ are closed and do not intersect. One of the sets A^+ and A^- is allowed to be empty. Without loss of generality we suppose that $A^+ \neq \emptyset$. For any point $\zeta \in A^+$, we have

$$\operatorname{Im} \frac{\zeta f'(\zeta, t')}{f(\zeta, t')} > 0. \tag{4.13}$$

The argument $\arg \frac{\zeta f'(\zeta,t')}{f(\zeta,t')}$ attains its maximum on $\zeta \in \partial U$ at the points of A^+. Therefore,

$$\frac{\partial}{\partial \theta} \arg \frac{e^{i\theta} f'(e^{i\theta},t')}{f(e^{i\theta},t')} = 0, \quad \zeta = e^{i\theta} \in A^+.$$

The argument $\arg \frac{re^{i\theta} f'(re^{i\theta},t')}{f(re^{i\theta},t')}$, $e^{i\theta} \in A^+$ attains its maximum on $r \in [0,1]$ at $r = 1$. Hence,

$$\frac{\partial}{\partial r} \arg \frac{re^{i\theta} f'(re^{i\theta},t')}{f(re^{i\theta},t')}\bigg|_{r=1} \geq 0.$$

We calculate

$$\operatorname{Re}\left[1 + \frac{\zeta f''(\zeta,t')}{f'(\zeta,t')} - \frac{\zeta f'(\zeta,t')}{f(\zeta,t')}\right] = 0, \tag{4.14}$$

$$\operatorname{Im}\left[1 + \frac{\zeta f''(\zeta,t')}{f'(\zeta,t')} - \frac{\zeta f'(\zeta,t')}{f(\zeta,t')}\right] \geq 0, \tag{4.15}$$

where $\zeta \in A^+$.

Let us represent the derivative

$$\frac{\partial}{\partial t} \arg \frac{\zeta f'(\zeta,t)}{f(\zeta,t)} = \operatorname{Im} \frac{\partial}{\partial t} \log \frac{f'(\zeta,t)}{f(\zeta,t)} = \operatorname{Im}\left(\frac{\frac{\partial}{\partial t} f'(\zeta,t)}{f'(\zeta,t)} - \frac{\frac{\partial}{\partial t} f(\zeta,t)}{f(\zeta,t)}\right). \tag{4.16}$$

Now we differentiate the Polubarinova-Galin equation (1.16) with respect to θ as

$$\operatorname{Im}\left(\overline{f'(\zeta,t)}\frac{\partial}{\partial t} f'(\zeta,t) - \overline{\zeta f'(\zeta,t)}\dot{f}(\zeta,t) - \overline{\zeta^2 f''(\zeta,t)}\dot{f}(\zeta,t)\right) = 0, \quad \zeta = e^{i\theta}. \tag{4.17}$$

This equality is equivalent to

$$\begin{aligned} |f'(\zeta,t)|^2 \operatorname{Im}\left(\frac{\frac{\partial}{\partial t} f'(\zeta,t)}{f'(\zeta,t)} - \frac{\frac{\partial}{\partial t} f(\zeta,t)}{f(\zeta,t)}\right) \\ = \operatorname{Im}\left[\overline{\zeta f'(\zeta,t)}\dot{f}(\zeta,t)\left(\overline{\left(\frac{\zeta f''(\zeta,t)}{f'(\zeta,t)}\right)} - \frac{\zeta f'(\zeta,t)}{f(\zeta,t)} + 1\right)\right]. \end{aligned}$$

Substituting (1.16) and (4.14) in the latter expression we have

$$\frac{\partial}{\partial t} \arg \frac{\zeta f'(\zeta,t)}{f(\zeta,t)}\bigg|_{\zeta \in A^+, t=t'} = \frac{Q}{|f'(\zeta,t')|^2} \operatorname{Im}\left(\frac{\zeta f'(\zeta,t')}{f(\zeta,t')} + \frac{\zeta f''(\zeta,t')}{f'(\zeta,t')}\right).$$

The right-hand side of this equality is continuous on A^+ and strictly negative because of (4.13), (4.15). Therefore,

$$\max_{\zeta \in A^+} \frac{\partial}{\partial t} \arg \frac{\zeta f'(\zeta,t)}{f(\zeta,t)}\bigg|_{t=t'} = -\delta < 0.$$

There exists a neighbourhood $A^+(\delta)$ on the unit circle of A^+ such that $A^+(\delta)$ and A^- do not intersect and

$$\frac{\partial}{\partial t}\arg\frac{\zeta f'(\zeta,t)}{f(\zeta,t)}\Big|_{\zeta\in A^+(\delta),t=t'} < -\frac{\delta}{2}.$$

There is a positive number σ such that

$$\max_{\zeta\in\partial U\setminus A^+(\delta)}\arg\frac{\zeta f'(\zeta,t)}{f(\zeta,t)}\Big|_{t=t'} = \frac{\alpha\pi}{2}-\sigma.$$

We choose such $s>0$ that

(i) $t'+s<t_0$;

(ii) $\displaystyle\frac{\partial}{\partial t}\arg\frac{\zeta f'(\zeta,t)}{f(\zeta,t)}\Big|_{\zeta\in A^+(\delta)} < 0, \quad t\in[t',t'+s]$;

(iii) $\displaystyle\max_{\zeta\in\partial U\setminus A^+(\delta)}\arg\frac{\zeta f'(\zeta,t)}{f(\zeta,t)} \le \frac{\alpha\pi}{2}-\frac{\sigma}{2}, \quad t\in[t',t'+s]$.

The condition (ii) implies that

$$\arg\frac{\zeta f'(\zeta,t)}{f(\zeta,t)} < \frac{\alpha\pi}{2}, \quad t\in(t',t'+s],\ \zeta\in A^+(\delta).$$

Thus, the condition (iii) yields

$$\alpha^+(t) := \max_{\zeta\in\partial U}\arg\frac{\zeta f'(\zeta,t)}{f(\zeta,t)} < \frac{\alpha\pi}{2} = \alpha(t'), \quad \text{for all } t\in(t',t'+s].$$

This means that $\alpha^+(t)$ is strictly decreasing in $[0,t_0)$.

If the set $A^-\neq\emptyset$, then we can define the function

$$\alpha^-(t) := -\min_{\zeta\in\partial U}\arg\frac{\zeta f'(\zeta,t)}{f(\zeta,t)}.$$

Similar argumentation shows that $\alpha^-(t)$ is strictly decreasing.

If $A^-=\emptyset$ (or $A^+=\emptyset$), then $\alpha(t)=\alpha^+(t)$ (or $=\alpha^-(t)$) for $t\in[t',t'+s]$, s sufficiently small.

We set the function $\alpha(t)=\max\{\alpha^+(t),\alpha^-(t)\}$ in the case $A^+\neq\emptyset$ and $A^-\neq\emptyset$. The so-defined function $\alpha(t)$ is strictly decreasing, which ends the proof. □

Directional convex dynamics

We proceed with the class $C_{\mathbb{R}}$ and the dynamics under injection ($Q < 0$).

Theorem 4.3.3. *If the initial domain* $\Omega(0)$ *is convex in the direction of* $\mathbb{R}$, *then the family of domains* $\Omega(t)$ *(in sequel, the family of univalent functions* $f(\zeta,t)$*) preserves this property as long as the classical solution of the Hele-Shaw problem exists and the level lines of the function* $f(\zeta,t)$ *also are convex in the direction of* $\mathbb{R}$ *in a neighbourhood* $|\zeta| \in (1-\varepsilon, 1]$.

Remark 4.3.4. *The last requirement is fulfilled always if the initial domain* Ω_0 *is symmetric with respect to the imaginary axis (Fejér and Szegö* [112]*).*

Proof. Let us again consider a critical map $f \in C_{\mathbb{R}}$, such that the image of U under the map $\zeta f'(\zeta,t)$, $|\zeta| \leq 1$ touches the imaginary axis. In other words, there exists $\zeta_0 = e^{i\theta_0}$, which satisfies the equality $\arg \zeta_0 f'(\zeta_0, 0) = \frac{\pi}{2}$ (or $-\frac{\pi}{2}$) at the initial instant $t = 0$ and for any $\varepsilon > 0$ there is such $t > 0$ and $\theta \in (\theta_0-\varepsilon, \theta_0+\varepsilon)$ that $\arg e^{i\theta} f'(e^{i\theta}, t) \geq \frac{\pi}{2}$ (or $\leq -\frac{\pi}{2}$). Of course, $\arg e^{i\varphi} f'(e^{i\varphi}, 0) = \frac{\pi}{2}$ and $\arg e^{i\psi} f'(e^{i\psi}, 0) = -\frac{\pi}{2}$. In this case the curve $e^{i\theta} f'(e^{i\theta}, t)$ has an intersection with the imaginary axis at the points φ, ψ. Then, $\frac{\partial}{\partial\theta} \arg e^{i\theta} f'(e^{i\theta}, t) \geq 0$, or $\mathrm{Re}\,(1 + \zeta f''(\zeta,0)/f'(\zeta,0)) > 0$ for $\zeta = re^{i\theta}$ for all $r \sim 1$, $r \neq 1$, and $\theta \sim \varphi$ or ψ. Thus, the level lines of the function f are of positive curvature, therefore, due to the argument of continuity, they are still of positive curvature locally in time $t > 0$ and reachable by horizontal rays. So we suppose that $\arg \zeta_0 \equiv \theta_0 \neq \varphi$, or ψ on the smooth boundary of $f(U,0) = \Omega_0$. Let us assume $\theta_0 \in [0,\varphi) \cup (\psi, 2\pi]$. For other locations of θ_0 the proof is similar. For definiteness, we put

$$\arg \zeta_0 f'(\zeta_0, 0) = \frac{\pi}{2}. \tag{4.18}$$

Since $f'(e^{i\theta}, t) \neq 0$, we have that $\mathrm{Im}\, \zeta_0 f'(\zeta_0, 0) > 0$. Since ζ_0 is a critical point and the image of the unit disk U under the mapping $\zeta f'(\zeta, 0)$ touches the imaginary axis at the point $\zeta_0 = e^{i\theta_0}$, we deduce that

$$\begin{aligned} \frac{\partial}{\partial\theta} \arg e^{i\theta} f'(e^{i\theta}, 0)\big|_{\theta=\theta_0} &= 0, \\ \frac{\partial}{\partial r} \arg e^{i\theta_0} f'(re^{i\theta_0}, 0)\big|_{r=1} &\geq 0. \end{aligned}$$

We calculate

$$\mathrm{Re}\left[1 + \frac{\zeta_0 f''(\zeta_0, 0)}{f'(\zeta_0, 0)}\right] = 0, \tag{4.19}$$

$$\mathrm{Im}\left[1 + \frac{\zeta_0 f''(\zeta_0, 0)}{f'(\zeta_0, 0)}\right] \geq 0. \tag{4.20}$$

Let us show that in (4.20) the equality sign is never attained. If it were so, we would conclude that $(\zeta f'(\zeta, 0))'\big|_{\zeta=\zeta_0} = 0$ because of (4.18). This means that the function

Re $\zeta f'(\zeta, 0)$ admits both signs in a neighbourhood of ζ_0 in U. This contradicts the condition that the level lines preserve the property to be convex in the direction of $\mathbb{R}$. Therefore, there is a strict inequality in (4.20).

Then we derive

$$\frac{\partial}{\partial t} \arg \zeta f'(\zeta, t) = \text{Im} \frac{\partial}{\partial t} \log f'(\zeta, t) = \text{Im} \frac{\frac{\partial}{\partial t} f'(\zeta, t)}{f'(\zeta, t)}. \tag{4.21}$$

Now we use the result of differentiating (4.11) and come to the equality

$$|f'(\zeta, t)|^2 \text{Im} \frac{\frac{\partial}{\partial t} f'(\zeta, t)}{f'(\zeta, t)} = \text{Im} \left(\overline{\zeta f'(\zeta, t)} \dot{f}(\zeta, t)\right) \left(\overline{\frac{\zeta f''(\zeta, t)}{f'(\zeta, t)}} - 1\right).$$

Substituting (4.31) and (4.19) in the latter expression we finally have

$$\frac{\partial}{\partial t} \arg \zeta f'(\zeta, t)\bigg|_{\zeta = e^{i\theta_0}} = \frac{Q}{2\pi |f'(e^{i\theta_0}, 0)|^2} \text{Im} \left(1 + \frac{e^{i\theta_0} f''(e^{i\theta_0}, 0)}{f'(e^{i\theta_0}, 0)}\right).$$

The right-hand side of this equality is strictly negative because of (4.20) and the remark thereafter. Hence, $\arg e^{i\theta} f'(e^{i\theta}, t) < \frac{\pi}{2}$ for $t > 0$ close to 0 in a neighbourhood of θ_0. This contradicts the assumption that ζ_0 is a critical point and ends the proof for the class $C_{\mathbb{R}}$. □

Close-to-convex dynamics

Of course, a function from $C_{\mathbb{R}}$ is close-to-convex. But in general, a result analogous to the previous theorem for close-to-convex functions is not true [173], because the level lines of a close-to-convex function continues to be close-to-convex but may fail to be $C_{\mathbb{R}}$. Here we give an example to prove that the solutions of the inner problem do not preserve the property of the initial flow domain to be close-to-convex.

It is known [190] that close-to-convexity is equivalent to the following analytic assertion: for any θ_1, θ_2 such that $0 < \theta_2 - \theta_1 < 2\pi$, the inequality

$$\int_{\theta_1}^{\theta_2} \left[1 + \text{Re} \frac{re^{i\theta} f''(re^{i\theta})}{f'(re^{i\theta})}\right] d\theta \geq -\pi, \quad 0 < r \leq 1,$$

holds (the equality sign is possible when $r = 1$). Denote

$$H(\theta_1, \theta_2, f(\zeta, t)) = \int_{\theta_1}^{\theta_2} \left[1 + \text{Re} \left(\zeta \frac{\partial}{\partial \zeta} \log f'(\zeta, t)\right)\right] d\theta, \quad \zeta = re^{i\theta}.$$

Let $f(\zeta, t')$ be a critical close-to-convex mapping, i.e., there are θ_1 and θ_2, such that $0 < \theta_2 - \theta_1 < 2\pi$ and $H(\theta_1, \theta_2, f(e^{i\theta}, t')) = -\pi$. Without loss of generality we assume $t' = 0$. Therefore, the integrals $J_1(\theta) = H(\theta, \theta_2, f_0(e^{i\theta}))$ and $J_2(\theta) = H(\theta_1, \theta, f_0(e^{i\theta}))$, as differentiable functions of the first and the second argument of H respectively, have the local maxima $J_1(\theta)$ at the point θ_1 and $J_2(\theta)$ at the point θ_2, i.e., $J_1'(\theta_1) = 0$ and $J_2'(\theta_2) = 0$. This means

$$\operatorname{Re}\left(e^{i\theta_1}\frac{f_0''(e^{i\theta_1})}{f_0'(e^{i\theta_1})}\right) = \operatorname{Re}\left(e^{i\theta_2}\frac{f_0''(e^{i\theta_2})}{f_0'(e^{i\theta_2})}\right) = -1. \tag{4.22}$$

The function $J_3(r) = H(\theta_1, \theta_2, f_0(re^{i\theta}))$ locally decreases in $r \in (0,1]$ in the neighbourhood of 1^-, and J_3 is differentiable in this semi-interval. Hence,

$$J_3'(1^-) = \operatorname{Im}\int\limits_{\theta_1}^{\theta_2} d\left(\frac{e^{i\theta}f_0''(e^{i\theta})}{f_0'(e^{i\theta})}\right) = \operatorname{Im}\left(\frac{e^{i\theta_2}f_0''(e^{i\theta_2})}{f_0'(e^{i\theta_2})} - \frac{e^{i\theta_1}f_0''(e^{i\theta_1})}{f_0'(e^{i\theta_1})}\right) \le 0. \tag{4.23}$$

Checking the sign of

$$\frac{\partial H(\theta_1, \theta_2, f_0(e^{i\theta}))}{\partial t}$$

at the point $t = 0^+$ we come to the decision about close-to-convexity. If it is negative, then there is a neighbourhood $(0, \varepsilon)$ where $H(\theta_1, \theta_2, f(\zeta, t)) < -\pi$, that contradicts the condition of close-to-convexity.

As in preceding subsections we deduce from the Polubarinova-Galin equation

$$\frac{\partial}{\partial t}H(\theta_1, \theta_2, f(e^{i\theta}, t))\bigg|_{t=0} = \frac{-Q}{|f'(e^{i\theta},0)|^2}\operatorname{Im}\frac{e^{i\theta}f''(e^{i\theta},0)}{f'(e^{i\theta},0)}\Bigg]_{\theta_1}^{\theta_2}. \tag{4.24}$$

We have $Q < 0$ for injection and consider an example of critical map

$$f_0(\zeta) = \int\limits_0^{\zeta} \exp\left[-\frac{1}{2\pi}\int\limits_0^{2\pi}(\gamma(\theta) - \theta - \frac{\pi}{2})S(\theta, \zeta)d\theta\right]d\zeta,$$

where $S(\theta, \zeta)$ is the Schwarz-Poisson kernel

$$S(\theta, \zeta) = \frac{e^{i\theta} + \zeta}{e^{i\theta} - \zeta},$$

$\gamma(\theta) = \frac{3}{2}\pi(1 + \sin(\frac{\alpha}{\pi}(\theta - \pi)))$, $\alpha = \pi - \arcsin\frac{2}{3}$, $\theta_2 = -\theta_1 = \pi - \frac{\pi^2}{2\alpha}$. This function satisfies the condition $f_0(\bar{\zeta}) = \overline{f_0(\zeta)}$, hence $|f_0'(e^{i\theta_1})| = |f_0'(e^{i\theta_2})|$. By (4.23) we obtain that the right-hand side in (4.24) is not positive. So it suffices to prove that $\operatorname{Im}\frac{e^{i\theta}f_0''(e^{i\theta})}{f_0'(e^{i\theta})}\Big|_{\theta_1}^{\theta_2} \ne 0$.

Calculate, integrating by parts

$$\operatorname{Im} \frac{e^{i\theta} f_0''(e^{i\theta})}{f_0'(e^{i\theta})}\Bigg|_{\theta_1}^{\theta_2} = \frac{-3\alpha^2}{2\pi^2} \int\limits_0^{2\pi} \sin(\frac{\alpha}{\pi}(\theta - \pi)) \log\left[1 + \cos(\theta + \frac{\pi^2}{2\alpha})\right] d\theta. \qquad (4.25)$$

From the obvious inequality $\frac{\pi}{2} < \frac{\pi^2}{2\alpha} < \pi$ it easily follows that the right-hand side of (4.25) is strictly negative and remains negative for $f(\zeta, t)$ in some time interval $t \in [0, \varepsilon)$.

To complete the proof we show that $f_0(\zeta)$ is close-to-convex and univalent, verifying the condition Re $f_0'(\zeta) \geq 0$. This condition is equivalent to the inequality

$$-\frac{\pi}{2} \leq \operatorname{Re} \frac{1}{2\pi} \int\limits_0^{2\pi} (\gamma(\theta) - \theta - \frac{\pi}{2}) \frac{e^{i\theta} + \zeta}{e^{i\theta} - \zeta} d\theta \leq \frac{\pi}{2}.$$

The right-hand side inequality is equivalent to $\int_0^{2\pi} (\gamma(\theta) - \theta - \pi) P(\zeta, \theta) d\theta \leq 0$, where $P(\zeta, \theta) \equiv \operatorname{Re} S(\zeta, \theta)$ is the Poisson kernel. The sign is obviously verified. The left-hand side inequality can be considered analogously.

4.3.2 Dynamics with small surface tension

As we mentioned in Section 1.4.4, in most practical experiments a zero surface tension process never occurs in the three-dimensional case. A 2-D approximation of the 3-D effect is given by introducing surface tension in the 2-D case. At the same time the non zero surface tension model regularizes the ill-posed problem.

We recall from Section 1.4.4 that the governing equations for the nonzero surface tension model are

$$\Delta p = Q\delta_0, \quad \text{in} \quad z \in \Omega(t), \qquad (4.26)$$

$$p = \gamma\kappa(z), \quad \text{on} \quad z \in \Gamma(t), \qquad (4.27)$$

$$v_n = -\frac{\partial p}{\partial n}, \quad \text{on} \quad z \in \Gamma(t), \qquad (4.28)$$

where κ is the curvature of the boundary and γ is surface tension. The problem of the existence of the solution in the non-zero surface tension case was discussed in Section 1.4.4.

Now we obtain the equation for the free boundary using an auxiliary parametric univalent map. To derive it let us consider the complex potential $W(z, t)$, Re $W = p$. For each fixed t this is an analytic function defined in $\Omega(t)$ which solves the Dirichlet problem (4.26)–4.27 in the sense that its real part induces the same distribution as the solution of the problem (4.26–4.27). In the neighbourhood of the origin we have the expansion

$$W(z, t) = \frac{Q}{2\pi} \log z + w_0(z, t), \qquad (4.29)$$

where $w_0(z,t)$ is an analytic regular function in $\Omega(t)$.

To derive the equation for the free boundary $\Gamma(t)$ we use the same arguments as in Sections 1.4.1, 1.4.2 and consider the Riemann conformal univalent map $f(\zeta,t)$ from the unit disk $U=\{\zeta : |z|<1\}$ into the phase plane $f: U\to\Omega(t)$, $f(0,t)=0$, $f'(0,t)>0$. Then the moving boundary is parameterized by $\Gamma(t)=\{f(e^{i\theta},t), \theta\in[0,2\pi)\}$. The normal velocity v_n of $\Gamma(t)$ in the outward direction is given by $v_n=-\partial p/\partial n$. The normal outer vector is given by the formula

$$\mathbf{n}=\zeta\frac{f'}{|f'|},\quad \zeta\in\partial U.$$

Therefore, the normal velocity is obtained as

$$v_n=\mathbf{V}\cdot\mathbf{n}=-\mathrm{Re}\left(\frac{\partial W}{\partial z}\zeta\frac{f'}{|f'|}\right).$$

The superposition $(W\circ f)(\zeta,t)$ is an analytic function in the unit disk. Since the Laplacian is invariant under a conformal map, the solution to the Dirichlet problem (4.26–4.27) is given in terms of the ζ-plane as

$$(W\circ f)(\zeta,t)=\frac{Q}{2\pi}\log\zeta+\frac{\gamma}{2\pi}\int_0^{2\pi}\kappa(e^{i\theta},t)\frac{e^{i\theta}+\zeta}{e^{i\theta}-\zeta}d\theta+iC, \tag{4.30}$$

where

$$\kappa(e^{i\theta},t)=\frac{\mathrm{Re}\,\left(1+e^{i\theta}f''(e^{i\theta},t)/f'(e^{i\theta},t)\right)}{|f'(e^{i\theta},t)|},\quad \theta\in[0,2\pi).$$

We calculate

$$\frac{\partial\kappa(e^{i\theta},t)}{\partial\theta}=\frac{-\mathrm{Im}\;e^{2i\theta}S_f(e^{i\theta},t)}{|f'(e^{i\theta},t)|},$$

with the Schwarzian derivative (see e.g. [92], [270], [271])

$$S_f(\zeta)=\frac{\partial}{\partial\zeta}\left(\frac{f''(\zeta,t)}{f'(\zeta,t)}\right)-\frac{1}{2}\left(\frac{f''(\zeta,t)}{f'(\zeta,t)}\right)^2.$$

Differentiating (4.30) we deduce that

$$\zeta\frac{\partial W}{\partial z}f'(\zeta,t)=\frac{Q}{2\pi}+\frac{\gamma}{\pi}\int_0^{2\pi}\frac{\kappa(e^{i\theta})\zeta e^{i\theta}}{(e^{i\theta}-\zeta)^2}d\theta,\quad \zeta\in U.$$

Integrating by parts we obtain

$$\zeta\frac{\partial W}{\partial z}f'(\zeta,t)=\frac{Q}{2\pi}-\frac{\gamma}{2\pi i}\int_0^{2\pi}\frac{\mathrm{Im}\;e^{2i\theta}S_f(e^{i\theta},t)}{|f'(e^{i\theta},t)|}\frac{e^{i\theta}+\zeta}{e^{i\theta}-\zeta}d\theta.$$

On the other hand, we have $v_n = \mathrm{Re}\,\dot{f}\,\overline{\zeta f'/|f'|}$, and applying the Sokhotskiĭ-Plemelj formulae [242] we, finally, get

$$\mathrm{Re}\,\dot{f}(\zeta,t)\overline{\zeta f'(\zeta,t)} = -\frac{Q}{2\pi} - \gamma H\Big[i\mathrm{Im}\,\frac{\zeta^2 S_f(\zeta,t)}{|f'(\zeta,t)|}\Big](\theta), \tag{4.31}$$

$\zeta = e^{i\theta}$, where the Hilbert transform in (4.31) is of the form

$$\frac{1}{\pi}\mathbf{p.v.}_\theta \int_0^{2\pi} \frac{\psi(e^{i\theta'})d\theta'}{1 - e^{i(\theta-\theta')}} = H[\psi](\theta).$$

For $\gamma = 0$ we just have equation (1.16).

From (4.31) one can derive a Löwner-Kufarev type equation by the Schwarz-Poisson formula:

$$\dot{f} = -\zeta f' \frac{1}{2\pi}\int_0^{2\pi} \frac{1}{|f'(e^{i\theta},t)|^2}\Big(\frac{Q}{2\pi} + \gamma H\Big[i\mathrm{Im}\,\frac{e^{2i\theta}S_f(e^{i\theta},t)}{|f'(e^{i\theta},t)|}\Big](\theta)\Big)\frac{e^{i\theta}+\zeta}{e^{i\theta}-\zeta}d\theta, \tag{4.32}$$

where $\zeta \in U$.

An interesting question appears when $\gamma \to 0$. It turns out that the solution in the limiting γ-surface-tension case need not always be the corresponding zero surface tension solution (see the discussion in [310], [324], [330]). This means that starting with a domain $\Omega(0) = \Omega(0,\gamma)$ we come to the domain $\Omega(t,\gamma)$ at an instant t using surface tension γ and to the domain $\Omega(t)$ at the same instant t in the zero surface tension model. Then the domain $\lim\limits_{\gamma\to 0}\Omega(t,\gamma) = \Omega(t,0)$ is not necessarily the same as $\Omega(t)$ (see numerical evidence in [277]). If the boundary Γ is highly curved, then the condition (4.27) must be used even though γ is small. Obviously, the non-zero surface tension model never develops cusps. Thus, solutions and geometric behaviour of the free boundary for small γ are of particular interest.

4.3.3 Geometric properties in the presence of surface tension

We need the following technical lemma.

Lemma 4.3.5. *For the function $f : U \to \mathbb{C}$ which parameterizes the phase domain $\Omega(t)$ we have the equality*

$$\frac{\partial}{\partial\theta}H\Big[\frac{ie^{2i\theta}\mathrm{Im}\,S_f(e^{i\theta})}{|f'(e^{i\theta})|}\Big](\theta) = -H[iA](\theta)$$

with

$$A(\zeta) = \frac{\mathrm{Re}\,\Big(2\zeta^2 S_f(\zeta) + \zeta\Big[\Big(\frac{f''(\zeta)}{f'(\zeta)}\Big)'' - \frac{f''(\zeta)}{f'(\zeta)}\Big(\frac{f''(\zeta)}{f'(\zeta)}\Big)'\Big]\Big) + \mathrm{Im}\,\frac{\zeta f''(\zeta)}{f'(\zeta)}\,\mathrm{Im}\,\zeta^2 S_f(\zeta)}{|f'(\zeta)|}.$$

Proof. We denote

$$\Phi(\zeta) = \frac{i}{\pi}\int_0^{2\pi} \frac{\operatorname{Im} e^{2i\theta'} S_f(e^{i\theta'})}{|f'(e^{i\theta'},t)|}\frac{e^{i\theta'}}{e^{i\theta'}-\zeta}d\theta', \ \ \zeta \in U.$$

Then, by the Sokhotskiĭ-Plemelj formulae we deduce that

$$\lim_{\zeta\to(1-0)e^{i\theta}} \Phi(\zeta) = i\frac{\operatorname{Im} e^{2i\theta'} S_f(e^{i\theta'})}{|f'(e^{i\theta'},t)|} - H\Big[i\frac{\operatorname{Im} e^{2i\theta'} S_f(e^{i\theta'})}{|f'(e^{i\theta'},t)|}\Big](\theta).$$

We note that the second term in the above relation is real, which can be easily seen by the definition of Φ and the Schwarz integral formula. The left-hand side in (4.31) is differentiable in θ and that is why one can calculate the derivative in question as the limit

$$\frac{\partial}{\partial\theta}H\Big[i\frac{\operatorname{Im} e^{2i\theta'} S_f(e^{i\theta'})}{|f'(e^{i\theta'},t)|}\Big](\theta) = \operatorname{Im} \lim_{\zeta\to(1^-)e^{i\theta}} \zeta\Phi'(\zeta).$$

We calculate

$$\Phi'(\zeta) = \frac{i}{\pi}\int_0^{2\pi} \frac{\operatorname{Im} e^{2i\theta'} S_f(e^{i\theta'})}{|f'(e^{i\theta'},t)|}\frac{e^{i\theta'}}{(e^{i\theta'}-\zeta)^2}d\theta'.$$

Integration by parts leads to the equality

$$\zeta\Phi'(\zeta) = \frac{1}{\pi}\int_0^{2\pi} A(e^{i\theta'})\frac{\zeta\, d\theta'}{e^{i\theta'}-\zeta}.$$

Thus, we apply the Sokhotskiĭ-Plemelj formulae once again and get the assertion of Lemma 4.3.5. □

Theorem 4.3.6. *Let $Q < 0$ and the surface tension γ be sufficiently small. If the initial domain $\Omega(0)$ is strongly starlike of order α, then there exists $t = t(\gamma) \leq t_0$, such that the family of domains $\Omega(t)$ (in the sequel, the family of univalent functions $f(\zeta,t)$) preserves this property during the time $t \in [0, t(\gamma)]$.*

Remark 4.3.7. *For $\gamma = 0$ we have the result of Theorem* 4.3.2.

Proof. If we consider f in the closure of U, then the inequality sign in (4.3) can be replaced by ($\leq$) where equality can be attained for $|\zeta| = 1$.

We suppose that there exists a critical map $f \in S^*(\alpha)$ of exact order α, that is the image of U under the map $\zeta f'(\zeta,t)/f(\zeta,t)$, $|\zeta| \leq 1$ touches the boundary rays $l^{\pm}$ of the angle $\arg w \in [-\alpha\frac{\pi}{2}, \alpha\frac{\pi}{2}]$, say there exist such $t' \geq 0$ and $\zeta_0 = e^{i\theta_0}$, that

$$\arg\frac{\zeta_0 f'(\zeta_0,t')}{f(\zeta_0,t')} = \alpha\frac{\pi}{2} \quad (\text{or } -\alpha\tfrac{\pi}{2}), \tag{4.33}$$

and for any $\varepsilon > 0$ there is such $t > t'$ and $\theta \in (\theta_0 - \varepsilon, \theta_0 + \varepsilon)$ that

$$\arg \frac{e^{i\theta} f'(e^{i\theta}, t)}{f(e^{i\theta}, t)} \geq \alpha \frac{\pi}{2} \quad (\text{or} \leq -\alpha \tfrac{\pi}{2}).$$

For definiteness we put the sign (+) in (4.33). Without loss of generality, let us assume $t' = 0$. Since $f'(e^{i\theta}, t) \neq 0$, our assumption about the sign in (4.33) yields that

$$\operatorname{Im} \frac{\zeta_0 f'(\zeta_0, 0)}{f(\zeta_0, 0)} > 0, \tag{4.34}$$

(the negative case is considered similarly).

Since ζ_0 is a critical point and the image of the unit disk U under the mapping $\frac{\zeta f'(\zeta,0)}{f(\zeta,0)}$ touches the ray l^+ at the point $\zeta_0 = e^{i\theta_0}$, we deduce that

$$\begin{aligned} \frac{\partial}{\partial \theta} \arg \frac{e^{i\theta} f'(e^{i\theta}, 0)}{f(e^{i\theta}, 0)} \bigg|_{\theta=\theta_0} &= 0, \\ \frac{\partial}{\partial r} \arg \frac{re^{i\theta_0} f'(re^{i\theta_0}, 0)}{f(re^{i\theta}, 0)} \bigg|_{r=1} &\geq 0. \end{aligned}$$

We calculate

$$\operatorname{Re} \left[1 + \frac{\zeta_0 f''(\zeta_0, 0)}{f'(\zeta_0, 0)} - \frac{\zeta_0 f'(\zeta_0, 0)}{f(\zeta_0, 0)} \right] = 0, \tag{4.35}$$

$$\operatorname{Im} \left[1 + \frac{\zeta_0 f''(\zeta_0, 0)}{f'(\zeta_0, 0)} - \frac{\zeta_0 f'(\zeta_0, 0)}{f(\zeta_0, 0)} \right] \geq 0. \tag{4.36}$$

One can derive

$$\frac{\partial}{\partial t} \arg \frac{\zeta f'(\zeta, t)}{f(\zeta, t)} = \operatorname{Im} \frac{\partial}{\partial t} \log \frac{f'(\zeta, t)}{f(\zeta, t)} = \operatorname{Im} \left(\frac{\frac{\partial}{\partial t} f'(\zeta, t)}{f'(\zeta, t)} - \frac{\frac{\partial}{\partial t} f(\zeta, t)}{f(\zeta, t)} \right). \tag{4.37}$$

We now differentiate the equation (4.31) with respect to θ using Lemma 4.3.5,

$$\operatorname{Im} \left(\overline{f'(\zeta, t)} \frac{\partial}{\partial t} f'(\zeta, t) - \overline{\zeta f'(\zeta, t)} \dot{f}(\zeta, t) - \overline{\zeta^2 f''(\zeta, t)} \dot{f}(\zeta, t) \right) = -\gamma H[iA](\theta), \tag{4.38}$$

for $\zeta = e^{i\theta}$. This equality is equivalent to the following:

$$\begin{aligned} &|f'(\zeta, t)|^2 \operatorname{Im} \left(\frac{\frac{\partial}{\partial t} f'(\zeta, t)}{f'(\zeta, t)} - \frac{\frac{\partial}{\partial t} f(\zeta, t)}{f(\zeta, t)} \right) \\ &= \operatorname{Im} (\overline{\zeta f'(\zeta, t)} \dot{f}(\zeta, t)) \left(\overline{\frac{\zeta f''(\zeta, t)}{f'(\zeta, t)}} - \frac{\zeta f'(\zeta, t)}{f(\zeta, t)} + 1 \right) - \gamma H[iA](\theta). \end{aligned}$$

Substituting (4.31) and (4.35) in the latter expression we finally have

$$\frac{\partial}{\partial t}\arg\frac{\zeta f'(\zeta,t)}{f(\zeta,t)}\Big|_{\zeta=e^{i\theta_0},t=0}=\frac{1}{|f'(e^{i\theta_0},0)|^2}\left(\frac{Q}{2\pi}+\gamma H\left[i\mathrm{Im}\,\frac{e^{2i\theta}S_f(e^{i\theta},t)}{|f'(e^{i\theta},t)|}\right](\theta_0)\right)$$

$$\times\mathrm{Im}\left(\frac{e^{i\theta_0}f'(e^{i\theta_0},0)}{f(e^{i\theta_0},0)}+\frac{e^{i\theta_0}f''(e^{i\theta_0},0)}{f'(e^{i\theta_0},0)}\right)-\gamma\frac{H[iA(e^{i\theta})](\theta_0)}{|f'(e^{i\theta_0},0)|^2}. \tag{4.39}$$

The right-hand side of this equality is strictly negative for small γ because of (4.34), (4.36). Therefore,

$$\arg\frac{e^{i\theta}f'(e^{i\theta},t)}{f(e^{i\theta},t)}<\alpha\frac{\pi}{2}$$

for $t>0$ (close to 0) in some neighbourhood of θ_0. This contradicts the assumption that $\Omega(t)$ fails to be starlike for $t>0$ and ends the proof for the class $S^*(\alpha)$. □

Problem 4.3.8. *Let $\Omega(0)$ be starlike of order α (not strongly), $f(\zeta,0)\in S^*_\alpha$, and $Q<0$. Do the domains of the Hele-Shaw evolution $\Omega(t)$ remain starlike of the same order?*

Problem 4.3.9. *Let $\Omega(0)$ be convex. It is known that for $Q<0$ the Hele-Shaw evolution $\Omega(t)$ is not necessarily convex. Are there conditions that guarantee convexity of $\Omega(t)$ for all the time?*

Problem 4.3.10. *If $Q=0$ and $\gamma\neq 0$, then the driving mechanism is only surface tension. Let $\Omega(0)$ be convex. Are the domains $\Omega(t)$ convex for all t?*

4.3.4 Unbounded regions with bounded complement

We now consider a Hele-Shaw cell in which the fluid occupies a full neighbourhood of infinity, so that the complementary set is a finite bubble. Injection or suction is supposed to take place at the point of infinity. This model has various applications in the boundary problems of gas mechanics, problems of metal or polymer swamping, etc., where the air viscosity is neglected. More about this problem is found in [104], [178].

As usual we denote by $\Omega(t)$ the fluid domain, i.e., in this case the unbounded complement of the bubble in consideration (see Figure 4.4). Let p be the pressure in the domain $\Omega(t)$ occupied by the fluid. We construct the complex potential $W(z,t)$, $\mathrm{Re}\, W=p$. For each fixed t this is an analytic function defined in $\Omega(t)$ which solves the problem

$$\Delta p = 0, \quad \text{in} \quad z\in\Omega(t), \tag{4.40}$$

$$p = 0, \quad \text{on} \quad z\in\Gamma(t), \tag{4.41}$$

$$v_n = -\frac{\partial p}{\partial n}, \quad \text{on} \quad z\in\Gamma(t), \tag{4.42}$$

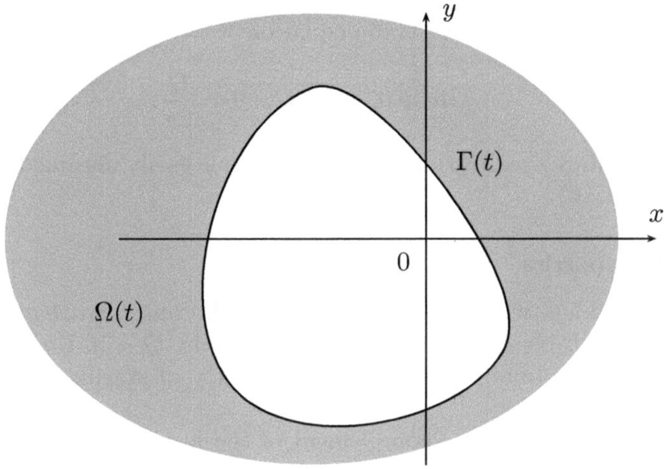

Figure 4.4: $\Omega(t)$ is the complement to a bounded simply connected bubble with the boundary $\Gamma(t)$ and the sink/source at infinity

normalized about infinity by

$$p \sim \frac{Q}{2\pi} \log |z|, \quad \text{as } |z| \to \infty,$$

where Q is the rate of bubble release caused by air extraction, $Q < 0$ in the case of a contracting bubble, $Q > 0$ otherwise. For this choice of Q the case of $Q < 0$ is stable whereas $Q > 0$ is not. Mathematical treatment for the case of a contracting bubble was presented in [104]. In particular, the problem of the limiting configuration was solved. It was proved that the moving boundary tends to a finite number of points which give the minimum to a certain potential. There an interesting problem was posed: to describe domains whose dynamics present only one limiting point. Howison [178] proved that a contracting elliptic bubble has a homothetic dynamics to a point (in particular, this is obvious for a circular one). Entov, Ètingof [104] (see also [333]) have shown that a contracting bubble which is convex at the initial instant preserves this property up to the moment when its boundary reduces to a point. These domains are called "simple" in [104].

Now our parametric domain is the exterior part of the unit disk, and there exists a unique conformal univalent map $F(\zeta, t)$ from the domain $U^* = \{\zeta : |\zeta| > 1\}$ into the phase plane $F : U^* \to \Omega(t)$, $F(\zeta, t) = a\zeta + a_0 + \frac{a_{-1}}{\zeta} + \cdots$, $a > 0$. By a shift we assume $0 \notin \text{closure}(F(U^*, t))$.

We repeat the calculations of the preceding subsections taking into account that the normal vector is calculated as

$$\mathbf{n} = -\zeta F'/|F'|, \quad |\zeta| = 1,$$

and come to the Polubarinova-Galin equation:

$$\mathrm{Re}\,[\dot F(\zeta,t)\overline{\zeta F'(\zeta,t)}] = \frac{Q}{2\pi}. \tag{4.43}$$

The Löwner-Kufarev realization of this equation is easily obtained by analogy with the equation (1.16).

Hereditary properties

We call the problem with injection at infinity the *outer problem.* In other words we consider the stable case of a contracting bubble, $Q<0$. Here we prove that a contracting starlike bubble preserves the property of starlikeness and directional convexity.

Let us suppose that the complement of the fluid domain contains the origin and is starlike with respect to the origin at the initial instant. Therefore, if a function $F(\zeta)=a\zeta+a_0+a_{-1}/\zeta+\cdots$ is defined outside of the unit disk, then the function $f(\zeta)=1/F(1/\zeta)$ is holomorphic in U. Then the equation (4.43) can be rewritten in terms of this holomorphic function as

$$\mathrm{Re}\;[\dot f(\zeta,t)\overline{\zeta\, f'(\zeta,t)}] = -\frac{Q|f(\zeta,t)|^4}{2\pi},\quad |\zeta|=1,\quad Q<0. \tag{4.44}$$

The condition of starlikeness Re $\zeta F'/F>0$, $|\zeta|>1$ is equivalent to Re $\zeta f'/f>0$, $|\zeta|<1$. We must control the sign of the functional $\frac{\partial}{\partial t}\arg \zeta f'/f$. Differentiating (4.44) with respect to θ we obtain

$$\frac{\partial}{\partial t}\arg\frac{\zeta f'(\zeta,t)}{f(\zeta,t)} = \mathrm{Im}\left(\frac{\dot f'}{f'}-\frac{\dot f}{f}\right)$$

$$=\frac{1}{|f'|^2}\mathrm{Im}\left(\dot f\,\overline{\zeta f'}\left(1-\frac{\zeta f'}{f}+\overline{\frac{\zeta f''}{f'}}\right)+4\frac{\zeta f'}{f}\frac{Q|f|^4}{2\pi}\right)$$

on the circle $|\zeta|=1$. At a critical point ζ_0 we have

$$\mathrm{Re}\,\frac{\zeta_0 f'(\zeta_0,0)}{f(\zeta_0,0)}=0,\quad \mathrm{Im}\,\frac{\zeta_0 f'(\zeta_0,0)}{f(\zeta_0,0)}>0,$$

$$\mathrm{Re}\left[1+\frac{\zeta_0 f''(\zeta_0,0)}{f'(\zeta_0,0)}-\frac{\zeta_0 f'(\zeta_0,0)}{f(\zeta_0,0)}\right]=0,$$

$$\mathrm{Im}\left[1+\frac{\zeta_0 f''(\zeta_0,0)}{f'(\zeta_0,0)}-\frac{\zeta_0 f'(\zeta_0,0)}{f(\zeta_0,0)}\right]\geq 0.$$

Finally, we have

$$\frac{\partial}{\partial t}\arg\frac{\zeta f'(\zeta,t)}{f(\zeta,t)} = \frac{Q|f|^4}{2\pi|f'|^2}\mathrm{Im}\left((1-\frac{\zeta f'}{f}+\frac{\zeta f''}{f'})+6\frac{\zeta f'}{f}\right). \tag{4.45}$$

The right-hand side of (4.45) is strictly negative due to the previous chain of inequalities. Therefore, we have the affirmative answer in the case of a contracting bubble.

Theorem 4.3.11. *Let $Q < 0$. If the domain of a contracting bubble $D_0 = \mathbb{C} \setminus \overline{\Omega_0}$ is starlike (or strongly starlike) with respect to a point $z_0 \in D_0$ at the initial instant, then the family of functions $f(\zeta, t)$ and the family of the domains $D(t) = \mathbb{C} \setminus \overline{\Omega(t)}$ preserve the same property as long as the solution exists and $z_0 \in D(t)$.*

In particular, a convex domain D_0 is starlike with respect to any point from D_0, and therefore, the convex dynamics is also preserved, as was proved earlier in [104], [333].

Remark 4.3.12. *Let us give a remark concerning the above result. It can be formulated as follows: if we find a point z_0 in D_0 with respect to which D_0 is starlike, then the domains $D(t)$ are also starlike with respect to the same point z_0 during the existence of the solution or up to the time when $z_0 \in \Gamma(t)$. This means that if D_0 is simple (in the terminology of* [104]*), z_0 is a limiting point in which $D(t)$ contracts, and D_0 is starlike with respect to z_0, then $D(t)$ remains starlike up to the instant when all air is removed (there exist non-convex simple domains, see* [104], [333]*).*

Similarly we establish the following.

Theorem 4.3.13. *If the initial domain is convex in the direction of $\mathbb{R}$ and symmetric with respect to the imaginary axis, then the contracting bubble preserves this property as long as the solution exists.*

Let us set up in the table the information about the dynamics for bounded and unbounded domains for the inner and outer stable (well-posed) problems known at the moment using special univalent functions. Here "yes" means that the property is preserved whereas "no" means that is not. For the outer problem we consider the complement to $\Omega(t)$.

Class of univalent functions	*Inner problem*	*Outer problem*
starlike (or strongly starlike)	yes	yes
convex	no	yes
close-to-convex	no	no
convex in the direction of $\mathbb{R}$ (with the condition for level lines)	yes	yes

For the ill-posed case fewer results are known. Injected air forms a bubble which grows as time increases. It was shown [178] that three kinds of behavior can occur. Firstly, the solution may cease to exist in finite time; secondly, the solution may exist for all time and the free boundary may have one or more limit points

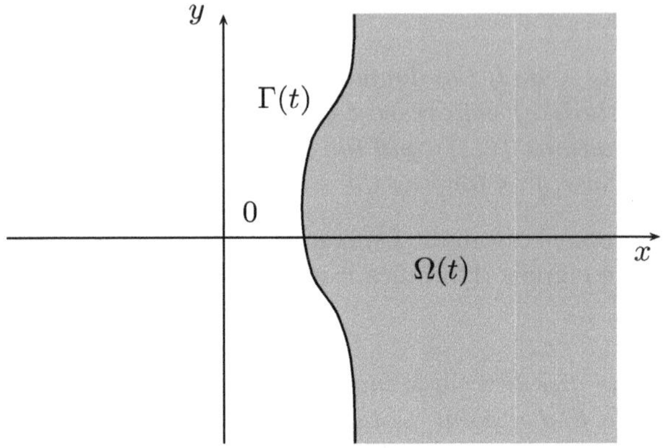

Figure 4.5: $\Omega(t)$ is an infinite domain with the boundary $\Gamma(t)$ extending to infinity and the sink/source at the infinity

as t tends to infinity; and thirdly, the bubble may exist for all time and fill the whole space as t tends to infinity. Making use of quadrature domains it has been proved that the only solutions of the third kind are those in which the bubble is always elliptical. The multidimensional case was treated in [82].

4.3.5 Unbounded regions with the boundary extending to infinity

This model corresponds to the moving fluid front which for definiteness we suppose to be located to the right. More precisely, we denote by $\Omega(t)$ a simply connected domain in the phase z-plane occupied by the moving fluid and its moving boundary $\Gamma(t) = \partial\Omega(t)$ contains the point at infinity. With $z = x + iy$ one can construct a parametrization $\Gamma(t)$ by the equation $\phi(x, y, t) \equiv \phi(z, t) = 0$, so that $\phi(\infty, t) \equiv 0$. Assuming a natural normalization for $\Gamma(t)$ close to ∞, we require that $\Gamma(t)$ is a vertical straight line near infinity (see Figure 4.5). The initial situation is represented at instant $t = 0$ as $\Omega(0) = \Omega_0$, and the boundary $\partial\Omega_0 = \Gamma(0) \equiv \Gamma_0$ is defined parametrically by an implicit function $\phi(x, y, 0) = 0$. We construct the complex potential $W(z, t)$, Re $W = p$, where p is, as usual, a pressure field in $\Omega(t)$. For each fixed t the potential W is an analytic function defined in $\Omega(t)$ which solves the problem

$$\Delta p = 0, \quad \text{in} \quad z \in \Omega(t), \tag{4.46}$$

$$p = 0, \quad \text{on} \quad z \in \Gamma(t), \tag{4.47}$$

$$v_n = -\frac{\partial p}{\partial n}, \quad \text{on} \quad z \in \Gamma(t). \tag{4.48}$$

We assume that the velocity tends to a constant value Q as $x \to \infty$, that is positive when fluid is removed to the right and negative otherwise. In terms of the potential p we have $p(x,y,t)/x \to -Q$ as $x \to \infty$ for any t fixed. The problem of the existence was discussed in Subsection 1.4.4. It is noteworthy that for this case the local solvability and uniqueness was proved by Kimura [195] in presence of surface tension.

We consider the auxiliary parametric complex ζ-plane, $\zeta = \xi + i\eta$. The Riemann Mapping Theorem yields that there exists a conformal univalent map $f(\zeta,t)$ of the right half-plane $H^+ = \{\zeta : \text{Re }\zeta > 0\}$ into the phase plane $f : H^+ \to \Omega(t)$. The half-plane H^+ is a natural parametric domain for Ω. The function $f(\zeta,0) = f_0(\zeta)$ produces a parametrization of Γ_0. The smoothness of the boundary $\Gamma(t)$ and its behavior in the neighbourhood of ∞ allows us to assume the normalization $f(\zeta,\cdot) = a\zeta + a_0 + \frac{a_{-1}}{\zeta} + \cdots$, $\zeta \sim \infty$, $a > 0$, i.e., the function f has an analytic continuation on the imaginary axis ∂H^+ near ∞. Thus, the moving boundary is parameterized by $\Gamma(t) = f(\partial H^+,t)$. The normal velocity v_n of $\Gamma(t)$ in the outward direction is given by $v_n = -\partial p/\partial n$. The normal exterior vector is given by the formula

$$\mathbf{n} = -\frac{\partial f}{\partial \zeta}\left|\frac{\partial f}{\partial \zeta}\right|^{-1}, \quad \zeta \in \partial H^+.$$

The harmonic function p is a linear one. The normalization about infinity implies that $W \circ f = -Q\zeta$ and the Polubarinova-Galin equation is of the form

$$\text{Re }\left(\dot{f}(\zeta,t)\overline{f'(\zeta,t)}\right) = Q, \text{ Re }\zeta = 0. \tag{4.49}$$

The application of the Schwarz integral formula enables us to deduce a Löwner–Kufarev type equation in the right-hand half-plane

$$\frac{\partial f}{\partial t} = -\frac{1}{\pi}\frac{\partial f}{\partial \zeta}\int\limits_{-\infty}^{\infty}\frac{Q}{|f'(i\eta',t)|^2}\frac{d\eta'}{i\eta'-\zeta}, \quad \zeta \in H^+, \tag{4.50}$$

with the initial condition $f(\zeta,0) = f_0(\zeta)$. Taking into account surface tension this equation becomes

$$\text{Re }\left(\dot{f}(\zeta,t)\overline{f'(\zeta,t)}\right) = Q + \gamma H\Big[\frac{i\text{Im }S_f}{|f'|}\Big](\eta), \quad \text{Re }\zeta = 0,$$

with the Hilbert transform defined as

$$\frac{1}{\pi i}\mathbf{p.v.}_\eta\int\limits_{-\infty}^{\infty}\frac{\psi(i\eta')d\eta'}{\eta'-\eta} = H[\psi](\eta).$$

Hereditary properties

We are going to prove that if the initial interface possesses the property to be convex of order α in the negative direction ($H^-_{\mathbb{R}}(\alpha)$), then the free boundary remains convex in the negative direction of the same order in so far as the solution to the Hele-Shaw problem exists in the case $Q < 0$ (the liquid moves to the left). An important remark is that the level lines of a function from $H^-_{\mathbb{R}}(\alpha)$ remain convex in the negative direction.

We are going to prove the following statement.

Theorem 4.3.14. *Let $Q < 0$ and $\Omega(0)$ be a domain convex in the negative direction of order (α). Let the solutions to the equation* (4.49) *exist during the time $t \in [0, t_0]$. Then, for all $t \in [0, t_0]$ the family of functions $f(\zeta, t)$ and the family of domains $\Omega(t)$ preserve the same property of convexity.*

Proof. Let us again suppose the contrary. In other words, there exists $\zeta_0 = e^{i\theta_0}$ that satisfies the equality

$$\arg f'(\zeta_0, 0) = \alpha\frac{\pi}{2} \quad (\text{or } -\alpha\tfrac{\pi}{2}) \tag{4.51}$$

at the initial instant $t = 0$ and for any $\varepsilon > 0$ there is such $t > 0$ and $\theta \in (\theta_0 - \varepsilon, \theta_0 + \varepsilon)$ that $\arg f'(e^{i\theta}, t) \geq \alpha\frac{\pi}{2}$ (or $\leq -\alpha\frac{\pi}{2}$). For definiteness we put the sign (+) in (4.51). This means that the mapping $f(\zeta, 0)$ is critical for the property of convexity in the negative direction. Since the free boundary $\Gamma(t)$ tends to a vertical line as $\eta \to \pm\infty$, we can consider finite critical points $\zeta_0 = i\eta_0$ and the set of such critical points lies in the compact subset on the imaginary axis.

$\Gamma(t)$ is analytic, the function f is analytically extendable onto the imaginary axis, and the derivative $f'(\zeta_0, t) \neq 0$. Suppose that $\operatorname{Im} f'(\zeta_0, 0) > 0$ (for $\operatorname{Im} f'(\zeta_0, 0) < 0$ the proof is similar). Now we show that $f''(\zeta_0, 0) \neq 0$. If not, then the point ζ_0 is a branch point of the function $f'(\zeta, 0)$ and in a neighbourhood of this point in H^+ the quantity $\arg f'(\zeta, 0) - \alpha\pi/2$ admits both positive or negative values. This contradicts the assumption that the function $f(\zeta, 0)$ is convex in the negative direction.

The image of the right half-plane H^+ under the map $f'(\zeta, 0)$ touches the ray $\arg w = \alpha\frac{\pi}{2}$ at the point $f'(\zeta_0, 0)$. Thus, the following statements are true:

$$\frac{\partial}{\partial\eta} \arg f'(i\eta, 0)\Big|_{\eta=\eta_0} = 0, \quad \frac{\partial}{\partial\xi} \arg f'(\xi + i\eta_0, 0)\Big|_{\xi=0} \leq 0.$$

Calculation of the left-hand sides of these formulae leads to the following:

$$\operatorname{Re} \frac{f''(i\eta_0, 0)}{f'(i\eta_0, 0)} = 0, \quad \operatorname{Im} \frac{f''(i\eta_0, 0)}{f'(i\eta_0, 0)} < 0. \tag{4.52}$$

We differentiate (4.49) with respect to η and at the point η_0 using (4.52) we obtain

$$|f'(i\eta,0)|^2 \mathrm{Im}\,\frac{\dot f'(i\eta,0)}{f'(i\eta,0)} = \mathrm{Im}\ \dot f(i\eta,0)\overline{f''(i\eta,0)}$$
$$= \mathrm{Im}\ \dot f(i\eta,0)\overline{f'(i\eta,0)}\,\overline{\frac{f''(i\eta,0)}{f'(i\eta,0)}}. \qquad (4.53)$$

From (4.52) and (4.53) it follows that

$$\frac{\partial}{\partial t}\arg f'(i\eta_0,t)\Big|_{t=0} = -Q\mathrm{Im}\,\frac{f''(i\eta_0,0)}{f'(i\eta_0,0)}. \qquad (4.54)$$

The inequality in (4.52) with $Q<0$ implies that the right-hand side of (4.54) is strictly negative for $t>0$ close to 0 and the inequality $\arg f'(i\eta_0,t)<\alpha\frac{\pi}{2}$ holds in some neighbourhood of $i\eta_0$. This contradicts the assumption that $\Omega(t)$ fails to be convex in the negative direction for $t\geq 0$ and ends the proof. □

Finally, in this section we should say that there are several other processes that involve planar dynamics in Hele-Shaw cells. Let us refer the reader, for example, to the papers [39], [110], [176], [181], [257], [273] and the references therein. We mention here a 600-paper bibliography of free and moving boundary problems for Hele-Shaw and Stokes flow since 1898 up to 1998 collected by Gillow and Howison [125]. Let us mention also a recent work [164] where the authors study the Hele-Shaw flow on hyperbolic surfaces.

Most of the results presented in this section are found in [151], [173], [275], [337], [334], [336].

4.4 Infinite life-span of starlike dynamics

In this section we prove precisely that, starting with a starlike bounded analytic phase domain Ω_0, the Hele-Shaw chain of subordinating domains $\Omega(t)$, $\Omega_0=\Omega(0)$, exists for all time under injection at the point of starlikeness. Suppose that at the initial time the phase domain Ω_0 occupied by the fluid is simply connected and bounded by a smooth analytic curve Γ_0.

In Section 4.3.1 we proved that, starting with a phase domain Ω_0 which is strongly starlike of order α and bounded by an analytic curve, we obtain a subordination chain of domains $\Omega(t)$ (and functions $f(\zeta,t)$) strongly starlike of order $\alpha(t)$ with a decreasing order $\alpha(t)$.

In this section we shall first prove that if the strong solution to (1.16) exists during the time interval $[0,t_0)$, then the limiting function $\lim\limits_{t\to t_0-0} f(\zeta,t)\equiv f(\zeta,t_0)$ is analytic in some neighbourhood of the unit disk U. Here the limit is taken with respect to the uniform convergence on compacts of the unit disk U. It exists because $f(\zeta,t)$ is a subordination chain and due to the Carathéodory Kernel Theorem. Then, we shall give the main result about the infinite life-span, see also [151]. Let us normalize the injection rate by taking $Q=-1$.

Theorem 4.4.1. *Let the strong solution to* (1.16) *with* $Q = -1$ *exist during the time interval* $[0, t_0)$, $0 < t_0 < \infty$, $\Omega(t) = f(U, t)$, *and let the initial function* $f(\zeta, 0)$ *be analytic and univalent in a neighbourhood of the closure of the unit disk* U. *Then the function* $f(\zeta, t)$ *is analytic in* $U_{R(t)}$, *where the radius of analyticity* $R(t) > 1$ *is a nondecreasing function in* $t \in [0, t_0]$. *The function* $f(\zeta, t)$ *is univalent in* U, *and possibly* $f(\zeta, t_0)$ *has a vanishing derivative at some points of the unit circle* ∂U *or is not univalent on* ∂U. *It follows that* $\Omega(t_0) \equiv f(U, t_0)$ *is a simply connected domain with an analytic boundary* $\partial\Omega(t_0)$ *with possible analytic singularities in the form of finitely many cusps and double points. In the case there are no singularities the strong solution can be extended to some time interval* $[0, t_0 + \varepsilon)$.

Proof. By the Carathéodory Kernel Theorem the domain

$$\Omega(t_0) = \bigcup_{t \in [0, t_0)} \Omega(t)$$

is just the same as in the formulation of the theorem and $\Omega(t_0)$ is a simply connected domain. It follows from Proposition 3.3.1 that $\Omega(t_0)$ is also the same as the domain at time t_0 for the weak solution.

We note also that since the normal velocity on the boundary never vanishes, we have the strict monotonicity of the subordination chain of domains:

$$\overline{\Omega(s)} \subset \Omega(t) \quad \text{for } s < t \text{ and } s, t \in (0, t_0). \tag{4.55}$$

Letting $t \to t_0$ we see that (4.55) holds for $t = t_0$, i.e., for $\Omega(t_0)$ as well.

The strong solution exists in the time interval $t \in [0, t_0)$ and coincides with the weak one. Therefore, the statements about $f(\zeta, t_0)$ and $\partial\Omega(t_0)$ follows directly from Theorem 3.4.1.

Let us prove the existence of the extension of the solution to the time interval $[0, t_0 + \varepsilon)$ when there are no singularities on $\partial\Omega(t_0)$. Construct the subordination chain of mappings $f_2(\zeta, t)$ satisfying the Polubarinova-Galin equation (1.16) with the initial data $f_2(\zeta, 0) \equiv f(\zeta, t_0)$. The strong solution exists and is unique locally in time, say $t \in [0, \varepsilon)$. Moreover, we have $\lim\limits_{t \to t_0 - 0} f(\zeta, t) = \lim\limits_{t \to 0+0} f_2(\zeta, t) = f(\zeta, t_0)$ and $\lim\limits_{t \to t_0 - 0} f'(\zeta, t) = \lim\limits_{t \to 0+0} f_2'(\zeta, t) = f'(\zeta, t_0)$ locally uniformly in $U_{1+\eta}$. We recall equation (1.17) (with $Q = -1$):

$$\dot{f}(\zeta, t) = \zeta f'(\zeta, t) \frac{1}{4\pi^2} \int_0^{2\pi} \frac{1}{|f'(e^{i\theta}, t)|^2} \frac{e^{i\theta} + \zeta}{e^{i\theta} - \zeta} d\theta, \quad t \in [0, t_0), \ |\zeta| < 1.$$

A similar equation is valid for the chain $f_2(\zeta, t)$ in the time interval $[0, \varepsilon)$. Taking the limit in the above equation as $t \to t_0 - 0$ we observe that there exists the one-sided limit $\dot{f}(\zeta, t_0 - 0)$. Similarly, there exists the one-sided limit $\dot{f}_2(\zeta, 0^+)$ and they are equal. Let us define $f(\zeta, t) \equiv f_2(\zeta, t - t_0)$ in the interval $t \in [t_0, t_0 + \varepsilon)$. Above observations yield that the so-extended function is continuous in the interval $t \in$

$[0, t_0+\varepsilon)$, analytic, univalent and starlike in some neighbourhood of $\bar{U}$. Moreover, it is differentiable at the point $t = t_0$, and being extended onto the unit circle, satisfies the equation (1.16). Thus, it is a unique strong solution in the interval $t \in [0, t_0 + \varepsilon)$. This finishes the proof of the theorem. □

Lemma 4.4.2. *Under the assumptions of the previous theorem, if Ω_0 is starlike ($f_0 \in S^*$) then the limiting domain $\Omega(t_0)$ has no singularities on the boundary.*

Proof. The function $f(\zeta, t)$ belongs to the class $S^*(\alpha(t))$ with $\alpha(t) < 1$ for any $t \in (0, t_0)$ due to Theorem 4.3.2, where $\alpha(t)$ strictly decreases with respect to t.

Define the limiting function $f(\zeta, t_0) = \lim_{t \to t_0 - 0} f(\zeta, t)$, where the limit is taken locally uniformly in U. The function $f(\zeta, t_0)$ is univalent, strongly starlike of order $\alpha(t_0) = \lim_{t \to t_0 - 0} \alpha(t) < 1$. According to the geometric characterization of the class $S^*(\alpha(t_0))$, the boundary of the domain $\Omega(t_0) = f(U, t_0)$ is reachable by the radial external angles $\pi(1 - \alpha(t_0))$, which implies that there is no cusp or a double point on the boundary of $\Omega(t_0)$. This completes the proof. □

Theorem 4.4.3. *Starting with a starlike phase domain Ω_0 with an analytic boundary, the life-span of the strong Hele-Shaw starlike dynamics $\Omega(t)$ is infinite.*

Proof. Indeed, if the strong solution exists during the finite interval $t \in [0, t_0)$ and does not in $t \in [t_0, t_0 + \varepsilon)$ for any $\varepsilon > 0$, then this contradicts Theorem 4.4.1 and Lemma 4.4.2. □

Problem 4.4.4. *Of course, any result on life-span of the strong solution would be welcomed.*

- *Are there other geometric conditions that guarantee infinite life-span?*
- *Given some geometric characteristics of the initial domain can we estimate life-span of the strong solution?*
- *Can we estimate life-span of polynomial solutions of degree n depending on n?*

4.5 Solidification and melting in potential flows

Another free boundary problem which we consider in this book is the problem of pattern formation in a forced hydrodynamic flow. The Ivantsov problem of dendritic solidification [182] and the Saffman-Taylor problem of viscous fingering [293] present a basis for a mathematical treatment of two-dimensional solidification/melting in a forced potential flow. Such a problem arises, for example, in models of artificial freezing and thawing of flows in porous media (see [14], [75], [126], [133], [204], [227], [228]). The behavior of a solution to our problem has common features with solutions to the one-phase zero surface tension Hele-Shaw problem, melting corresponds to the stable case of the injection into the Hele-Shaw cell, and crystallization to the unstable case of suction. Mathematically, the problem that appears for the complex potential of the unfrozen flow is governed by Darcy's law which takes into account additional equations for the temperature field. One of the typical features of this problem is that there is not, in general, uniqueness of the solution. At the same time the existence can be proved in a usual way.

Let us formulate the governing equations. In the exterior part $\Omega(t)$ of the crystal cross-section we introduce the complex coordinate $z = x + iy$ and the complex flow potential $W = \varphi + i\psi$, where φ is the velocity potential and ψ is the stream function. We consider a dimensionless model such that $\varphi = p$ refers to the pressure and gravity is neglected. Let us denote the temperature field by $\theta(z) \equiv \theta(x, y)$ in $\Omega(t)$ and suppose that the phase transition is taken under the temperature $\theta = 0$ on $\Gamma(t) = \partial\Omega(t)$. A suitable scaling leads to the condition that $\theta(x, y) = \pm 1$ if $x \to -\infty$, where we suppose that the fluid moves to the right and (+) corresponds to melting and (-) to crystallization with supercooling. Within strong Hele-Shaw assumptions, the mathematical model is described by the following equations [14], [75], [70], [133], [204]:

$$\begin{cases} \nabla \cdot \mathbf{V} = 0, & \mathbf{V} = \nabla\varphi, & Pe\,\mathbf{V} \cdot \nabla\theta = \Delta\theta, & z \in \Omega(t); \\ \theta = 0, & v_n = -\frac{\partial \theta}{\partial n}, & \frac{\partial \varphi}{\partial n} = 0, & z \in \Gamma(t). \end{cases} \tag{4.56}$$

In this system the Péclet number Pe is a measure of the intensity of heat transfer by convection compared with conduction. We note that this model is time-reversible [70]. In fact, reversing the sign of the temperature changes only the sign condition for $\lim_{x\to-\infty} \theta$ and for the kinematic boundary condition which are both reversed.

4.5.1 Close-to-parabolic semi-infinite crystal

Let us specify the shape of the initial crystal. In this subsection we suppose that the initial melting crystal is approximately parabolic when $x, y \in \Gamma(t)$, $x \to \infty$. Let us note that if $\Gamma(0)$ admits such a normalization, then $\Gamma(t)$ is of the same normalization for $t > 0$. We add to (4.56) the initial conditions

$$\lim_{x\to-\infty} \theta = 1, \quad \lim_{y\to\pm\infty} \frac{\partial\theta}{\partial y} = 0.$$

The Boussinesq transformation [42] applied to the convective heat transfer equation (4.56) leads to uncoupling of the problem and permits us to apply analytic univalent functions. Uncoupling means that the initial problem (4.56) may be split into two independent tasks ([204], [227], [228]), the first of which is the problem of heat exchange, the second refers to the free boundary nature of the problem. In fact, the Boussinesq transformation is equivalent to the existence of a conformal univalent map from the phase domain $\Omega(t)$ onto the plane of the complex potential $W = \varphi + i\psi$. Under this transformation the boundary of the crystal cross-section is mapped onto the slit directed along the positive real axis $\psi = 0$, $\varphi > 0$ in the W-plane. Thus, the problem admits the form:

$$Pe\frac{\partial \theta}{\partial \varphi} = \Delta\theta, \quad W \in D, \tag{4.57}$$

where $D = \{\mathbb{C} \setminus [0, \infty)\}$. The boundary conditions are

$$\lim_{\varphi\to-\infty} \theta = 1, \quad \lim_{\psi\to\pm\infty} \frac{\partial\theta}{\partial\psi} = 0, \quad \theta = 0, \quad W \in \partial D. \tag{4.58}$$

Now let us introduce the auxiliary parametric complex ζ-plane, $\zeta = \xi + i\eta$. The Riemann Mapping Theorem yields that there exists a conformal univalent map $f(\zeta, t)$ of the left half-plane $H^- = \{\zeta : \text{Re}\,\zeta < 0\}$ onto the phase domain $f : H^- \to \Omega(t)$. The parabolic shape of $\Gamma(t)$ implies the normalization $f(\zeta, \cdot) = -\zeta^2 + a_1\zeta + a_0 + a_{-1}/\zeta + \cdots$. Fortunately, for the problem (4.57)–4.58 the method of separating variables is applicable. First we introduce the map $H^- \to D$ given by $W = -\zeta^2$. Then we are looking for a similarity solution $\theta = g(\xi)$. Elementary calculation leads to the relations

$$\left|\frac{\partial \theta}{\partial \psi}\right| = \frac{\sigma}{\sqrt{\varphi}} \quad \text{on the slit } \partial D,$$

where $\sigma = \sqrt{Pe/\pi}$, and

$$\left|\frac{\partial \theta}{\partial n}\right| = \frac{\sigma}{\sqrt{\varphi}}|W'(z)|.$$

The unit normal outer vector to the moving interface is $\mathbf{n} = f'(i\eta, t)/|f'(i\eta, t)|$, and the normal velocity, hence, is of the form

$$v_n = \text{Re}\ \dot{f}(i\eta, t)\overline{\frac{f'(i\eta, t)}{|f'(i\eta, t)|}}, \quad \eta \in (-\infty, \infty).$$

Besides, we have

$$\pm i\left|\frac{\partial \theta}{\partial \psi}\right|\overline{W'(z)} = -v_n\hat{n}, \quad \psi = \pm 0, \ \varphi > 0.$$

Changing variables, $W = -\zeta^2$ we come to the Polubarinova-Galin type equation for the free boundary:

$$\text{Re}\ \dot{f}(i\eta, t)\overline{f'(i\eta, t)} = -2\sigma, \quad \xi = 0, \eta \in (-\infty, \infty). \tag{4.59}$$

Using the same argumentation as in the above subsections we prove the following statement.

Theorem 4.5.1. *If the initial crystal interface possesses the property to be convex of order α in the negative direction ($H^-_{\mathbb{R}}(\alpha)$), then the free boundary remains convex in the negative direction of the same order as long as the solution to the problem (4.57)–(4.58) (or (4.59)) exists ($\sigma > 0$ and the liquid moves to the right).*

The result for $\alpha = 1$ is obtained in [205].

4.6 Geometry of weak solutions

Some of the results on geometry of solutions to Hele-Shaw flow problems are most easily discussed in terms of weak solutions, in fact, they will really be results on the geometry of domains obtained by partial balayage (see Section 3.5). We recall that the weak solution $\Omega(t)$ of the one-point injection Hele-Shaw problem with $Q = -1$ is expressed as

$$\operatorname{Bal}(\chi_{\Omega(0)} + t\delta_0, 1) = \chi_{\Omega(t)}$$

in terms of balayage. What will count in the results on geometry is just the support, or even the convex hull of the support, of the measure $\mu = \chi_{\Omega(0)} + t\delta_0$.

4.6.1 Starlikeness of the weak solution

We already proved via conformal mappings that starting with a domain $\Omega(0)$ which is starlike with respect to the origin, the injection at the origin gives a strong solution with the starlikeness preserved. In the weak formulation the preservation of starlikeness is even easier to show. The following result is originally due to Di Benedetto and Friedman [81].

Theorem 4.6.1. *Let* $\operatorname{Bal}(\chi_{\Omega(0)} + t\delta_0) = \chi_{\Omega(t)}$, *where* $\Omega(0)$ *is starlike with respect to the origin. Then also* $\Omega(t)$ *is starlike with respect to the origin for* $t > 0$.

Proof. We write

$$\chi_{\Omega(0)} + t\delta_0 + \Delta u = \chi_{\Omega(t)}$$

with $u \geq 0$, $u = 0$ outside $\Omega(t)$, as usual. Then in terms of polar coordinates

$$\frac{1}{r}\frac{\partial}{\partial r}\left(r\frac{\partial u}{\partial r}\right) + \frac{1}{r^2}\frac{\partial^2 u}{\partial \theta^2} = 1 - \chi_{\Omega(0)}$$

in $\Omega(t) \setminus \{0\}$. Multiplying by r^2 and then applying $\frac{1}{r}\frac{\partial}{\partial r}$ to both parts gives

$$\frac{1}{r}\frac{\partial}{\partial r}\left(r\frac{\partial}{\partial r}\left(r\frac{\partial u}{\partial r}\right)\right) + \frac{1}{r^2}\frac{\partial^2}{\partial \theta^2}\left(r\frac{\partial u}{\partial r}\right) = \frac{1}{r}\frac{\partial}{\partial r}\left(r^2(1 - \chi_{\Omega(0)})\right).$$

Here the left member is $\Delta(r\frac{\partial u}{\partial r})$ and the right member is non-negative due to the starlikeness of $\Omega(0)$. Thus $r\frac{\partial u}{\partial r}$ is subharmonic in $\Omega(t)\setminus\{0\}$, and $r\frac{\partial u}{\partial r} = 0$ on $\partial\Omega(t)$.

At the origin u has the positive singularity $u \sim -\frac{t}{2\pi}\log|z|$, hence $r\frac{\partial u}{\partial r} < 0$ near the origin, and therefore, in all $\Omega(t)\setminus\{0\}$, from which the starlikeness follows. $\square$

4.6.2 The inner normal theorem

Next we show that $\Omega(t)$ has very good properties outside the convex hull of $\Omega(0)$, e.g., that there are natural bounds on the curvature of $\partial\Omega(t)$. So we consider a weak solution

$$\chi_{\Omega(t)} = \mathrm{Bal}\,(\chi_{\Omega(0)} + t\delta_0, 1),$$

or more generally,

$$\chi_{\Omega(t)} = \mathrm{Bal}\,(\chi_{\Omega(0)} + \nu(t), 1),$$

for any measure $\nu(t) \geq 0$ which vanishes outside $\Omega(0)$. Let $K = \mathrm{conv}\,\overline{\Omega(0)}$ be the closed convex hull of $\Omega(0)$.

Theorem 4.6.2. [146], [149], [150] *Under the above assumptions $\Omega = \Omega(t)$ has the following properties:*

(i) *$\partial\Omega \setminus K$ is smooth analytic;*

(ii) *for any $z \in \partial\Omega \setminus K$ the inward normal ray N_z from z intersects K (if $\Omega(0)$ is connected, then it has to intersect $\overline{\Omega(0)}$ itself);*

(iii) *the normal rays N_z in (ii) do not intersect each other before they reach K;*

(iv) *Ω can be expressed as a union of disks with centers on $K \cap \Omega$:*

$$\Omega = \bigcup_{a \in K \cap \Omega} U_{r(a)}(a)$$

for suitable $r(a) > 0$.

Proof. Set $\mu = \chi_{\Omega(0)} + \nu(t)$ and write

$$\chi_\Omega = \mathrm{Bal}\,(\mu, 1) = \mu + \Delta u,$$

where u is the smallest function satisfying $u \geq 0$, $\Delta u \leq 1 - \mu$.

We first assume that K lies in the lower half-plane $K \subset \{y \leq 0\}$, i.e., that $\Omega(0) \subset \{y < 0\}$, and we shall study the geometry of

$$\Omega^+ = \Omega \cap \{y > 0\}.$$

Let u^* be the reflection of u with respect to the real axis, i.e., $u^*(x+iy) = u(x-iy)$, and set

$$v = u - \inf(u, u^*) = (u - u^*)_+.$$

Since $\Delta u \leq 1 - \mu \leq 1$ everywhere, we have $\Delta u^* \leq 1$. Therefore, $\Delta \inf(u, u^*) \leq 1$ everywhere. Indeed, the infimum of two superharmonic functions is again superharmonic, so

$$\Delta \inf(u, u^*) - 1 = \Delta(\inf(u, u^*) - \frac{1}{4}|z|^2) = \Delta \inf(u - \frac{1}{4}|z|^2, u^* - \frac{1}{4}|z|^2) \leq 0.$$

Since $\Delta u = 1$ in Ω^+ it follows that $\Delta v \geq 0$ in Ω^+. Moreover, $v = 0$ on $\partial(\Omega^+)$.

The maximum principle now shows that $v \leq 0$ in Ω^+. This means that $u \leq u^*$ in Ω^+, i.e., that u is smaller (or at least not larger) at any point in the upper half-plane, than in the reflected point in the lower half-plane. On the real axis this gives

$$\frac{\partial u}{\partial y}(x, 0) \leq 0, \tag{4.60}$$

and in general it shows that the reflection of Ω^+ in the real axis is contained in Ω:

$$(\Omega^+)^* \subset \Omega. \tag{4.61}$$

Now since $\nabla u = 0$ on $(\partial\Omega)^+$ and $\Delta(\frac{\partial u}{\partial y}) = \frac{\partial}{\partial y}\Delta u = 0$ in Ω^+ we can apply the maximum principle again, now to $\frac{\partial u}{\partial y}$, to obtain that $\frac{\partial u}{\partial y} \leq 0$ in Ω^+. This inequality is everywhere strict because if we had equality at some point, then it would follow that $u = 0$ in a whole component of Ω^+; and this is impossible because $1 = \chi_\Omega = \mu + \Delta u = \Delta u$ in Ω^+.

The conclusion now is that $u(x+iy)$ is a strictly decreasing function of $y > 0$ in Ω^+. Therefore, since $u = 0$ outside Ω, every vertical line L in the upper half-plane intersects Ω^+ in at most one segment ($L \setminus \Omega^+$ is connected). It follows that $(\partial\Omega)^+$ is a graph of a function, say

$$(\partial\Omega)^+ = \{z = x + iy,\ y = g(x)\}.$$

The domain of definition of the function g may consist of more than one interval. It follows from the general regularity theory (e.g., [46], [297]) that g is real analytic.

Next, with K still in the lower half-plane we shall obtain a similar convexity statement, but for semicircles instead of vertical lines. Let (r, θ) be the polar coordinates. In the proof of Theorem 4.6.1 we studied

$$r\frac{\partial u}{\partial r} = x\frac{\partial u}{\partial x} + y\frac{\partial u}{\partial y},$$

here we shall study

$$\frac{\partial u}{\partial \theta} = -y\frac{\partial u}{\partial x} + x\frac{\partial u}{\partial y}, \tag{4.62}$$

in Ω^+. Since $\Delta u = 1$ in Ω^+ and the coefficients in

$$\Delta u = \frac{1}{r}\frac{\partial}{\partial r}\left(r\frac{\partial u}{\partial r}\right) + \frac{1}{r^2}\frac{\partial^2 u}{\partial \theta^2}$$

do not depend on θ, $\frac{\partial u}{\partial \theta}$ is harmonic,

$$\Delta\frac{\partial u}{\partial \theta} = \frac{\partial}{\partial \theta}\Delta u = 0 \quad \text{in } \Omega^+.$$

As to the boundary values of $\frac{\partial u}{\partial \theta}$ on $\partial(\Omega^+)$, we have

$$\frac{\partial u}{\partial \theta} = 0 \quad \text{on } (\partial\Omega)^+.$$

By (4.60), (4.62) we have that

$$\frac{\partial u}{\partial \theta} \leq 0 \quad \text{for } x > 0,$$

$$\frac{\partial u}{\partial \theta} \geq 0 \quad \text{for } x < 0,$$

on the real axis.

Now consider a circular arc

$$C_R = \{z = re^{i\theta}, r = R, 0 < \theta < \pi\}$$

in the upper half-plane. We shall prove that $C_R \backslash \Omega^+$ consists of at most one segment (more precisely, that it is connected), and we shall argue by contradiction.

So suppose $C_R \setminus \Omega^+$ has at least two components. Then there are points $z_1 = Re^{i\theta_1}$ and $z_2 = Re^{i\theta_2}$ with $0 < \theta_1 < \theta_2 < \pi$, such that $z_1, z_2 \in (\partial\Omega)^+$, and $z = Re^{i\theta} \in \Omega^+$ for all $\theta_1 < \theta < \theta_2$. Since $u(z_1) = u(z_2) = 0$ and $u(z) > 0$, we have, integrating along C_R,

$$\int_{z_1}^{z} \frac{\partial u}{\partial \theta} d\theta = u(z) - u(z_1) > 0,$$

$$\int_{z}^{z_2} \frac{\partial u}{\partial \theta} d\theta = u(z_2) - u(z) < 0.$$

Therefore, there must be points $z = Re^{i\theta}$ with $\theta_1 < \theta < \theta_2$ arbitrarily close to θ_1 for which $\frac{\partial u}{\partial \theta} > 0$. Similarly, there must be points $z = Re^{i\theta}$ with $\theta_1 < \theta < \theta_2$ arbitrarily close to θ_2 for which $\frac{\partial u}{\partial \theta} < 0$.

Now we apply the maximum principle. Every component of $\{\frac{\partial u}{\partial \theta} > 0\}$ must reach some part of the negative real axis, because we know that $\frac{\partial u}{\partial \theta} \leq 0$ on all other possible parts of the boundary of that component. Similarly, every component of $\{\frac{\partial u}{\partial \theta} < 0\}$ must reach some part of the positive real axis. But it is obviously topologically impossible to have components of $\{\frac{\partial u}{\partial \theta} > 0\}$ stretching from points arbitrarily close to z_1 (the rightmost end point of the described component of C_R) to the negative real axis, and simultaneously, components of $\{\frac{\partial u}{\partial \theta} > 0\}$ stretching from points arbitrarily close to z_2 on the positive real axis.

This contradiction shows that $C_R \setminus \Omega^+$ actually is connected. The same reasoning applies with the center of the polar coordinates at any point of the real axis. Thus for any semicircle C in the upper half-plane with its center on the real axis, $C \setminus \Omega^+$ is connected. Together with the first part of the proof, saying that $L \setminus \Omega^+$

is connected for any vertical semiline L, what we have proved can be expressed by saying that the complement of Ω^+ in the upper half-plane is convex with respect to the Poincaré metric in the upper half-plane.

Next, as for Euclidean convexity, $C^+ \setminus \Omega^+$ being convex (in the Poincaré metric) implies that it is the intersection of Poincaré half-planes, or that, turning to the complements, Ω^+ is the union of such. This means that

$$\Omega^+ = \bigcup_{a\in\mathbb{R}} U_{r(a)}(a) \tag{4.63}$$

for suitable radii $r(a) \geq 0$.

Let $z \in (\partial\Omega)^+$ and let N_z be the inward normal ray at z. Since $(\partial\Omega)^+$ is a graph of a function, N_z intersects the real axis at a point $p(z)$, which we call the *foot point* of the normal. In terms of the equation $y = g(x)$ for $(\partial\Omega)^+$ we have $z = x + ig(x)$ and one easily computes $p(z) = x + g(x)g'(x)$. The fact (4.63) that Ω^+ is a union of semidisks $(U_{r(a)}(a))^+$ implies that actually

$$(U_{|z-p(z)|}(p(z)))^+ \subset \Omega^+. \tag{4.64}$$

Let $z_1 \neq z_2$. Assume that two normals N_{z_1} and N_{z_2} intersect in $\mathbb{C}^+$, say $w \in N_{z_1} \cap N_{z_2}$, where $w \in \mathbb{C}^+$. We may assume that $|z_1 - w| \leq |z_2 - w|$, for example. Then z_1 is contained in the closure of the disk with the center at w and the radius $|z_2 - w|$, and therefore, in the interior of the larger disk

$$U_{|z_2-p(z_2)|}(p(z_2)).$$

But $z_1 \in (\partial\Omega)^+$, so this disk must contain points outside $\overline{\Omega^+}$ as well. This contradicts (4.64), and we conclude that N_{z_1} and N_{z_2} actually cannot intersect in $\mathbb{C}^+$.

One easily sees that the inner ball property (4.63), or the fact that the inner normals do not intersect in $\mathbb{C}^+$, is equivalent to the foot point $p(z) = x+g(x)g'(x)$ being a monotone increasing function of x.

So far we have assumed that $K \subset \{y < 0\}$. Adapting the results we have obtained to all half-planes containing K easily gives the statements of the theorem. We give some details below.

(i) We already know that $\partial\Omega \setminus K$ is analytic but with possible singularities. But since near any point of $\partial\Omega \setminus K$, $\partial\Omega$ will be a graph seen from several different angles (choosing different half-planes containing K) there can be no singular points.

(ii) If the inward normal N_z at a point $z \in \partial\Omega \setminus K$ did not intersect K we could find a half-plane $H \supset K$ which does not meet N_z, contradicting that $\partial\Omega \setminus H$ is a graph when seen from H.

(iii) Similarly, if $w \in N_{z_1} \cap N_{z_2}$, $w \in \Omega \setminus K$, $z_1, z_2 \in \partial\Omega \setminus K$, $z_1 \neq z_2$, we choose a half-plane $H \supset K$, such that $w \notin \bar{H}$. By (ii), both N_{z_1} and N_{z_2} intersect

K. This shows that $z_1 \notin \bar{H}$, $z_2 \notin \bar{H}$, and we are back in the situation with $H = \{y < 0\}$, i.e., we have a contradiction. Thus N_{z_1} and N_{z_2} do not meet before they reach K.

(iv) From what we already did it follows that

$$\Omega \setminus K = \left(\bigcup_{a \in \partial K} U_{r(a)}(a) \right) \setminus K$$

for suitable $r(a) \geq 0$. Also, by (4.61),

$$\bigcup_{a \in \partial K} U_{r(a)}(a) \subset \Omega.$$

Obviously, $r(a) = 0$ for $a \in \partial K \setminus \Omega$. The points from $\Omega \cap K$ can be trivially covered by the disks $U_{r(a)}(a) \subset \Omega$ with $a \in \Omega \cap K$. Now (iv) follows, and the proof of the theorem is complete.

□

4.6.3 Distance to the boundary (revisited)

Now we discuss the distances from points in the initial domain $\Omega(0)$ to points on the boundary of $\Omega(t)$. The following theorem is due to Sakai [300].

Theorem 4.6.3. *Let $\mu \geq 0$ be a measure with a support in the disk U_R, $R > 0$. Define $r(\mu)$ by*

$$\pi(r(\mu))^2 = \iint d\mu,$$

and let Ω be the saturated set (3.15) *for* Bal $(\mu, 1)$. *Then*

$$\Omega \subset U_{r(\mu)+R},$$

and if $r(\mu) \geq 2R$ we moreover have

$$U_{r(\mu)-R} \subset \Omega.$$

Proof. First we recall from (3.20) that

$$\mathrm{Bal}\,(\mu, 1) = \chi_\Omega + \mu\chi_{\mathbb{C}\setminus\Omega}.$$

From this it follows that

$$\iint_{U_R} \varphi d\mu \leq \iint_{\Omega} \varphi d\sigma_z + \iint_{\mathbb{C}\setminus\Omega} \varphi d\mu$$

for all functions φ in $\mathbb{C}$ which are integrable and subharmonic in Ω. In particular, taking $\varphi = \pm 1$, $|\Omega| \leq \iint d\mu$.

The upper bound $\Omega \subset U_{R+r(\mu)}$ is actually a direct consequence of the Inner Normal Theorem. By that theorem Ω is a union of disks with centers in $\overline{U_R}$, so if Ω contained points outside $U_{R+r(\mu)}$, then it would contain a disk of radius greater than $r(\mu)$, which is impossible because $|\Omega| \leq \iint d\mu = |U_{r(\mu)}|$.

Next assume $r(\mu) \geq 2R$ and let $z \notin \Omega$. We shall show that $|z| \geq r(\mu) - R$, hence that $U_{r(\mu)-R} \subset \Omega$. By a rotation we may assume that $z = \rho \geq 0$.

If $0 < \rho < R$ we choose

$$\varphi(\zeta) = G(\zeta, \rho) + c,$$

where $G(\zeta,\rho) = \frac{1}{2\pi}\log|\frac{R^2-\rho\zeta}{R(\zeta-\rho)}|$ is Green's function of U_R and $c > 0$. Since $\rho \notin \Omega$, φ is subharmonic in Ω. The level lines of $G(\cdot,\rho)$ and φ are circles, so

$$\{\zeta : \varphi(\zeta) > 0\} = \{\zeta : G(\zeta,\rho) > -c\} = U_t(a)$$

for some disk $U_t(a)$, which contains U_R because $c > 0$.

Now choose c so that

$$\iint_{U_t(a)} G(\zeta,\rho)\, d\sigma = 0.$$

By a straightforward computation (see section 6 in [300]), this gives $R < t < 2R$. Thus, using that $U_t(a)$ is exactly the set where φ is positive we get

$$c\pi r(\mu)^2 = c\iint_{U_R} d\mu \leq \iint_{U_R} (G(\zeta,\rho)+c)\, d\mu(\zeta) = \iint_{U_R} \varphi\, d\mu$$

$$\leq \iint_\Omega \varphi\, d\sigma + \iint_{\mathbb{C}\setminus\Omega} \varphi d\mu \leq \iint_{U_t(a)} \varphi\, d\sigma = \iint_{U_t(a)} c\, d\sigma = c\pi t^2.$$

Hence $r(\mu) \leq t < 2R$, contrary to our assumption in the beginning. Thus there is no $0 < \rho < R$ with $\rho \notin \Omega$. So $U_R \subset \Omega$ and $\mathrm{Bal}\,(\mu, 1) = \chi_\Omega$.

If $\rho \geq R$ then we take instead φ to be the Poisson kernel for the disk $U_{R+\rho}(-R)$ and with the pole at ρ:

$$\varphi(\zeta) = \frac{(R+\rho)^2 - |\zeta+R|^2}{|(R+\rho)-(\zeta+R)|^2} = \frac{(R+\rho)^2 - |\zeta+R|^2}{|\zeta-\rho|^2}.$$

Since $\rho \notin \Omega$, φ is subharmonic in Ω. Also in this case the level lines of φ are circles. The circle where $\varphi = 1$ passes through the center $-R$ of $U_{R+\rho}(-R)$ and through the pole $\zeta = \rho$, hence $\varphi \geq 1$ inside that circle, in particular $\varphi \geq 1$ in $U_R(0)$. Moreover, $U_{R+\rho}(-R)$ is exactly the set where $\varphi \geq 0$. All this, combined with the mean value property of φ in $U_{R+\rho}(-R)$, gives that

$$\pi r(\mu)^2 \iint_{U_R} d\mu \leq \iint_{U_R} \varphi\, d\mu \leq \iint_\Omega \varphi\, d\sigma$$

$$\leq \iint_{U_{R+\rho}(-R)} \varphi \, d\sigma = |U_{R+\rho}(-R)| \, \varphi(-R) = \pi(R+\rho)^2.$$

Thus $\rho \leq R - r(\mu)$ as required. □

Corollary 4.6.4. *In the one point injection Hele-Shaw case* $\mu = \chi_{\Omega(0)} + t\delta_a$. *This gives, if* $a \in U_R$, $\Omega(0) \subset U_R$, *that*

$$\Omega(t) \subset U_{\sqrt{(|\Omega(0)|+t)/\pi}+R}.$$

If, in addition, $t \geq 4\pi R^2 - |\Omega(0)|$, *then*

$$U_{\sqrt{(|\Omega(0)|+t)/\pi}-R} \subset \Omega(t).$$

Chapter 5

Capacities and Isoperimetric Inequalities

Isoperimetric inequalities have been known since antiquity. The simplest version of an isoperimetric theorem reads in two equivalent forms:

- Among all planar shapes with the same perimeter the circle has the largest area.
- Among all planar shapes with the same area the circle has the shortest perimeter.

This is the solution of what is sometimes known as Dido's problem because of the story that Queen Dido of Tyre bargained for some land bounded on one side by the (straight) Mediterranean coast and agreed to pay a fixed sum for as much land as could be enclosed by a bull's hide. Both statements can be expressed in a more algebraic form which indeed underlines the fact that they are equivalent. Denote the perimeter and area of a planar shape by L and A, respectively. Then, $4\pi A \leq L^2$. The equality only holds for a circle. In higher dimensional spaces, for example, if S is a surface area while V a volume of a three-dimensional body, then $36\pi V^2 \leq S^3$ (see, e.g., [59], [90]).

Pappus of Alexandria (ca 300 A.D.) wrote: *bees, then, know just this fact which is useful to them, that the hexagon is greater than the square and the triangle and will hold more honey for the same expenditure of material in constructing each. But we, claiming a greater share of wisdom than the bees, will investigate a somewhat wider problem, namely that, of all equilateral and equiangular plane figures having the same perimeter, that which has the greater number of angles is always greater, and the greatest of them all is the circle having its perimeter equal to them.*

Probably, the most representative work on isoperimetric inequalities in various aspects of mathematical physics is the famous monograph [265] written by

George Pólya (1887–1985) and **Gabor Szegö** (1895–1985) (see also [240]). By an isoperimetric inequality we mean an inequality that links a measure of volume with a measure of its boundary. We shall be concerned mainly with the following related question: how is the area of the phase domain controlled by the capacity of its boundary (or conformal radius of the domain)?

5.1 Conformal invariants and capacities

We start by giving some background on quantities we are going to use in isoperimetric inequalities. These quantities are moduli, reduced moduli and capacities.

5.1.1 Modulus of a family of curves

The notion of the *modulus of a family of curves* goes back to early works of Grötzsch [139], [140]. Later, Ahlfors and Beurling [5], [6] introduced the notion of *extremal length* (the reciprocal of the *modulus*) which stimulated the active development of the method of extremal lengths. Major contributions to the subject were made by Jenkins [183], [184], Strebel [315] and Ohtsuka [254] who connected the modulus problem with the problem of the extremal partitioning of a Riemann surface and proved the existence of the extremal metric by Schiffer's variations.

Let Ω be a domain in $\mathbb{C}$ and $\rho(z)$ be a real-valued, Borel measurable, non-negative function in $L^2(\Omega)$. Let this function define a differential metric ρ on Ω by $\rho := \rho(z)|dz|$.

Let γ be a locally rectifiable curve in Ω. The integral

$$l_\rho(\gamma) := \int_\gamma \rho(z)|dz|, \tag{5.1}$$

is said to be the ρ - *length* of γ. If $\rho(z) \equiv 1$ almost everywhere in Ω, then the 1 - length of any rectifiable $\gamma \subset \Omega$ coincides with its Euclidian length. The integral

$$A_\rho(\Omega) := \iint_\Omega \rho^2(z) d\sigma_z, \qquad d\sigma_z = dxdy, \tag{5.2}$$

is called the ρ-*area* of Ω.

Let Γ be a family of curves γ in Ω. Denote by

$$L_\rho(\Gamma) := \inf_{\gamma\in\Gamma} l_\rho(\gamma)$$

the ρ-length of the family Γ. Then, the quantity

$$m(\Omega, \Gamma) = \inf_\rho \frac{A_\rho(\Omega)}{L_\rho^2(\Gamma)}$$

is said to be the *modulus* of the family Γ in Ω where the infimum is taken over all metrics ρ in Ω.

Another equivalent and suitable (in a view of further applications) definition of the modulus can be formulated as follows. Denote by P the family of all *admissible* (for Γ) metrics in Ω, that is, metrics $\rho \in P$ that satisfy the additional condition $l_\rho(\gamma) \geq 1$ for all $\gamma \in \Gamma$. If $P \neq \emptyset$, then we can define the modulus as

$$m(\Omega, \Gamma) = \inf_{\rho \in P} A_\rho(\Omega).$$

If there is a metric ρ^*, such that $m(\Omega,\Gamma) = A_{\rho^*}(\Omega)$, then this metric is called *extremal.*

Two main properties of the modulus are its conformal invariance and the uniqueness of the extremal metric (if it exists). More precisely, let Γ be a family of curves in a domain $\Omega \in \overline{\mathbb{C}}$, and let $w = f(z)$ be a conformal map of Ω onto $\widetilde{\Omega} \in \overline{\mathbb{C}}$. If $\widetilde{\Gamma} := f(\Gamma)$, then

$$m(\Omega, \Gamma) = m(\widetilde{\Omega}, \widetilde{\Gamma}).$$

Let ρ_1 and ρ_2 be two extremal metrics for the modulus $m(\Omega,\Gamma)$. Then, $\rho^* := \rho_1 = \rho_2$ almost everywhere. Moreover, $L_{\rho^*}(\Gamma) = 1$.

The property of monotonicity reads as follows. If $\Gamma_1 \subset \Gamma_2$ in Ω, then $m(\Omega,\Gamma_1) \leq m(\Omega,\Gamma_2)$.

Example 5.1.1. *Let Ω be a rectangle $\{z = x + iy : 0 < x < a, 0 < y < b\}$ and Γ be the family of curves in Ω that connect the opposite horizontal sides of Ω. Then, $m(\Omega,\Gamma) = a/b$.*

Example 5.1.2. *Let Ω be an annulus $\{z = re^{i\theta} : 1 < r < R, 0 < \theta \leq 2\pi\}$ and Γ be the family of curves in Ω that separate the opposite boundary components of Ω. Then, $m(\Omega,\Gamma) = \frac{1}{2\pi}\log R$.*

Example 5.1.3. *Let Ω be an annulus $\{z = re^{i\theta} : 1 < r < R, 0 < \theta \leq 2\pi\}$ and Γ be the family of curves in Ω that connect the two boundary components of Ω. Then, $m(\Omega,\Gamma) = \frac{2\pi}{\log R}$.*

For more information see, e.g., [5], [184], [254], [335].

5.1.2 Reduced modulus and capacity

Let $\Omega \subset \overline{\mathbb{C}}$ be a simply connected hyperbolic domain, $a \in \Omega$, $|a| < \infty$. We consider the doubly connected domain $\Omega_\varepsilon = \Omega \setminus U(a,\varepsilon)$ for a sufficiently small $\varepsilon > 0$. The quantity

$$M(\Omega, a) := \lim_{\varepsilon \to 0} \left(M(\Omega_\varepsilon) + \frac{1}{2\pi} \log \varepsilon \right)$$

is said to be the *reduced modulus* of the circular domain Ω with respect to the point a, where $M(\Omega_\varepsilon)$ is the modulus of the doubly connected domain Ω_ε with respect to the family of curves that separate its boundary components.

Let a simply connected hyperbolic domain Ω have the conformal radius $R(\Omega, a)$ with respect to a fixed point $a \in \Omega$. Then, the quantity $M(\Omega, a)$ exists, is finite, and is equal to $\frac{1}{2\pi} \log R(\Omega, a)$, see [184], [335]. An immediate corollary when $|a| < \infty$ says that if $f(z)$ is a conformal map of Ω, such that $|f(a)| < \infty$, then $M(f(\Omega), f(a)) = M(\Omega, a) + \frac{1}{2\pi} \log |f'(a)|$.

Now, we define the reduced modulus $M(\Omega, \infty)$ of a simply connected domain Ω, $\infty \in \Omega$ with respect to infinity as the reduced modulus of the image of Ω under the map $1/z$ with respect to the origin,

$$M(\Omega, \infty) = -\frac{1}{2\pi} \log R(\Omega, \infty).$$

So, if Ω is a simply connected hyperbolic domain, $a \in \Omega$, $|a| < \infty$, and $f(z) = a_{-1}/(z-a) + a_0 + a_1(z-a) + \cdots$ is a conformal map from Ω, then $M(f(\Omega), \infty) = M(\Omega, a) - \frac{1}{2\pi} \log |a_{-1}|$.

We give all further definitions only for compact sets, however, there are generalizations to the Borel sets as well. Denote by $\mathrm{Lip}\,(\Omega)$ the class of functions $u(z) : \Omega \to \mathbb{R}$ satisfying the Lipschitz condition in Ω, i.e., for every function $u \in \mathrm{Lip}\,(\Omega)$ there is a constant c such that for any two points $z_1, z_2 \in \Omega$ the inequality $|u(z_1) - u(z_2)| \leq c|z_1 - z_2|$ holds. In the case $\infty \in \Omega$ the continuity of $u(z)$ at ∞ is required. Functions from $\mathrm{Lip}\,(\overline{\mathbb{C}})$ are absolutely continuous on lines which are parallel to the axes and the integral

$$I(u) := \iint\limits_{\overline{\mathbb{C}}} |\nabla u(z)|^2 d\sigma_z$$

exists.

An ordered pair of disjoint compact sets K_1, K_2 is called a *condenser* $C = \{K_1, K_2\}$ with the *field* $\overline{\mathbb{C}} \setminus \{K_1 \cup K_2\}$. The *capacity of a condenser* C is the quantity

$$\mathrm{cap}\ C := \inf I(u)$$

as u ranges over the class $\mathrm{Lip}\,(\overline{\mathbb{C}})$ and $0 \leq u(z) \leq 1$ whenever $z \in \overline{\mathbb{C}}$, $u(z) \equiv 0$ in K_1, $u(z) \equiv 1$ in K_2.

A condenser C is said to be *admissible* if there exists a continuous real-valued in $\overline{\mathbb{C}}$ function $\omega(z)$, $0 \leq \omega(z) \leq 1$ which is harmonic in $\overline{\mathbb{C}} \setminus \{K_1 \cup K_2\}$ and $\omega(z) = 0$ for $z \in K_1$, $\omega(z) = 1$ for $z \in K_2$. This function is said to be a *potential* . The Dirichlet principle yields that in the definition of capacity equality appears only in the case of an admissible condenser and $u(z) \equiv \omega(z)$ almost everywhere for the potential function ω. Then by Green's formula

$$\mathrm{cap}\ C = \iint\limits_{\overline{\mathbb{C}} \setminus \{K_1 \cup K_2\}} |\nabla \omega|^2 \, d\sigma = \int\limits_{\partial K_2} \left|\frac{\partial \omega}{\partial n}\right| ds.$$

Obviously, the capacity is a conformal invariant, that is, if C_f is a condenser $\overline{\mathbb{C}} \setminus f(\overline{\mathbb{C}} \setminus \{K_1 \cup K_2\})$ for a conformal map f in $\overline{\mathbb{C}} \setminus \{K_1 \cup K_2\}$, then $\mathrm{cap}\ C = \mathrm{cap}\ C_f$.

If K_1 and K_2 are two disjoint continua, then we can construct the conformal map $w = f(z)$ of the doubly connected domain $\overline{\mathbb{C}} \setminus \{K_1 \cup K_2\}$ onto an annulus $1 < |w| < R$ and the potential function for the condenser $C = \{K_1, K_2\}$ is

$$\omega(z) = \frac{\log \frac{R}{|f(z)|}}{\log R}, \quad z \in \overline{\mathbb{C}} \setminus \{K_1 \cup K_2\},$$

$\omega(z) \equiv 0$ in K_1, $\omega(z) \equiv 1$ in K_2. Therefore, cap $C = 2\pi/\log R$.

Let $C = \{K_1, K_2\}$ and $C_k = \{K_1^k, K_2^k\}$, $k = 1, \dots, n$ be such condensers that all C_k have non-intersecting fields and

$$K_1 \subset \bigcap_{k=1}^{n} K_1^k, \quad K_2 \subset \bigcap_{k=1}^{n} K_2^k.$$

From the definition of capacity and from the Dirichlet principle one can derive the inequality (Grötzsch's Lemma)

$$\frac{1}{\text{cap } C} \geq \sum_{k=1}^{n} \frac{1}{\text{cap } C_k}. \tag{5.3}$$

(possibly with equality, see, e.g., [87]).

Let K be a compact set in $\mathbb{C}$. We consider condensers of special type $C_R = \{|z| \geq R, K\}$ for R large. If $C_{R_1,R_2} = \{|z| \leq R_1\} \cup \{|z| \geq R_2\}$ for $R_1 < R_2$, then the inequality (5.3) implies

$$\frac{1}{\text{cap } C_{R_2}} \geq \frac{1}{\text{cap } C_{R_1}} + \frac{1}{2\pi} \log \frac{R_2}{R_1}.$$

Therefore, the function $\frac{1}{\text{cap } C_R} - \frac{1}{2\pi} \log R$ increases with increasing R and the limit

$$\text{cap } K = \lim_{R \to \infty} R \exp\left(-\frac{2\pi}{\text{cap } C_R}\right) \tag{5.4}$$

exists and is said to be the *logarithmic capacity* of the compact set $K \subset \mathbb{C}$. Equality (5.4) is also known as Pfluger's theorem (see e.g. [270], Theorem 9.17).

Next we briefly summarize the definition and some properties of the logarithmic capacity of a compact set $K \subset \mathbb{C}$ following Fekete. For $n = 2, 3, \dots$ we consider

$$\Delta_n(K) = \max_{z_1, \cdots, z_n \in K} \prod_{1 \leq k < j \leq n}^{n} |z_k - z_j|.$$

The maximum exists and is attained for so-called Fekete points $z_{nk} \in \partial K$, $k = 1, \dots, n$. The value Δ_n is equal to the Vandermonde determinant

$$\Delta_n(K) = \left| \det_{k=1,\dots,n} (1\ z_{nk}\ \cdots\ z_{nk}^{n-1}) \right|.$$

Then, the limit

$$\operatorname{cap} K = \lim_{n\to\infty} (\Delta_n(K))^{\frac{2}{n(n-1)}}$$

exists (see [270]), is known as the *transfinite diameter*, and is equal to the logarithmic capacity (see also [265], [292]).

Let K be a continuum (a closed connected set containing at least two points) in $\overline{\mathbb{C}}$ and $\Omega = \overline{\mathbb{C}} \setminus K$. Then, from the definition of the logarithmic capacity and the reduced modulus it is clear that $\operatorname{cap} K = \operatorname{cap} \partial K = \exp(-2\pi\, m(\Omega, \infty))$.

It is well known that for $K = [0, 1]$ the capacity is given by $\operatorname{cap} K = 1/4$.

If we have a condenser $C^{(h)} = \{K_1, K_2\}$ of the special type $K_1 \subset U$, $K_2 = \overline{\mathbb{C}} \setminus U$, then $\operatorname{cap} C^{(h)}$ is said to be the hyperbolic capacity of K_1 and $\operatorname{cap}^{(h)} K_1 = \operatorname{cap} C^{(h)}$. One can define also $\operatorname{cap}^{(h)} K$ by means of the *hyperbolic transfinite diameter*. Set

$$\Delta_n^{(h)}(K) = \max_{z_1,\cdots,z_n\in D} \prod_{1\leq k<j\leq n}^{n} \left|\frac{z_k - z_j}{1 - z_k\overline{z_j}}\right|.$$

Then,

$$\operatorname{cap}^{(h)} K = \lim_{n\to\infty} (\Delta_n^{(h)}(K))^{\frac{2}{n(n-1)}}.$$

Finally, if K is a continuum in U and $\Omega = U \setminus K$, then

$$\operatorname{cap}^{(h)} K = \operatorname{cap}^{(h)} \partial K = \exp(-2\pi M(\Omega)),$$

where $M(\Omega)$ is the modulus of the doubly connected domain Ω with respect to the family of separating curves.

5.1.3 Integral means and the radius-area problem

We consider the zero surface tension Hele-Shaw model with injection through a source at the origin and with a bounded initial phase domain Ω_0.

Let f be a univalent function $f(\zeta) = a\zeta + a_2\zeta^2 + \cdots$, $a > 0$, defined in the unit disk U and let S be the area of $f(U)$. An obvious inequality, which we call the *radius-area estimate*, is $S \geq \pi a^2$. It follows from the formula for the area $S = \pi(a^2 + \sum_{k=2}^{\infty} k|a_k|^2)$. Equality is attained for a trivial map $f(\zeta) = a\zeta$. There is no upper estimate of S, namely, the area can be infinite. As for the perimeter, the classical result by Pólya and Schiffer [266] states that $L \geq 2\pi a$. The upper estimate is ∞ in general and $8R(f(U), \infty)$ for convex domains.

Let $S(t)$ be the area of the domain $\Omega(t)$ in the Hele-Shaw dynamics under the conditions of Section 1.4.2. A simple application of Green's theorem implies that the rate of the area change is expressed as $\dot{S} = -Q$ under injection ($Q < 0$). From (1.17) we deduce that

$$\dot{a} = -a\frac{1}{4\pi^2}\int_0^{2\pi} \frac{Q}{|f'(e^{i\theta}, t)|^2} d\theta \geq -a\frac{1}{4\pi^2}\int_0^{2\pi} \operatorname{Re}\frac{Q}{[f'(e^{i\theta}, t)]^2} d\theta = \frac{-Q}{2\pi a} = \frac{\dot{S}}{2\pi a},$$

where $a = f'(0,t)$. In other words the area rate is controlled by the rate of the conformal radius of the domain $\Omega(t)$ with respect to the origin: $\dot{S} \leq 2\pi a\dot{a}$. Equality in the above inequalities is attained for $\Omega(t) = \{z : |z| < a(t)\}$.

The lower bound for $\dot{S}$ in terms of a is much more difficult. One must estimate the integral mean

$$\int_0^{2\pi} \frac{1}{|f'(e^{i\theta},t)|^2} d\theta \tag{5.5}$$

from above. The fact that cusps may develop shows that there is no uniform estimate with respect to t. But one can estimate (5.5) under some geometric constraints on the domain $\Omega(t)$ at an instant t. For example, assume that the domain $\Omega(t)$ is convex. The function f is also convex and, thus, $\frac{1}{2}$-starlike, i.e., Re $\zeta f'(\zeta,t)/f(\zeta,t) > 1/2$. Moreover the Koebe covering theorem for the convex functions says that $|f(\zeta,t)| \geq a/2$. This implies the estimate

$$\int_0^{2\pi} \frac{1}{|f'(e^{i\theta},t)|^2} d\theta < \frac{32\pi}{a^2}, \quad \text{and so } \dot{S} > \frac{\pi a}{8}\dot{a}.$$

Let us give a precise estimate for this integral mean in the case of convex functions f. If f is convex, then the function $g(\zeta) \equiv \zeta f'(\zeta)$ is starlike in U and the function $h(\zeta) \equiv 1/g(1/\zeta) = \frac{1}{a}(\zeta + c_0 + c_1/\zeta + \cdots)$ is starlike in the complement U^* of the closure of U. The function $h(\zeta)$ is univalent, bounded in U, and has no zeros in the closure $\bar{U}$. Therefore,

$$\begin{aligned} \frac{1}{2\pi}\int_0^{2\pi} \frac{d\theta}{|f'(e^{i\theta})|^2} = \frac{1}{2\pi}\int_0^{2\pi} |h(e^{i\theta})|^2 d\theta &= \frac{1}{a^2}(1 + |c_0|^2 + \sum_{k=1}^{\infty} |c_k|^2) \leq \\ &\leq \frac{1}{a^2}(1 + |c_0|^2 + \sum_{k=1}^{\infty} k|c_k|^2). \end{aligned} \tag{5.6}$$

We have $|c_0| \leq 2$, and by the Area Theorem (see e.g. [127], [132], [270]) the right-hand side of (5.6) is $\leq 6/a^2$. This estimate is sharp. Finally, for domains $\Omega(t)$ which are convex at an instant t we have

$$\dot{S} \geq \frac{\pi a}{3}\dot{a}.$$

Problem 5.1.4. *If the domain $\Omega(t)$ is convex, then during the time of the existence of the solution of* (1.16) *convexity may be lost at the next instant. It is better to find a geometric condition that is preserved during some time interval of the existence of the solution of* (1.16) *and that permits us to estimate the integral mean* (5.5) *from above.*

Let a univalent function f be defined in U, have the non-vanishing finite angular derivatives almost everywhere at the unit circle, and the boundary of $f(U)$ be reachable by outer angles $> \pi/2$. Then $1/f'$ is from the Hardy class H^2. Generally, we operate with univalent functions with analytic boundaries of $f(U)$. Of course, we can consider domains with angles on the boundary and weak solutions. For example, the following theorem was proved in [275].

Theorem 5.1.5. *Let a univalent map $z = f(\zeta) = \zeta + a_2\zeta^2 + \cdots$ be α-convex in U. Then the angular derivative of f exists almost everywhere on the unit circle and*

$$\frac{1}{2\pi}\int_0^{2\pi} \frac{1}{|f'(e^{i\theta})|^2} d\theta \leq \frac{2^{8(1-\alpha)}}{\pi} \mathbf{B}(\frac{5}{2} - 2\alpha, \frac{5}{2} - 2\alpha),$$

where $\mathbf{B}(\cdot,\cdot)$ stands for the Euler Beta-function. The inequality is sharp. In particular,

$$\frac{1}{2\pi}\int_0^{2\pi} \frac{1}{|f'(e^{i\theta})|^2} d\theta \leq \frac{4^{1-4\alpha}}{2\pi} \frac{(3-4\alpha)(1-4\alpha)}{(1-\alpha)(1-2\alpha)} \mathbf{B}(\frac{1}{2} - 2\alpha, \frac{1}{2} - 2\alpha)$$

for $0 \leq \alpha < 1/4$.

Proof. If a function f is α-convex in U, then the analytic function $g(z) \equiv zf'(z)$ is α-starlike (S^*_α, see Section 4.2). Functions from S^*_α admit the following known integral representation,

$$g(z) \in S^*_\alpha \Leftrightarrow g(z) = z\ \exp\{-2(1-\alpha)\int_{-\pi}^{\pi} \log(1 - e^{i\theta}z) d\mu(\theta)\},$$

where $\mu(\theta)$ is a non-decreasing function of $\theta \in [-\pi, \pi]$ and $\int_{-\pi}^{\pi} d\mu(\theta) = 1$.

If $\mu(\theta)$ is a piecewise constant function, then we have a set of complex-valued functions $g_n(z)$ that admit the representation

$$g_n(z) = \frac{z}{\prod_{k=1}^{n}(1 - e^{i\theta_k}z)^{2(1-\alpha)\beta_k}} \in S^*_\alpha, \quad \theta_k \in [-\pi, \pi], \quad \beta_k \geq 0, \quad \sum_{k=1}^{n} \beta_k = 1. \tag{5.7}$$

Using known properties of Stieltjes' integral and Vitali's theorem it is easy to show that the set of functions (5.7) is dense in S^*_α, i.e., for every function $g(z) \in S^*_\alpha$ there exists a sequence $\{g_n(z)\}$ satisfying (5.7) that locally uniformly converges to $g(z)$ in U. Therefore, we need to prove our result only for $g(z) = g_n(z)$.

Let us present a chain of inequalities

$$\begin{aligned}
\frac{1}{2\pi}\int_0^{2\pi}\frac{1}{|g_n(e^{i\theta})|^2}d\theta &= \frac{1}{2\pi}\int_0^{2\pi}\prod_{k=1}^{n}|1-e^{i(\theta-\theta_k)}|^{4(1-\alpha)\beta_k}d\theta\\
&\leq \frac{1}{2\pi}\int_0^{2\pi}\sum_{k=1}^{n}\beta_k|1-e^{i(\theta-\theta_k)}|^{4(1-\alpha)}d\theta\\
&= \frac{1}{2\pi}\sum_{k=1}^{n}\beta_k\int_0^{2\pi}|1-e^{i(\theta-\theta_k)}|^{4(1-\alpha)}d\theta\\
&= \frac{1}{2\pi}\int_0^{2\pi}|1-e^{i\theta}|^{4(1-\alpha)}d\theta\\
&= \frac{4^{1-\alpha}}{2\pi}\int_0^{2\pi}(1-\cos\theta)^{2(1-\alpha)}d\theta\\
&= \frac{2^{8(1-\alpha)}}{\pi}\mathbf{B}(\tfrac{5}{2}-2\alpha,\tfrac{5}{2}-2\alpha).
\end{aligned}$$

The last assertion of the theorem follows from the formulae of reduction of the Beta-function. □

We summarize the results of this section in the following theorem.

Theorem 5.1.6. *Let $\Omega(t)$ be a phase domain occupied by a fluid injected through the origin, let the area of $\Omega(t)$ be $S(t)$, and $a(t)$ be the conformal radius of $\Omega(t)$ with respect to the origin. Then $\dot S \leq 2\pi a\dot a$. If, moreover, $\Omega(t)$ is α-convex at an instant t, then*

$$\frac{2\pi^2 a\dot a}{2^{8(1-\alpha)}\mathbf{B}(\frac{5}{2}-2\alpha,\frac{5}{2}-2\alpha)} \leq \dot S \leq 2\pi a\dot a.$$

In the case of a contracting bubble we have a similar estimate $\dot S \geq 2\pi a\dot a$, where $S(t)$ means the area of the bubble and $a = \operatorname{cap}\Gamma(t)$. The good thing is that the outer Hele-Shaw problem preserves the convex dynamics. More about estimates for integral means can be found, e.g., in [271].

5.2 Hele-Shaw cells with obstacles

Recent studies of Robin's function and Robin's capacity [93]–[98] showed their connections with several problems of potential theoretic nature as well as extremal length and minimal energy considerations. Our goal is to give another physical interpretation that contains the Hele-Shaw problem with an obstacle inside. We shall connect the rate of area change of the phase domain with the rate of change of Robin's reduced modulus of the free boundary.

5.2.1 Robin's capacity and Robin's reduced modulus

Let Ω be a finitely connected domain in $\overline{\mathbb{C}}$, and A be an arbitrary closed set of the boundary $\partial\Omega$. Let us denote by B the complementary part of $\partial\Omega$, so that

$\partial\Omega = A \cup B$. For a fixed finite point $z_0 \in \Omega$, the complex *Robin's function* $R(z, z_0)$ is defined by the following requirements:

- $R(z, z_0)$ is analytic in Ω except at the point z_0 where R has a logarithmic singularity: $R(z, z_0) = \frac{1}{2\pi}\log(z - z_0) + w_0(z)$, where $w_0(z)$ is a regular function in Ω;
- Re $R(z, z_0) = 0$ for all $z \in A$, while $\frac{\partial \text{Re } R}{\partial n}(z, z_0) = 0$ for all $z \in B$.

For $z_0 = \infty$ the definition is modified by requiring $R(z, \infty) - \frac{1}{2\pi}\log z$ to be regular in a neighbourhood of infinity. The real part of this function $R_{re}(z, z_0) =$ Re $R(z, z_0)$ is the classical Robin's function that was studied deeply in [93]–[98]. The main property which we use here is its conformal invariance. For basic properties of Robin's function we refer the reader to [93].

Let us define the *Robin's reduced modulus* $M_\Omega(A, z_0)$ of the set A with respect to the domain Ω and the point $z_0 \in \Omega$ as

$$M_\Omega(A, z_0) = \lim_{r \to 0} R_{re}(z, z_0) + \frac{1}{2\pi}\log r, \quad |z - z_0| = r$$

in the case of a finite z_0, and

$$M_\Omega(A, \infty) = \lim_{r \to \infty} R_{re}(z, \infty) - \frac{1}{2\pi}\log r, \quad |z| = r$$

otherwise. We note that $\delta_\Omega(A) := \exp(2\pi M_\Omega(A, \infty))$ is Robin's capacity of the set A with respect to Ω. In particular, if $B = \partial\Omega \setminus A = \emptyset$, then Robin's capacity coincides with the usual logarithmic capacity $d(A)$ and Robin's reduced modulus is exactly the reduced modulus of the domain Ω with respect to the finite point z_0.

Another description of Robin's capacity and Robin's reduced modulus is provided by means of the modulus of a family of curves.

Let $C_r(z_0) = \{z : |z - z_0| = r\}$ and $C_r = C_r(0)$. For r sufficiently small and a finite $z_0 \in \Omega$ let us consider the family Γ of curves that connect the set A with $C_r(z_0)$. Then the limit

$$\lim_{r \to 0} \frac{1}{m(\Omega, \Gamma)} + \frac{1}{2\pi}\log r$$

exists and is exactly Robin's reduced modulus $M_\Omega(A, z_0)$. Analogously, for $z_0 = \infty \in \Omega$ and r sufficiently large we define Γ to be the family of rectifiable arcs that connect A with C_r. Then the limit

$$\lim_{r \to \infty} \frac{1}{m(\Omega, \Gamma)} - \frac{1}{2\pi}\log r$$

exists and is Robin's reduced modulus $M_\Omega(A, \infty)$. From this definition it follows that Robin's modulus is changed under a conformal map $f : \Omega \to \Omega'$ by the following rule: for finite points $w_0 = f(z_0)$

$$w = f(z) = w_0 + a(z - z_0) + \cdots, \quad M_{\Omega'}(f(A), w_0) = M_\Omega(A, z_0) + \frac{1}{2\pi}\log |a|,$$

for infinite points ($z_0 = w_0 = \infty$):

$$w = f(z) = az + a_0 + \frac{a_{-1}}{z} + \cdots, \quad M_{\Omega'}(f(A), \infty) = M_\Omega(A, \infty) - \frac{1}{2\pi} \log |a|,$$

and for the mixed case ($z_0 = \infty$, w_0 finite):

$$w = f(z) = w_0 + \frac{a}{z} + \cdots, \quad M_{\Omega'}(f(A), w_0) = M_\Omega(A, \infty) + \frac{1}{2\pi} \log |a|.$$

Let us mention some results about distortion of Robin's capacity under an "admissible" conformal map $f(z) = z + a_0 + \frac{a_{-1}}{z} + \cdots$. Ch. Pommerenke [271], [268], [270] proved that for an arbitrary closed set A on the unit circle the sharp estimate $d(f(A)) \geq (d(A))^2$ holds. Later on, P. Duren and M. M. Schiffer [95] generalized this result to an arbitrary multiply connected domain giving Robin's interpretation to the inequality $d(f(A)) \geq \delta_\Omega(A)$, which is sharp. Let us give two elementary examples of Robin's capacity and Robin's reduced modulus.

Examples.

- Let $\bar{U}$ be the closed unit disc and A be an arc on the boundary which subtends an angle 2α at the center. Then $d(A) = \sin\frac{\alpha}{2}$, while $\delta_{\bar{U}}(A) = \left(\sin\frac{\alpha}{2}\right)^2$. Besides, $\delta_{\bar{U}}(A) + \delta_{\bar{U}}(B) = 1 = d(A \cup B)$ (see [95]).
- Let $U' = U \setminus (-1, -r]$, $r \in (0,1)$, and A be the unit circle. Then Robin's reduced modulus $M_{U'}(A, 0) = 0$ of the set A with respect to U', whereas the usual reduced modulus of the domain U' with respect to the origin is $\frac{1}{2\pi} \log \frac{4r}{(1+r)^2} < 0$. To see this we use the standard Pick function $\zeta = \varphi(w)$,

$$\begin{aligned} \varphi(w) &= \frac{\left(\beta(1-w) - \sqrt{\beta^2(1-w)^2 + 4w}\right)^2}{4w} \\ &= \frac{1}{\beta^2} w + \cdots \\ &= \frac{4w}{\left(\beta(1-w) + \sqrt{\beta^2(1-w)^2 + 4w}\right)^2}, \end{aligned}$$

$\beta \geq 1$, that maps the unit disk U onto U' with

$$r = \frac{1}{\beta + \sqrt{\beta^2 - 1}} = \beta - \sqrt{\beta^2 - 1}.$$

The arc $\gamma = \{e^{i\theta}, \theta \in [\pi - \alpha, \pi + \alpha]\}$, $\cos(\alpha/2) = 1/\beta$, is mapped onto the slit $[-1, -r]$. Robin's modulus of γ is $M_U(\gamma, 0) = \frac{1}{2\pi} \log \beta^2$. Therefore, making use of the formula of the modulus transformation we have $M_{U'}(A, 0) = \frac{1}{2\pi} \log 1/\beta^2 + \frac{1}{2\pi} \log \beta^2 = 0$.

Sometimes the notion of reduced modulus of the domain Ω with respect to the point z_0 is replaced by the notion of conformal radius. These concepts are linked by the formula

$$M_\Omega(z_0) = M(\Omega, z_0) = \frac{1}{2\pi} \log R(\Omega, z_0).$$

Similarly we define Robin's radius of the set A with respect to the domain Ω and the point z_0 as

$$M_\Omega(A, z_0) = \frac{1}{2\pi} \log R_\Omega(\Gamma, z_0).$$

5.2.2 A problem with an obstacle

Let a viscous fluid be injected into a Hele-Shaw cell through a point well, producing a simply connected evolution until it meets a straight wall. Then it starts sliding along the wall. We denote by $\Omega(t)$ the bounded plane domain in the phase z-complex plane occupied by the moving fluid at instant t. The source is located at the origin and is of strength $-Q$, $Q < 0$. The unique force in consideration is the dimensionless pressure p scaled so that 0 corresponds to the atmospheric pressure. The initial moment we choose to be when the fluid reaches the wall. This stationary infinite straight wall is placed in the Hele-Shaw cell so that $\partial\Omega(t)$ splits into two parts: $\Gamma(t)$ is the free boundary and $\Pi(t)$ is the complementary arc on the wall. The potential function solves the mixed boundary value problem

$$\Delta p = Q\delta_0(z), \quad \text{in} \quad z \in \Omega(t), \tag{5.8}$$

$$p = 0, \quad v_n = -\frac{\partial p}{\partial n} \quad \text{on} \quad z \in \Gamma(t), \tag{5.9}$$

$$v_n = 0, \quad \text{on} \quad z \in \Pi(t). \tag{5.10}$$

The complex potential is exactly given by Robin's function as $W = QR(z, 0)$ (of course R depends on t).

Richardson [286], [286], considered a similar problem in a wedge assuming circular initial evolution.

We consider the case when the boundary $\Pi(t)$ is an interval during the time of consideration. Observation of the velocities at the contact point between $\Gamma(t)$ and $\Pi(t)$ suggests the contact angle to be $\pi/2$. To derive the equation for the free boundary $\Gamma(t)$ we involve an auxiliary parametric complex ζ-plane, $\zeta = \xi + i\eta$. The Riemann Mapping Theorem yields that there exists a unique conformal univalent map $f(\zeta, t)$ from the unit disk minus a radial slit $U' = U \setminus (-1, -r(t)]$, $r(t) \in (0, 1]$, $U = \{\zeta : |\zeta| < 1\}$, onto the phase domain $f : U' \to \Omega(t)$, $f(0, t) = 0$, $a(t) = f'(0, t) > 0$, so that $\Gamma(t) = \{f(e^{i\theta}), \theta \in (-\pi, \pi)\}$, and $\Pi(t) = \{f(\zeta), \zeta \in (-1 - 0i, -r(t) - 0i) \cup (-r(t) + 0i, -1 + 0i)\}$. The point $\zeta = r(t)$ corresponds to a stagnation point at $\Pi(t)$. The function $f(\zeta, 0) = f_0(\zeta)$ produces a parametrization of $\partial\Omega(0) = \overline{\Gamma(0) \cup \Pi(0)}$. The moving boundary is parameterized by $\Gamma(t) = f(\partial U, t)$. The normal outer vector is given by the formula

$$n = \zeta \frac{f'}{|f'|}, \quad \zeta = e^{i\theta}, \quad \theta \in (-\pi, \pi),$$

and $n = -1$ on $\Pi(t)$. Therefore, the normal velocity at the free boundary is obtained as

$$v_n = \mathbf{V} \cdot \mathbf{n} = \begin{cases} -\mathrm{Re}\left(W'\zeta \frac{f'}{|f'|}\right), & \text{for } z \in \Gamma,\ \zeta = e^{i\theta},\ \theta \in (-\pi, \pi), \\ \mathrm{Re}\, W', & \text{for } z \in \Pi(t). \end{cases}$$

The superposition $(W \circ f)(\zeta, t) \equiv Q\, R \circ f(\zeta, t)$ is $-Q$ times Robin's function of the domain U' because of the conformal invariance. The set A for the function $R \circ f$ is the unit circle and the set B is the radial segment $[-1, -r(t)]$. Robin's function for U' with the chosen A and B is simply $\frac{1}{2\pi} \log \zeta$. Hence, $W'f' = \frac{Q}{2\pi\zeta}$. On the other hand, we have $v_n = \mathrm{Re}\, \dot{f}\, \overline{\zeta f'/|f'|}$, for $\zeta = e^{i\theta}$, $\theta \in (-\pi, \pi)$ and

$$\mathrm{Re}\, \frac{Q}{2\pi\zeta f'} = \mathrm{Re}\, W' = \mathrm{Re}\, \dot{f} = 0$$

on the wall. This implies that $\dot{f}$ and $1/\zeta f'$ are imaginary or $\dot{f}/\zeta f'$ is real. Finally we deduce the Polubarinova-Galin type equations

$$\mathrm{Re}\, [\dot{f}(\zeta, t)\overline{\zeta f'(\zeta, t)}] = \frac{-Q}{2\pi}, \quad |\zeta| = 1,\ \arg \zeta \in (-\pi, \pi), \tag{5.11}$$

$$\mathrm{Im}\, [\dot{f}(\zeta, t)\overline{\zeta f'(\zeta, t)}] = 0 \quad \text{on the radial slit } [-1, -r(t)]. \tag{5.12}$$

The length of the radial slit $1-r(t)$ is such that the conformal radius of the domain $\Omega(t)$ with respect to the origin is equal to $4r(t)a(t)(1 + r(t))^{-2}$.

If $\Pi(t)$ is the union of intervals, then the function $f(\zeta, t)$ maps the unit disk minus several slits onto the phase domain. Each slit corresponds to a connected component of $\Pi(t)$.

Now we apply Robin's reduced modulus to estimate the area growth of the phase domain of the injecting fluid in the Hele-Shaw problem. First of all, let us recall that the boundary $\partial\Omega(t)$ of the domain occupied by viscous fluid contains a free part $\Gamma(t)$ and the solid part $\Pi(t)$. The fluid is injected through the origin $0 \in \Omega(t)$. The parametric function $f(\zeta, t)$ maps $U' = U \setminus (-1, -r(t)]$ onto $\Omega(t)$ and satisfies the equations (5.11)–(5.12).

A simple application of Green's theorem implies that the rate of the area change is expressed as $\dot{S} = -Q$, where $S(t)$ is the Euclidean area of $\Omega(t)$. For injection we have $Q < 0$.

Let the Pick function $\zeta = \varphi(w, t)$ map U onto U',

$$\varphi(w, t) = \frac{4r(t)}{(1 + r(t))^2} w + b_2 w^2 + \cdots \equiv bw + b_2 w^2 + \cdots, \cdots b = 1/\beta^2$$

so that the arc $\{e^{i\theta}, \theta \in (-\alpha(t), \alpha(t))\}$ is mapped onto $\partial U \setminus \{-1\}$ and the arc $\{e^{i\theta}, \theta \in (\alpha(t), 2\pi - \alpha(t))\}$ is mapped onto the radial slit $(-1, -r(t))$.

Set the analytic function

$$\Phi(w,t) = \frac{\dot{f}\circ\varphi}{\varphi(f'\circ\varphi)}(w,t),$$

defined in U. The mixed boundary value problem (5.11)–(5.12) can be reformulated as the Riemann-Hilbert problem for the analytic function Φ as

$$\text{Re } \Phi(e^{i\theta},t) = \frac{-Q}{2\pi|(f'\circ\varphi)(e^{i\theta},t)|^2}, \qquad \theta \in (-\alpha,\alpha),$$

$$\text{Im } \Phi(e^{i\theta},t) = 0, \quad \theta \in (\alpha, 2\pi-\alpha),$$

with bounded values of $|\lim_{\theta\to\pm\alpha^\pm} \Phi(e^{i\theta},t)|$. The solution to this problem is given by the integral representation

$$\begin{aligned}\Phi(w,t) &= \frac{1}{2\pi}\frac{\sqrt{(w-e^{i\alpha})(w-e^{-i\alpha})}}{w+1} \\ &\times \int_0^{2\pi} \frac{e^{i\theta}+1}{\sqrt{(e^{i\theta}-e^{i\alpha})(e^{i\theta}-e^{-i\alpha})}} h(e^{i\theta},t)\frac{e^{i\theta}+w}{e^{i\theta}-w}d\theta,\end{aligned}$$

where the branch of the root is chosen so that $\sqrt{1}=1$, and

$$h(e^{i\theta},t) = \frac{-Q}{2\pi|(f'\circ\varphi)(e^{i\theta},t)|^2}, \qquad \theta \in (-\alpha,\alpha),$$

and vanishes in the complementary arc of ∂U, see, e.g., [116]. We deduce that

$$\frac{\dot{a}}{a} = \frac{1}{2\pi}\int_0^{2\pi} \frac{\sqrt{2}\cos\frac{\theta}{2}}{\sqrt{\cos\theta - \cos\alpha}} h(e^{i\theta},t)d\theta.$$

Obviously,

$$\frac{\sqrt{2}\cos\frac{\theta}{2}}{\sqrt{\cos\theta-\cos\alpha}} \geq \frac{1}{\sin\frac{\alpha}{2}} = \beta, \quad \text{for } \theta \in (-\alpha,\alpha).$$

Therefore,

$$\frac{\dot{a}}{a} \geq \frac{-Q\beta}{4\pi^2}\int_0^{2\pi} \chi_{[-\alpha(t),\alpha(t)]}\frac{1}{|f'(\varphi(e^{i\theta},t),t)|^2}d\theta,$$

where $\chi_{[-\alpha(t),\alpha(t)]}$ is the characteristic function of the segment $[-\alpha(t),\alpha(t)]$, and $\varphi(e^{i\alpha(t)},t) = -1$. The Hölder inequality implies

$$\frac{\dot{a}}{a} \geq \frac{-Q\beta}{8\pi^3}\left(\int_0^{2\pi} \chi_{[-\alpha(t),\alpha(t)]}\frac{1}{|f'(\varphi(e^{i\theta},t),t)|}d\theta\right)^2.$$

In its turn,

$$\frac{-Q\beta}{8\pi^3}\left(\int_0^{2\pi}\chi_{[-\alpha(t),\alpha(t)]}\frac{1}{|f'(\varphi(e^{i\theta},t),t)|}d\theta\right)^2$$

$$\geq\frac{-Q\beta}{8\pi^3}\left(\int_0^{2\pi}\chi_{[-\alpha(t),\alpha(t)]}\mathrm{Re}\,\Big[\frac{1}{f'(\varphi(e^{i\theta},t),t)}\Big]d\theta\right)^2.$$

But $\mathrm{Re}\, f'(\varphi(e^{i\theta},t),t)=0$ for $\theta\in(\alpha(t),2\pi-\alpha(t))$. Therefore,

$$\frac{-Q\beta}{8\pi^3}\left(\int_0^{2\pi}\chi_{[-\alpha(t),\alpha(t)]}\mathrm{Re}\,\Big[\frac{1}{f'(\varphi(e^{i\theta},t),t)}\Big]d\theta\right)^2$$

$$=\frac{-Q\beta}{8\pi^3}\left(\int_0^{2\pi}\mathrm{Re}\,\Big[\frac{1}{f'(\varphi(e^{i\theta},t),t)}\Big]d\theta\right)^2=\frac{Q\beta}{2\pi a^2}.$$

Finally, we have an estimate $2\pi\dot{a}a\sin\frac{\alpha}{2}\geq\dot{S}$, where $a=f'(0,t)$. The conformal radius $R(\Omega(t),0)$ of the domain $\Omega(t)$ is just $ab=a/\beta^2=a\sin^2\frac{\alpha}{2}$. Robin's radius of the arc $\Gamma(t)$ is $R_{\Omega(t)}(\Gamma(t),0)=a$. Therefore, we have our main isoperimetric inequality

$$\dot{S}\leq 2\pi\dot{R}_{\Omega(t)}(\Gamma(t),0)\sqrt{R_{\Omega(t)}(\Gamma(t),0)\;R(\Omega(t),0)}.$$

In other words this means that the rate of area change of the phase domain $\Omega(t)$ is controlled by Robin's radius of the free boundary as well as by its conformal radius.

Finally, let us remark that a general case of disconnected boundary component $\Pi(t)$ can be treated in the same way. The solution to the corresponding Riemann-Hilbert problem yields more complicated formulations, so we have considered only the simplest case.

5.3 Isoperimetric inequality for a corner flow

In this section we shall obtain an analogue of the right-hand side estimate given in Theorem 5.1.2 for the corner flow. In Section 2.2 we already considered such flows and derived the governing equations for the conformal map that parameterizes the phase domain. Here we use a slightly different parametrization that fits better for our concrete purpose.

Similarly to the model analysed in Section 2.2 we consider a Hele-Shaw cell where the viscous fluid occupies a simply connected domain $\Omega(t)$ in the phase z-plane whose boundary $\Gamma(t)$ at an instant t consists of two walls $\Gamma_1(t)$ and $\Gamma_2(t)$ of the corner and a free interface $\Gamma_3(t)$ between them. The inviscid fluid (or air) fills the complement of $\Omega(t)$. The simplifying assumption of constant pressure at the interface between the fluids means that the surface tension effect is neglected. We let the positive real axis x contain one of the walls and fix the angle between walls as $\alpha \in (0, 2\pi)$. The motion of the boundary $\Gamma_3(t)$ is due to injection of strength $Q > 0$ through the vertex of the corner placed at the origin. The initial domain $\Omega(0)$ fills the vertex. In our model we consider the local behavior of $\Gamma_3(t)$ and agree that $\Gamma_3(t)$ is connected. At the wall-fluid contact points where Γ_1 or Γ_2 join with Γ_3 the velocity vector is directed along the walls that implies that Γ_1 and Γ_2 are perpendicular to Γ_3 at these points.

As before, the pressure field p satisfies the Laplacian equation and the boundary conditions split into the free boundary condition (given on Γ_3) for pressure and the wall conditions for pressure's normal derivative. The potential p behaves near the origin as

$$p \sim -\frac{Q}{\alpha} \log |z|, \quad \text{as } |z| \to 0.$$

Let us consider an auxiliary parametric complex ζ-plane, $\zeta = \xi + i\eta$. We set $D = \{\zeta : |\zeta| < 1, 0 < \arg \zeta < \pi\}$, $D_3 = \{z : z = e^{i\theta}, \theta \in (0, \pi)\}$, $D_1 = \{z : z = -r, r \in (0,1)\}$, $D_2 = \{z : z = r, r \in (0,1)\}$, $\partial D = D_1 \cup D_2 \cup D_3$. Construct a conformal univalent time-dependent map $z = f(\zeta, t)$, $f : D \to \Omega(t)$, such that being continued onto ∂D, $f(0,t) \equiv 0$, and the circular arc D_3 of ∂D is mapped onto Γ_3. This map has the expansion $f(\zeta, t) = \zeta^{\alpha/\pi} \sum_{k=0}^{\infty} a_k(t) \zeta^k$ near the origin, and $a_0(t) > 0$. The function f parameterizes the boundary of the domain $\Omega(t)$ by $\Gamma_j = \{z : z = f(\zeta, t), \zeta \in D_j\}$, $j = 1, 2, 3$.

Using standard steps of Section 2.2 we arrive at the free boundary condition expressed in terms of the function f as

$$\operatorname{Re} (\dot{f} \, \overline{\zeta f'}) = \frac{Q}{\pi}, \quad \text{for } \zeta \in D_3. \tag{5.13}$$

The wall conditions imply that

$$\operatorname{Im} (\dot{f} e^{-i\alpha}) = 0 \quad \text{for } \zeta \in D_1; \quad \operatorname{Im} (\dot{f}) = 0 \quad \text{for } \zeta \in D_2. \tag{5.14}$$

We note that the derivative of the mapping function $f'(\zeta, t)$ satisfies the following conditions at D_1 and D_2,

$$\arg(\zeta f'(\zeta, t)) = \alpha \quad \text{for } \zeta \in D_1,$$

and

$$\operatorname{Im} (\zeta f'(\zeta, t)) = 0 \quad \text{for } \zeta \in D_2.$$

Hence, we can rewrite the conditions (5.13)–(5.14) as a mixed boundary value problem for the analytic function

$$\Phi(\zeta, t) := \frac{\dot{f}(\zeta, t)}{\zeta f'(\zeta, t)},$$

given by

$$\operatorname{Re}(\Phi(\zeta,t)) = \frac{Q}{\pi|f'(\zeta,t)|^2}, \quad \text{for } \zeta \in D_3, \tag{5.15}$$

$$\operatorname{Im}(\Phi(\zeta,t)) = 0, \quad \text{for } \zeta \in D_1 \cup D_2. \tag{5.16}$$

First, we solve the mixed boundary value problem (5.15)–(5.16). Making use of an auxiliary Joukowski transform

$$\omega(\zeta) := \frac{1}{2}\left(\zeta + \frac{1}{\zeta}\right), \quad \text{or } \zeta(\omega) := \omega - \sqrt{\omega^2 - 1},$$

we reduce this problem to a Riemann-Hilbert problem in the upper ω-half-plane. Applying the Keldysh-Sedov formula (see, e.g., [116]) for the analytic function $\Phi(\zeta(\omega),t)$ which is bounded at ± 1, we get

$$\Phi(\zeta(\omega),t) = \frac{\sqrt{\omega^2-1}}{\pi i}\int_{-1}^{1} \frac{Q}{\pi|f'(\zeta(\tau),t)|^2\sqrt{\tau^2-1}}\frac{d\tau}{\tau-\omega}.$$

The analytic function in the right-hand side is defined in $\mathbb{C} \setminus [-1,1]$, therefore, choosing a suitable branch of the root we can calculate

$$\lim_{\omega\to\infty} \Phi(\zeta(\omega),t) = \frac{1}{\pi}\int_{-1}^{1} \frac{Q}{\pi|f'(\zeta(\tau),t)|^2\sqrt{1-\tau^2}}d\tau.$$

Secondly, we return to the variable ζ and obtain

$$\frac{\dot a_0(t)}{a_0(t)} = \frac{\alpha Q}{\pi^3}\int_0^{\pi} \frac{1}{|f'(e^{i\theta},t)|^2}d\theta.$$

Certainly,

$$\frac{\dot a_0(t)}{a_0(t)} \geq \frac{\alpha Q}{\pi^3}\operatorname{Im}\int_0^{\pi} \frac{e^{2i\alpha\theta/\pi}}{(e^{i\theta}f'(e^{i\theta},t))^2} ie^{i\theta}\frac{d\theta}{e^{i\theta}},$$

or

$$\frac{\dot a_0(t)}{a_0(t)} \geq \frac{\alpha Q}{\pi^3}\operatorname{Im}\int_{D_1\cup D_2\cup D_3} \frac{\zeta^{2\alpha/\pi}}{(\zeta f'(\zeta,t))^2}\frac{d\zeta}{\zeta}.$$

The function

$$\frac{\zeta^{2\alpha/\pi}}{(\zeta f'(\zeta,t))^2}$$

is analytic about the origin and symmetric with respect to the real axis. Hence, we can take a small circle $S_\varepsilon = \{\zeta : |\zeta| = \varepsilon\}$ and write

$$\begin{aligned}\operatorname{Im} \int\limits_{D_1 \cup D_2 \cup D_3} \frac{\zeta^{2\alpha/\pi}}{(\zeta f'(\zeta,t))^2} \frac{d\zeta}{\zeta} &= \operatorname{Im} \int\limits_{D_3} \frac{\zeta^{2\alpha/\pi}}{(\zeta f'(\zeta,t))^2} \frac{d\zeta}{\zeta} \\ &= \frac{1}{2} \operatorname{Im} \int\limits_{S_\varepsilon} \frac{\zeta^{2\alpha/\pi}}{(\zeta f'(\zeta,t))^2} \frac{d\zeta}{\zeta} \\ &= \frac{\pi^3}{\alpha^2 a_0^2}.\end{aligned}$$

So we have the inequality $Q \leq \alpha a_0 \dot{a}_0$. The constant Q corresponds to the rate of the area growth. However, one can obtain this directly using Green's theorem. In fact, if $S(t)$ means the area of $\Omega(t)$ and $\Gamma = \Gamma_1 \cup \Gamma_2 \cup \Gamma_3$, then

$$S(t) = \frac{1}{2} \operatorname{Im} \int\limits_{\Gamma} \bar{z} dz = \frac{1}{2} \operatorname{Im} \int\limits_{0}^{\pi} \bar{f} f' i e^{i\theta} d\theta.$$

Therefore,

$$\dot{S} = \frac{1}{2} \operatorname{Im} \int\limits_{0}^{\pi} \dot{\bar{f}} f' i e^{i\theta} d\theta + \frac{1}{2} \operatorname{Im} \int\limits_{0}^{\pi} \bar{f} \dot{f}' i e^{i\theta} d\theta.$$

Integrating by parts the second term and using (5.13) we come to the equality $\dot{S} = Q$. Finally, we obtain the desirable inequality

$$\dot{S} \leq \alpha a_0 \dot{a}_0, \tag{5.17}$$

which is known for $\alpha = 2\pi$ (see Theorem 5.1.2).

To make inequality (5.17) isoperimetric we interpret a_0 as a certain entity related to the free boundary Γ_3. Let D be a hyperbolic simply connected domain in $\mathbb{C}$ with three finite fixed boundary points z_1, z_2, and a on its piecewise smooth boundary. Denote by D_ε the domain $D \setminus U(a, \varepsilon)$ for a sufficiently small ε, where $U(a, \varepsilon) = \{z : |z - a| < \varepsilon\}$. Denote by $M(D_\varepsilon)$ the modulus of the family of arcs in D_ε joining the boundary arc of $U(a, \varepsilon)$ that lies in the circumference $|z - a| = \varepsilon$ with the leg of the triangle D which is opposite to a (we choose a unique arc of the circle so that it can be connected in D_ε with the leg (z_1, z_2) for any $\varepsilon \to 0$). If the limit

$$M_\Delta(D, a) = \lim_{\varepsilon \to 0} \left(\frac{1}{M(D_\varepsilon)} + \frac{1}{\varphi_a} \log \varepsilon \right)$$

exists, where $\varphi_a = \sup \Delta_a$ is the inner angle and Δ_a are the Stolz angles inscribed in D at a, then it is called the *reduced modulus of the triangle D*. The conditions for the reduced modulus to exist are found in [272], [313], [335]. It turns out that the reduced modulus exists if D is conformal at a. Let there exist a conformal map $f(z)$ of the triangle D onto a triangle D' such that there is an angular limit $f(a)$ (see definitions in [271]) with the inner angle ψ_a at the vertex $f(a)$. If the function f has the angular finite non zero derivative $f'(a)$, then $\varphi_a = \psi_{f(a)}$ and the reduced modulus of D exists and changes [313], [335] according to the rule

$$M_\Delta(f(D), f(a)) = M_\Delta(D, a) + \frac{1}{\psi_a} \log |f'(a)|.$$

If we suppose, moreover, that f has the expansion

$$f(z) = w_1 + (z-a)^{\psi_a/\varphi_a}(c_1 + c_2(z-a) + \cdots)$$

in a neighborhood of the point a, then the reduced modulus of D changes according to the rule

$$M_\Delta(f(D), f(a)) = M_\Delta(D, a) + \frac{1}{\psi_a} \log |c_1|.$$

Similarly to the connection between the conformal radius and the usual reduced modulus of a simply connected domain with respect to an inner point, we introduce the conformal triangle radius $R_\Delta(D, a)$ as

$$R_\Delta(D, a) = \exp[\varphi_a M_\Delta(D, a)].$$

The conformal triangle radius of the half-disk $\{|z| < 1\} \cap \{\operatorname{Im} z > 0\}$ with marked vertices $0, \pm 1$ with respect to the origin is 1. The phase domain Ω is conformal at the origin. Using this interpretation we rewrite inequality (5.17) as

$$\dot{S} \leq \alpha \; R_\Delta(\Omega(t), 0)\dot{R}_\Delta(\Omega(t), 0).$$

This is the isoperimetric inequality we were looking for.

5.4 Melting of a bounded crystal

In Section 3.4 have already discussed the governing equation for a melting crystal. In this section we consider a bounded initial crystal that is melting in forced flow. The fluid moves to the right and there are two stagnating points on the interface of the crystal. The governing equations are the same (4.56) with the initial conditions

$$\lim_{x\to\pm\infty} \theta = 1, \qquad \lim_{y\to\pm\infty} \frac{\partial \theta}{\partial y} = 0.$$

The Boussinesq transformation applied to the convective heat transfer equation (4.56) leads to uncoupling of the problem. There is a conformal univalent map from the phase domain $\Omega(t)$ onto the plane of the complex potential $W = \varphi + i\psi$. Under this transformation the boundary of the crystal cross-section is mapped into the slit along the positive real axis $\psi = 0$, $\varphi \in [-2a, 2, a]$ in the W-plane. Thus, the problem admits the form:

$$Pe\frac{\partial\theta}{\partial\varphi} = \Delta\theta, \quad W \in D, \tag{5.18}$$

where $D = \{\mathbb{C} \setminus [-2a, 2a]\}$. The boundary conditions are

$$\lim_{\varphi\to\pm\infty} \theta = 1, \quad \lim_{\psi\to\pm\infty} \frac{\partial\theta}{\partial\psi} = 0, \quad \theta = 0, \quad W \in \partial D. \tag{5.19}$$

We introduce the auxiliary parametric complex ζ-plane, $\zeta = \xi + i\eta$. The Riemann Mapping Theorem yields that there exists a conformal univalent map $f(\zeta, t)$ of the exterior part U^* of the unit disk U onto the phase domain $f : U^* \to \Omega(t)$, normalized by $f(\zeta, \cdot) = a\zeta + a_0 + a_{-1}/\zeta + \cdots$. This problem does not admit separation of variables as in the previous case. In [204] it was shown that the heat flux density at the slit on the plane of the complex potential $W = \varphi + i\psi$ can be expressed as $|\partial\theta/\partial\psi| = (4a^2 - \varphi)^{-1/2}\mu(\varphi/2a)$. The Joukowski function $W = a(\zeta+1/\zeta)$ permits us finally to come to the Polubarinova-Galin type equation for the free boundary:

$$\text{Re }[\dot{f}(\zeta,t)\overline{\zeta f'(\zeta,t)}] = -\mu(\cos\theta), \quad \zeta = e^{i\theta}. \tag{5.20}$$

The function μ in (5.20) satisfies the integral equation [204] for a crystal that admits reflection with respect to the real axis and for small Péclet numbers:

$$\int_{-1}^{1} \frac{\mu(\xi)}{\sqrt{1-\xi^2}} \ln\left|\frac{\varphi}{2a} - \xi\right| d\xi = \pi - \frac{Q}{2}\ln\frac{ae^{\gamma}\, Pe}{2},$$

where $\varphi \in [-2a, 2a]$, γ is Euler's constant, and

$$Q = 2\int_{-1}^{1} \frac{\mu(\xi)}{\sqrt{1-\xi^2}} d\xi = \int_{0}^{2\pi} \mu(\cos\theta)d\theta$$

is the total heat flux. Obviously, the sign of the function μ is connected with the sign of the normal velocity. Therefore, for a melting crystal we have $\mu(\cos\theta) \geq 0$ for all $\theta \in [0, 2\pi)$. A simple applications of Green's Theorem yields that the rate of the area change of the nucleus $\dot{S}$ is exactly equal to the total heat flux taken with (-): $\dot{S} = -Q$. In fact, we have

$$2S(t) = -\int_{\Gamma(t)} \text{Im }(f d\bar{f}) = \text{Re}\int_0^{2\pi} f(e^{i\theta}, t)\overline{\frac{\partial f(\zeta,t)}{\partial\zeta}\Big|_{\zeta=e^{i\theta}}} e^{-i\theta}d\theta.$$

Then,

$$2\frac{dS}{dt} = -\int_0^{2\pi} \mu(\cos\theta)d\theta - \text{Im}\int_0^{2\pi} f\frac{\partial}{\partial t}\left(\frac{\partial f}{\partial\theta}\right)d\theta.$$

Integrating the last term by parts we obtain that $\dot{S} = -\int_0^{2\pi} \mu(\cos\theta)d\theta = -Q$.

The equation (5.20) implies

$$\dot{a} = \frac{-a}{2\pi}\int_0^{2\pi} \frac{\mu(\cos\theta)}{|f'(e^{i\theta},t)|^2}d\theta.$$

Since we have the inequality $|f'(e^{i\theta},t)| \leq 2a$ for functions that map U^* onto a convex domain, the radius-area estimate $\dot{S} \geq 8\pi a\dot{a}$ can be given.

Theorem 5.4.1. *If the initial nucleus is convex, then locally in time we have the estimate* $\dot{S} \geq 8\pi a\dot{a}$ *where* $a = \operatorname{cap}\Gamma(t)$.

Problem 5.4.2. *Give an analogous estimate without the assumption about convexity.*

Problem 5.4.3. *Give an estimate of* $\dot{a}$ *or of the above integral mean depending on the total heat flux* Q.

Chapter 6

General Evolution Equations

Let us consider the solutions to the Polubarinova-Galin equation (1.16) in the case of injection (with $Q < 0$). The fluid is advancing in the normal direction and the solutions form subordination chains of conformal univalent maps (and corresponding chains of hyperbolic univalent domains). This particular case of subordination chains has been considered in the preceding chapters. The existence theorem makes it natural to assume that at least the initial domain Ω_0 of the Hele-Shaw dynamics $\Omega(t)$ is bounded by a smooth analytic curve. A closed Jordan curve is called a *quasicircle* (*quasidisk*) if it is an image of the unit circle (disk) under a quasiconformal homeomorphism of $\overline{\mathbb{C}}$. A piece-wise smooth Jordan curve bounds a quasidisk if and only if it has no cusps, i.e., no corners of angle 0 or 2π. So all domains $\Omega(t)$, $t \in [0, t_0)$ in a Hele-Shaw evolution are quasidisks until a cusp or a double point (Theorem 4.4.1) occurs on the boundary $\Omega(t_0)$.

This chapter is devoted to general subordination dynamics that corresponds to the Löwner-Kufarev equation. We construct a parametric method for conformal maps that admit quasiconformal extensions and, in particular, such that the associated quasidisks are bounded by smooth Jordan curves. Some applications to Hele-Shaw flows of viscous fluids are given.

As usual, U denotes the unit disk and $S^1 = \partial U$. By S we denote the class of all holomorphic univalent functions in U normalized by $f(\zeta) = \zeta + a_2\zeta^2 + \cdots$, $\zeta \in U$, and by Σ, the class of all univalent meromorphic functions in U^* normalized by $f(\zeta) = \zeta + c_0 + \dfrac{c_1}{\zeta} + \cdots$, $\zeta \in U^*$, Σ_0 stands for all functions from Σ with $c_0 = 0$. These classes have been one of the principal objects of research in complex analysis for a long time. The most challenging problem for the class S, posed by Bieberbach[1] in 1916 [38], was finally solved in 1984 by Louis de Branges

[1] Ludwig Georg Elias Moses Bieberbach, 1886–1982, was a student of Felix Klein at Göttingen. He received his doctorate in 1911. Bieberbach was appointed professor of mathematics in Basel in Switzerland, Frankfurt in Germany, and the University of Berlin where he held the chair of geometry and where he developed a reputation as an inspiring but rather disorganised teacher.

[43] who proved that $|a_n| \leq n$ for any $f \in S$ and that equality is attained only for the Koebe function $k(z) = z(1 - ze^{i\theta})^{-2}$, $\theta \in [0, 2\pi)$. The main tool of the proof turned out to be the parametric representation of a function from S by the Löwner homotopic deformation of the identity map given by the *Löwner differential equation*. This parametric method emerged almost 80 years ago in a seminal paper by Löwner [224]. Löwner studied a one-parameter semigroup of conformal one-slit maps of U. His main achievement was an infinitesimal description of the semi-flow of such maps by the Schwarz kernel that led him to what is now called the Löwner equation. This crucial result was later on generalized in several ways.

Attempts have been made to derive an equation that allowed one to describe a representation of the whole class S. Nowadays, it is rather difficult to follow the correct historic line of the development of the parametric method because in the middle of the 20th century a number of works dedicated to this general equation appeared independently. In particular, Kufarev [212] studied a one-parameter family of domains $\Omega(t)$, and regular functions $f(z,t)$ defined in $\Omega(t)$. He proved differentiability of $f(z,t)$ with respect to t for z in the Carathéodory kernel $\Omega(t_0)$ of $\Omega(t)$, and derived a generalization of the Löwner equation. Pommerenke [268] proposed to consider subordination chains of domains that led him to a general equation. We mention here also papers by Gutlyanskiĭ [155] and Goryainov [134] in this direction. One can learn more about this method in the monographs [13], [92], [270] (see also the references therein). Let us draw reader's attention to Goryainov's approach [134]. He suggested using a method of semigroups to derive several other parametric representations of classes of analytic maps and to apply it to study the dynamics of stochastic branching processes. This approach is based on the study of one-parameter semi-flows on semigroups of conformal maps and their infinitesimal descriptions by evolution equations (see also [309]).

In 1959 Shah Dao-Shing [307] suggested a parametric method for quasiconformal automorphisms of U. In another form this method appeared in the paper by Gehring and Reich [122], and then, in [219]. Later, Cheng Qi He [159] obtained an analogous equation for classes of quasiconformally extendable univalent functions (to be more precise, in terms of inverse functions). Unlike the parametric method for conformal maps, its analogue for quasiconformal maps did not receive so much attention.

Several attempts were launched to specialize the Löwner-Kufarev equation to obtain conformal maps that admit quasiconformal extensions (see [24], [25], [26], [156]).

The principal goal of this chapter is to study evolution equations for conformal maps with quasiconformal extensions. In particular, we are interested in maps smoothly extendable onto the unit circle. Our approach is based on the study of

He was converted to the views of the Nazis soon after Hitler came to power and energetically persecuted his Jewish colleagues (e.g., Landau and Schur). However, after the end of World War II in 1945 Ostrowski invited him to lecture at Basel University in 1949 respecting Bieberbach's contribution as a mathematician. It is interesting that de Branges became the first winner of the Ostrowski Prize for solving the Bieberbach conjecture.

flows on the universal Teichmüller space T and on the manifold Diff $S^1/\text{Rot } S^1$ embedded into T. Another question we are interested in is what a Hele-Shaw evolution looks like in the universal Teichmüller space.

6.1 The Löwner-Kufarev equation

We consider a *subordination chain* of simply connected hyperbolic domains $\Omega(t)$ in the Riemann sphere $\overline{\mathbb{C}}$, which is defined for $0 \leq t < t_0$. This means that $\Omega(s) \subset \Omega(t)$ when $s < t$. We suppose that all $\Omega(t)$ are unbounded with $\infty \in \Omega(t)$ for all t. By the Riemann Mapping Theorem we can construct a subordination chain of mappings $f(\zeta, t)$, $\zeta \in U^*$, where the function $f(\zeta,t) = \alpha(t)\zeta + a_0(t) + \dfrac{a_1(t)}{\zeta} + \cdots$ is a meromorphic univalent map of U^* onto $\Omega(t)$ for each fixed t. Pommerenke [268], [270] first introduced such chains in order to generalize Löwner's equation. His result says that given a subordination chain of domains $\Omega(t)$ with a differentiable decreasing real-valued coefficient $\alpha(t)$ (e^{-t} after suitable rescaling), there exists a regular analytic function

$$p(\zeta,t) = p_0(t) + \frac{p_1(t)}{\zeta} + \frac{p_2(t)}{\zeta^2} + \cdots, \qquad \zeta \in U^*,$$

such that $\operatorname{Re} p(\zeta,t) > 0$ for $\zeta \in U^*$ and

$$\dot{f}(\zeta,t) = -\zeta f'(\zeta,t) p(\zeta,t), \tag{6.1}$$

for almost all $t \in [0, t_0)$. The coefficient $\alpha(t) = \alpha(0)\exp(-\int_0^t p_0(\tau)d\tau)$ is the conformal radius of $\Omega(t)$. A reciprocal statement is also true. This equation is known nowadays as the Löwner-Kufarev equation due to the contributions by Löwner [224] and Kufarev [212].

Geometrically, it readily corresponds to the normal motion of the boundary $\partial\Omega(t)$. Indeed, supposing an analytic boundary $\partial\Omega(t)$, the normal vector in the outward direction is $\mathbf{n} = -\zeta f'(\zeta,t)/|f'(\zeta,t)|$, $|\zeta| = 1$, and defining $p(\zeta,t)$ as $(-\dot{f}(\zeta,t))/(\zeta f'(\zeta,t))$ the normal velocity v_n is given by

$$v_n = -\operatorname{Re}\left(\frac{\dot{f}(\zeta,t)\overline{\zeta f'(\zeta,t)}}{|f'(\zeta,t)|}\right) = \operatorname{Re}\left[p(\zeta,t)|f'(\zeta,t)|\right], \quad |\zeta| = 1,$$

and is positive. Therefore, $\operatorname{Re} p(\zeta,t) > 0$ as stated in (6.1). Of course, the general case of a nonanalytic boundary requires finer argumentation.

We consider two main **questions:**

- What does $p(\zeta,t)$ look like when $\partial\Omega(t)$ is a quasicircle?
- The same question in the case of a smooth $\partial\Omega(t)$.

Analogous problems can be posed for the flow in a Hele-Shaw cell under injection at infinity. Suppose that at the initial time the phase domain Ω_0 occupied by the fluid is simply connected and bounded by a smooth curve Γ_0. The model can be thought of as a receding air bubble in a viscous flow and was discussed in Section 4.3.4. The evolution of the phase domains $\Omega(t)$ is described by an auxiliary conformal mapping $f(\zeta,t)$ of U^* onto $\Omega(t)$, $\Omega(0)=\Omega_0$, normalized by $f(\zeta,t)=\alpha(t)\zeta+a_0(t)+\frac{a_1(t)}{\zeta}+\cdots$, $\alpha(t)>0$. Rescaling $Q=-1$ this mapping satisfies the Polubarinova-Galin equation

$$\operatorname{Re}\left[\dot f(\zeta,t)\overline{\zeta f'(\zeta,t)}\right]=-1,\quad \zeta=e^{i\theta}. \tag{6.2}$$

The corresponding Löwner-Kufarev type equation is

$$\dot f(\zeta,t)=\zeta f'(\zeta,t)\frac{1}{2\pi}\int\limits_0^{2\pi}\frac{1}{|f'(e^{i\theta},t)|^2}\frac{\zeta+e^{i\theta}}{\zeta-e^{i\theta}}d\theta, \tag{6.3}$$

where $\zeta\in U^*$.

The equation (6.2) is equivalent to the kinematic condition on the free boundary and, in particular, implies that the phase domains $\Omega(t)$ form a subordination chain. Unlike the classical Löwner-Kufarev equation (6.1), the equation (6.3) is even not quasilinear and the problem of the short-time existence and uniqueness of the solution is much more difficult. The function $p(\zeta,t)$ is not explicitly given as a function of ζ and t but depends on f in a nonlinear way. It is the integral operator in the right-hand side of (6.3). As discussed in Section 2.1.2, $\partial\Omega(t)$ remains a smooth (even analytic) boundary up to the time t_0 when possible cusps develop or when the domain $\Omega(t)$ is no longer simply connected. This means that $\Omega(t)$ may fail to be a quasidisk as $t\to t_0^-$. Quasidisks can be thought of as elements of the *universal Teichmüller space*, which we shall use as a general parametric space.

We ask the following **question**: given an initial smooth phase domain Ω_0, and the Hele-Shaw evolution $\Omega(t)$, what kind of evolution does it generate in the universal Teichmüller space?

A general scheme of the proposed investigation is shown in Figure 6.1

6.2 Quasiconformal maps and Teichmüller spaces

The theory of quasiconformal mapping emerged at the beginning of the twentieth century. At that time, quasiconformal maps arose by geometric reasons and their definition was based on the works by Grötzsch [139], [140] (who introduced so-called regular quasiconformal maps) and the notion of extremal length suggested by Ahlfors and Beurling, and also, as solutions of a special type of elliptic systems of differential equations in the works by Lavrentiev (see e.g. [218]). Important applications to various fields of mathematics, such as discrete group theory,

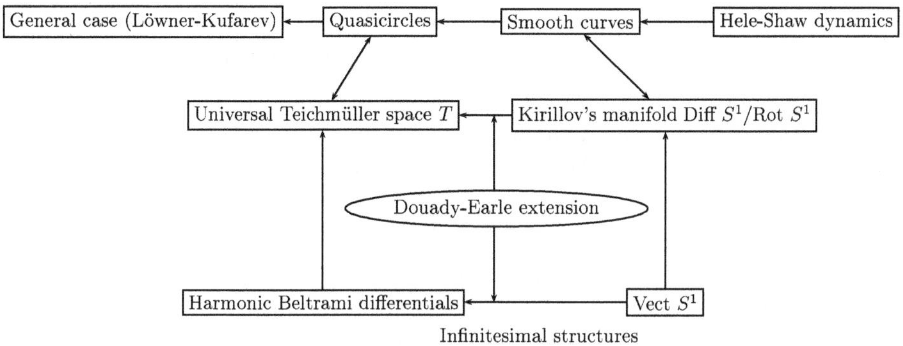

Figure 6.1: General scheme of investigation

mathematical physics, and complex differential geometry, have stimulated much development of the theory of quasiconformal mappings that, nowadays, is an important branch of Complex Analysis. Major contributions to this theory were made by Lavrentiev, Grötzsch, Ahlfors (who was one of the first Fields laureates (1936)), Bers, Teichmüller, Belinskiĭ, Volkovyskiĭ, in the past and many contributors recently.

At the mid-20th century it was established that the classical methods of geometric function theory could be extended to complex hyperbolic manifolds. The Teichmüller[2] spaces became the most important of them. In 1939 Teichmüller [327] proposed and partially realized an adventurous program of investigation in the moduli problem for Riemann surfaces. His main theorem asserts the existence and uniqueness of the extremal quasiconformal map between two compact Riemann surfaces of the same genus modulo an equivalence relation.

Teichmüller brought together the moduli problem, extremal quasiconformal maps, and relevant quadratic differentials on Riemann surfaces. This led him to the well-known theory of Teichmüller spaces. Later on, Teichmüller's ideas were thoroughly substantiated by Ahlfors, Bers [7] and other specialists.

6.2.1 Quasiconformal maps

Let D be a domain in $\overline{\mathbb{C}}$ (possibly equal to $\overline{\mathbb{C}}$) and $w = f(z)$ be a homeomorphism of D onto a domain $D' \subseteq \overline{\mathbb{C}}$. We define distributional derivatives as

$$f_{\bar{z}} := \frac{\partial f}{\partial \bar{z}} = \frac{1}{2}\left(\frac{\partial f}{\partial x} + i\frac{\partial f}{\partial y}\right); \; f_z := \frac{\partial f}{\partial z} = \frac{1}{2}\left(\frac{\partial f}{\partial x} - i\frac{\partial f}{\partial y}\right),$$

which are supposed to be locally square integrable on D, $z = x + iy$. A homeomorphism f is said to be quasiconformal in D if the complex-valued function

[2]Paul Julius Oswald Teichmüller, 1913–1943, a student of Bieberbach in Berlin was also an active member of the Nazi party. He died in heavy fighting along the river Dnieper, USSR.

$\mu_f(z) = f_{\bar{z}}/f_z$ satisfies an inequality $|\mu_f(z)| \leq k < 1$ almost everywhere in D. If $\|\mu_f\|_\infty = \operatorname{ess}\sup_{z\in D} |\mu_f(z)| \leq k < 1$, then the homeomorphism f is said to be K-quasiconformal, $K = (1+k)/(1-k)$. The function $\mu_f(z)$ is called its complex characteristic or *dilatation.* A quasiconformal map $w = f(z)$ is a homeomorphic distributional solution w of the Beltrami equation

$$w_{\bar{z}} = \mu_f(z)w_z, \tag{6.4}$$

for a given dilatation $\mu_f(z)$. This solution is unique up to a conformal homeomorphism. Imposing some standard conformal normalization (for instance, three boundary fixed points for a simply connected domain) implies the uniqueness of the solution to (6.4). Detailed descriptions of properties of quasiconformal maps can be easily found in [5], [31], [118], [121], [211].

To give a geometric definition of a quasiconformal map one can consider the notion of the modulus of a family of curves as a basis of the notion of quasiconformality. A sense preserving homeomorphism f of a domain D onto a domain D' is said to be a *K-quasiconformal map* if for any doubly connected hyperbolic domain $R \subset D$ the ratio $M(f(R))/M(R)$ is bounded and the following inequality is satisfied,

$$\frac{M(R)}{K} \leq M(f(R)) \leq KM(R), \tag{6.5}$$

where $M(R)$ is the modulus of the family of curves that separate the boundary components of R. We call inequality (6.5) the property of *quasi-invariance* of the modulus. A quasiconformal map is conformal if and only if $K = 1$ (or $k = 0$).

An important point to note here is the dependence of a quasiconformal map on its dilatation. We let the dilatation $\mu_f(z,t)$ depend on $z \in U$ and on a real or complex parameter t; $\mu_f(z,\cdot)$ is assumed to be a measurable function with respect to z, $\|\mu_f\|_\infty < 1$. If μ_f is n-differentiable with respect to z and the n-th derivative is Hölder continuous of order $\alpha \in (0,1)$, then a quasiconformal solution $f^\mu \in U_K$ to the equation (6.4) is $(n+1)$-differentiable and the $(n+1)$-th derivative satisfies the same Hölder condition [31], $n \geq 1$. Thus, one could expect that f^μ possesses a continuous derivative whenever μ is continuous. However, this not true, as is shown by the example $f(z) = z(1 - \log|z|)$, $f(0) = 0$, $z \in U$, where f is not even Lipschitz continuous at $z = 0$. Belinskiĭ [31] proved that a continuous μ produces a Hölder continuous f for any $0 < \alpha < 1$.

The dependence on the parameter t is much easier. If $\mu(\cdot,t)$ is a differentiable or continuous function with respect to t (for instance, holomorphic for complex t), then the same is true for f^μ.

6.2.2 The universal Teichmüller space

Let us consider the family $\mathcal{F}$ of all quasiconformal automorphisms of U. Every such map f satisfies the Beltrami equation $f_{\bar{\zeta}} = \mu_f(\zeta)f_\zeta$ in U in the distributional sense, where μ_f is a measurable essentially bounded function ($L^\infty(U)$) in

U, $\|\mu_f\| = \text{ess sup}_U |\mu_f(\zeta)|_\infty < 1$. Conversely, for each measurable Beltrami coefficient μ essentially bounded as above, there exists a quasiconformal automorphism of U, that satisfies the Beltrami equation, which is unique if provided with some conformal normalization, e.g., three-point normalization $f(\pm 1) = \pm 1$, $f(i) = i$. Two normalized maps f_1 and f_2 are said to be equivalent, $f_1 \sim f_2$, if being extended onto the unit circle S^1, the superposition $f_1 \circ f_2^{-1}$ restricted to S^1 is the identity map. The quotient set $\mathcal{F}/\sim$ is called *the universal Teichmüller space* T. It is a covering space for all Teichmüller spaces of analytically finite Riemann surfaces. By definition we have two realizations of T: as a set of equivalence classes of quasiconformal maps and, due to the relation between $\mathcal{F}/\sim$ and the unit ball $B \subset L^\infty(U)$, as a set of equivalence classes of corresponding Beltrami coefficients.

The normalized maps from $\mathcal{F}$ form a group $\mathcal{F}_0$ with respect to superposition and the maps that act identically on S^1 form a normal subgroup $\mathcal{I}$. Thus, T is the quotient of $T = \mathcal{F}_0/\mathcal{I}$.

If $g \in \mathcal{F}$, $f \in \mathcal{F}_0$, then there exists a Möbius transformation h, such that $h \circ f \circ g^{-1} \in \mathcal{F}_0$. Let us denote by $[f] \in T$ the equivalence class represented by $f \in \mathcal{F}_0$. Then, one defines *the universal modular group* $\mathcal{M}$, $\omega \in \mathcal{M}$, $\omega : T \to T$, by the formula $\omega([f]) = [h \circ f \circ g^{-1}]$. Its subgroup $\mathcal{M}_0$ of right translations on T is defined by $\omega_0([f]) = [f \circ g^{-1}]$, where $f, g \in \mathcal{F}_0$.

An important fact (see [221, Chapter III, Theorem 1.1]) is that there are real analytic mappings in any equivalence class $[f] \in T$.

Given a Beltrami coefficient $\mu \in B \subset L^\infty(U)$ let us extend it by zero into U^*. We normalize the corresponding quasiconformal map f, which is conformal in U^*, by $f(\zeta) = \zeta + a_1/\zeta + \dots$ about infinity. Then, two Beltrami coefficients μ and ν are equivalent if and only if the corresponding normalized mappings f^μ and f^ν map U^* onto one and the same domain in $\overline{\mathbb{C}}$. Thus, the universal Teichmüller space can be thought of as the family of all normalized conformal maps of U^* admitting quasiconformal extension. Moreover, any compact subset of T consists of conformal maps f of U^* that admit quasiconformal extension to U with $\|\mu_f\|_\infty \leq k < 1$ for some k.

As we mentioned above, a normalized conformal map $f \in [f] \in T$ defined in U^* can have a quasiconformal extension to U which is real analytic in U, but on the unit circle f may behave quite irregularly. For example, the resulting quasicircle $f(S^1)$ can have a Hausdorff dimension greater than 1.

Remark 6.2.1. *Given a bounded K-quasicircle Γ, $K = (1+k)/(1-k)$, in the plane let $N(\varepsilon, \Gamma)$ denote the minimal number of disks of radius $\varepsilon > 0$ that are needed to cover Γ. Let*

$$\beta(K) = \sup_\Gamma \limsup_{\varepsilon \to 0} \log N(\varepsilon, \Gamma)/\log(1/\varepsilon)$$

denote the supremum of the Minkowski dimension of curves Γ where Γ ranges over all bounded K-quasicircles. The Hausdorff dimension of Γ is bounded from above by $\beta(K)$ (see [27]). In [27] several explicit estimates for $\beta(K)$ were also established, e.g., $\beta(K) \leq 2 - cK^{-3.41}$.

Let us denote by $\Sigma_0^{qc} \subset \Sigma_0$ the class of those univalent conformal maps f defined in U^* which admit a quasiconformal extension to U, normalized by $f(\zeta) = \zeta + a_1/\zeta + \cdots$. Let $x, y \in T$ and $f, g \in \Sigma_0^{qc}$ be such that $\mu_f \in x$ and $\mu_g \in y$. Then, *the Teichmüller distance* $\tau(x,y)$ on T is defined as

$$\tau(x,y) = \inf_{\mu_f \in x,\ \mu_g \in y} \frac{1}{2} \log \frac{1 + \|\mu_{g \circ f^{-1}}\|_\infty}{1 - \|\mu_{g \circ f^{-1}}\|_\infty}.$$

For a given $x \in T$ we consider an extremal Beltrami coefficient μ^* such that $\|\mu^*\|_\infty = \inf_{\nu \in x} \|\nu\|_\infty$. Let us remark that μ^* need not be unique. A geodesic on T can be described in terms of the extremal coefficient μ^* as a continuous homomorphism $x_t : [0,1] \mapsto T$ such that $\tau(0, x_t) = t\tau(0, x_1)$. Due to the above remark the geodesic need not be unique as well.

We consider the Banach space $B(U)$ of all functions holomorphic in U equipped with the norm

$$\|\varphi\|_{B(U)} = \sup_{\zeta \in U} |\varphi(\zeta)|(1 - |\zeta|^2)^2.$$

For a function f in Σ the Schwarzian derivative

$$S_f(\zeta) = \frac{\partial}{\partial \zeta}\left(\frac{f''(\zeta)}{f'(\zeta)}\right) - \frac{1}{2}\left(\frac{f''(\zeta)}{f'(\zeta)}\right)^2$$

is defined and Nehari's [245] estimate $\|S_f(1/\zeta)\|_{B(U)} \leq 6$ holds. Given $x \in T$, $\mu \in x$ we construct the mapping $f^\mu \in \Sigma_0^{qc}$ and have the homeomorphic embedding $T \to B(U)$ by the Schwarzian derivative.

The universal Teichmüller space T is an analytic infinite dimensional Banach manifold modelled on $B(U)$. The Banach space $B(U)$ is an infinite dimensional vector space that can be thought of as the cotangent space to T at the initial point (represented by $\mu \equiv 0$). More rigorously, let the map f^μ be a quasiconformal homeomorphism of the unit disk U. It has a Fréchet derivative with respect to μ in a direction ν. Let us construct the variation of $f^{\tau\nu} \in \Sigma_0^{qc}$, $\mu = \tau\nu$, with respect to a small parameter τ:

$$f^{\tau\nu}(\zeta) = \zeta + \tau V(\zeta) + o(\tau), \quad \zeta \in U^*.$$

Taking the Schwarzian derivative in U^* we get

$$S_{f^{\tau\nu}} = \tau V'''(\zeta) + o(\tau), \quad \zeta \in U^*,$$

locally uniformly in U^*. Taking into account the normalization of the class Σ_0^{qc} we have (see, e.g., [221])

$$V(\zeta) = -\frac{1}{\pi} \iint_U \frac{\nu(w) d\sigma_w}{w - \zeta}, \quad V'''(\zeta) = -\frac{6}{\pi} \iint_U \frac{\nu(w) d\sigma_w}{(w - \zeta)^4}.$$

The integral formula implies $V'''(A(\zeta))A'(\zeta)^2 = V'''(\zeta)$ (subject to the relation for the Beltrami coefficient $\nu(A(\zeta))\overline{A'(\zeta)} = \nu(\zeta)A'(\zeta)$) for any Möbius transform A. Now let us change variables $\zeta \to 1/\bar{\zeta}$ and reduce the first variation to a holomorphic function in the unit disk by changing $f^{\tau\nu}(\zeta)$ to $g^{\tau\nu}(\zeta) \equiv \overline{f^{\tau\nu}(1/\bar{\zeta})}$. Setting $\Lambda_\nu(\zeta) = S_{g^{\tau\nu}}(\zeta)$ and $\dot{\Lambda}_\nu(\zeta) = \frac{1}{\zeta^4}\overline{V'''(1/\bar{\zeta})}$ we have (see, e.g., [119, Section 6.5, Theorem 5]) that

$$\Lambda_\nu(\zeta) - \tau\dot{\Lambda}_\nu(\zeta) = \frac{o(\tau)}{(1-|\zeta|^2)^2}.$$

So the operator $\dot{\Lambda}_\nu$ is the derivative of Λ_ν at the initial point of the universal Teichmüller space with respect to the norm of the Banach space $B(U)$. The reproducing property of the Bergman integral gives

$$\varphi(\zeta) = \frac{3}{\pi}\iint\limits_U \frac{\varphi(w)(1-|w|^2)^2 d\sigma_w}{(1-\bar{w}\zeta)^4}, \quad \varphi \in B(U). \tag{6.6}$$

The latter integral leads us to the so-called harmonic (Bers) Beltrami differential

$$\nu(\zeta) = \Lambda^*_\varphi(\zeta) \equiv -\frac{1}{2}\overline{\varphi(\zeta)}(1-|\zeta|^2)^2, \quad \zeta \in U.$$

Let us denote by $A(U)$ the Banach space of analytic functions with the finite L^1 norm in the unit disk. We have that $A(U) \hookrightarrow B(U)$ is a continuous inclusion (see, e.g., [244, Section 1.4.2]). On $L^\infty(U) \times A(U)$ one can define a coupling

$$\langle \mu, \varphi \rangle := \iint\limits_U \mu(\zeta)\varphi(\zeta)\, d\sigma_\zeta.$$

Denote by N the space of *locally trivial Beltrami coefficients*, which is the subspace of $L^\infty(U)$ that annihilates the operator $\langle \cdot, \varphi\rangle$ for all $\varphi \in A(U)$. Then, one can identify the tangent space to T at the initial point with the space $H := L^\infty(U)/N$. It is natural to relate it to a subspace of $L^\infty(U)$. The superposition $\dot{\Lambda}_\nu \circ \Lambda^*_\varphi$ acts identically on $A(U)$ due to (6.6). The space N is also the kernel of the operator $\dot{\Lambda}_\nu$. Thus, the operator Λ^* splits the exact sequence

$$0 \longrightarrow N \hookrightarrow L^\infty(U) \xrightarrow{\dot{\Lambda}_\nu} A(U) \longrightarrow 0.$$

Then, $H = \Lambda^*(A(U)) \cong L^\infty(U)/N$. The coupling $\langle \mu, \varphi\rangle$ defines $A(U)$ as a cotangent space. Let $A^2(U)$ denote the Banach space of analytic functions φ with the finite norm

$$\|\varphi\|_{A^2(U)} = \iint\limits_U |\varphi(\zeta)|^2(1-|\zeta|^2)^2 d\sigma_\zeta.$$

Then $A(U) \hookrightarrow A^2(U)$ and Petersson's Hermitian product [347] is defined on $A^2(U)$ as

$$(\varphi_1, \varphi_2) = \iint\limits_U \varphi_1(\zeta)\overline{\varphi_2(\zeta)}(1-|\zeta|^2)^2 d\sigma_\zeta.$$

The Kählerian Weil-Petersson metric $\{\nu_1, \nu_2\} := \langle \nu_1, \dot{\Lambda}_{\nu_2} \rangle$ can be defined on the tangent space to T and gives a Kählerian manifold structure to T. Defined as just a complex Banach manifold, T requires additional Hilbert manifold structure. The regorous results about the Weil-Petersson metric on T were obtained in [319]–[321].

The universal Teichmüller space is a smooth manifold on which a Lie group Diff T of real sense-preserving diffeomorphisms is defined. The tangent bundle is defined on T and is represented by the harmonic differentials from H translated to all points of T. We will consider tangent vectors from H at the initial point of T represented by the map $f(\zeta) \equiv \zeta$. The Weil-Petersson metric defines a Lie algebra of vector fields on T by the Poisson-Lie bracket $[\nu_1, \nu_2] = \{\nu_2, \nu_1\} - \{\nu_1, \nu_2\}$, where $\nu_1, \nu_2 \in H$. One can define the Poisson-Lie bracket at all other points of T by left translations from Diff T. To each element $[x]$ from Diff T an element x from T is associated as an image of the initial point. Therefore, a curve in Diff T generates a traced curve in T that can be realized by a one-parameter family of quasiconformal maps from Σ_0^{qc}.

For each tangent vector $\nu \in H$ there is a one-parameter semi-flow in Diff T and a corresponding flow $x^\tau \in T$ with the velocity vector ν. To make an explicit representation we use the variational formula for the subclass Σ_0^{qc} of Σ_0 of functions with quasiconformal extension (see, e.g., [221]) to $\overline{\mathbb{C}}$. If $f^\mu \in \Sigma_0^{qc}$, $\nu \in H$ and

$$\mu_f(\zeta, \tau) = \begin{cases} \tau\nu(\zeta) + o(\tau) & \text{if } \zeta \in U, \\ 0 & \text{if } \zeta \in U^*, \end{cases}$$

then the map

$$f^\mu(\zeta) = \zeta - \frac{\tau}{\pi} \iint\limits_U \frac{\nu(w) d\sigma_w}{w - \zeta} + o(\tau)$$

locally describes the semi-flow x^τ on T.

6.3 Diff S^1/Rot S^1 embedded into T

In this section we study a diffeomorphic embedding of the homogeneous manifold Diff S^1/Rot S^1 into the universal Teichmüller space T.

6.3.1 Homogeneous manifold Diff S^1/Rot S^1

We denote the Lie group of C^∞ sense preserving diffeomorphisms of the unit circle S^1 by Diff S^1. Each element of Diff S^1 is represented as $z = e^{i\phi(\theta)}$ with a monotone increasing, C^∞ real-valued function $\phi(\theta)$, such that $\phi(\theta + 2\pi) = \phi(\theta) + 2\pi$. The Lie algebra for Diff S^1 is identified with the Lie algebra Vect S^1 of smooth (C^∞) tangent vector fields to S^1 with the Poisson-Lie bracket given by

$$[\phi_1, \phi_2] = \phi_1\phi_2' - \phi_2\phi_1'.$$

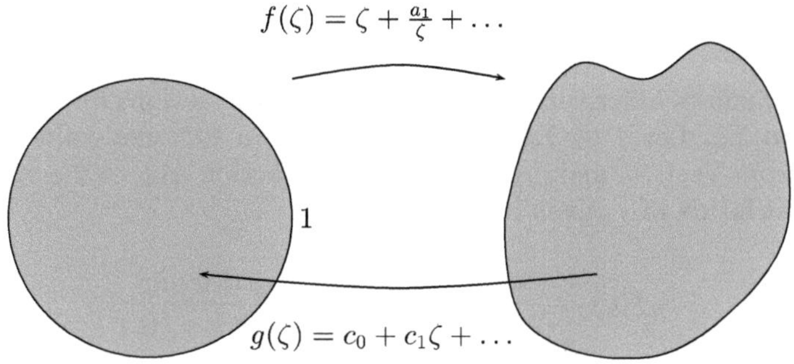

Figure 6.2: Representation of M

Fixing the trigonometric basis in Vect S^1 the commutator relations take the form

$$\begin{aligned}
[\cos n\theta, \cos m\theta] &= \frac{n-m}{2}\sin{(n+m)\theta} + \frac{n+m}{2}\sin{(n-m)\theta}, \\
[\sin n\theta, \sin m\theta] &= \frac{m-n}{2}\sin{(n+m)\theta} + \frac{n+m}{2}\sin{(n-m)\theta}, \\
[\sin n\theta, \cos m\theta] &= \frac{m-n}{2}\cos{(n+m)\theta} - \frac{n+m}{2}\cos{(n-m)\theta}.
\end{aligned}$$

There is no general theory of infinite dimensional Lie groups, an example of which is under consideration. The interest in this particular case comes first of all from string theory where the Virasoro algebra appears as the central extension of Vect S^1. The entire necessary background for the construction of the theory of unitary representations of Diff S^1 is found in the study of Kirillov's homogeneous Kählerian manifold M = Diff S^1/Rot S^1, where Rot S^1 denotes the group of rotations of S^1. The group Diff S^1 acts as a group of translations on the manifold M with Rot S^1 as a stabilizer. The Kählerian geometry of M was described by Kirillov and Yuriev in [199]. The manifold M admits several representations, in particular, in the space of smooth probability measures, symplectic realization in the space of quadratic differentials. We will use its analytic representation that is based on the class $\tilde{\Sigma}_0$ of functions from Σ_0 which, being extended onto the closure $\overline{U}^*$ of U^*, are supposed to be smooth on S^1. The class $\tilde{\Sigma}_0$ is dense in Σ_0 in the local uniform topology of U^*.

Let $\tilde{S}$ denote the class of all univalent holomorphic maps in the unit disk $g(\zeta) = c_0 + c_1\zeta + c_2\zeta^2 + \cdots$ which are smooth on S^1. Then, for each $f \in \tilde{\Sigma}_0$ we have $\infty \in f(U^*)$ and there is an *adjoint map* $g \in \tilde{S}$ such that $\overline{\mathbb{C}} \setminus f(U^*) = \overline{g(U)}$. The superposition $g^{-1} \circ f$ restricted to S^1 is in M (see Figure 6.2). Reciprocally, for each element of M there exist such f and g. A piece-wise smooth closed Jordan curve is a quasicircle if and only if it has no cusps. So any function f from $\tilde{\Sigma}_0$ has a quasiconformal extension to U. By this realization the manifold M is naturally embedded into the universal Teichmüller space T. Moreover, the Kählerian structure on M corresponds to the Kählerian structure on T given by the Weil-Petersson metric.

The Goluzin-Schiffer variational formulae lift the actions from the Lie algebra Vect S^1 onto $\tilde{\Sigma}_0$. Let $f \in \tilde{\Sigma}_0$ and let $d(e^{i\theta})$ be a C^∞ real-valued function in $\theta \in (0, 2\pi]$ from Vect S^1 making an infinitesimal action as $\theta \mapsto \theta + \tau d(e^{i\theta})$. Let us consider a variation of f given by

$$\delta_d f(\zeta) = \frac{-1}{2\pi i} \int\limits_{S^1} \left(\frac{wf'(w)}{f(w)} \right)^2 \frac{wd(w)dw}{f(w) - f(\zeta)}. \tag{6.7}$$

Kirillov and Yuriev [199], [200] have established that the variations $\delta_d f(\zeta)$ are closed with respect to the commutator and the induced Lie algebra is the same as Vect S^1. Moreover, Kirillov's result [201] states that there is the exponential map Vect $S^1 \to$ Diff S^1 such that the subgroup Rot S^1 coincides with the stabilizer of the map $f(\zeta) \equiv \zeta$ from $\tilde{\Sigma}_0$.

6.3.2 Douady-Earle extension

Let $\varphi \colon S^1 \to S^1$ be a circle quasisymmetric homeomorphism, i.e., a homeomorphism that possesses a quasiconformal extension into U (for a precise definition see, e.g., [221]). Then φ has infinitely many quasiconformal extensions into U, one of the most remarkable of which is the Beurling-Ahlfors[3] extension. In 1986 Douady and Earle [85] defined for any such $\varphi \colon S^1 \to S^1$ a conformally natural extension $h \colon \overline{U} \to \overline{U}$ from $\mathcal{F}$. The map h is a homeomorphism which is real analytic in the interior. The idea was to introduce the concept of a conformal barycenter of a measure on $S^1 = \partial U$. Douady and Earle proved that $w = h(\zeta) \in \mathcal{F}$ satisfies the functional equation

$$F(\zeta, w) \equiv \frac{1}{2\pi} \int\limits_{S^1} \left(\frac{\varphi(z) - w}{1 - \overline{w}\varphi(z)} \right) \frac{1 - |\zeta|^2}{|\zeta - z|^2} |dz| = 0. \tag{6.8}$$

An advantage of this extension is that if $\sigma, \tau \in$Möb(U), then the extension of $\sigma \circ \varphi \circ \tau$ is given by $\sigma \circ h \circ \tau$, which is not true for the Beurling-Ahlfors extension. The three-point boundary normalization of $\mathcal{F}_0$ can be always attained, and thus, the Douady-Earle extension is compatible with the definition of the universal Teichmüller space. Later, in 1988, another proof of Douady-Earle's result appeared in [220] where the authors worked with the inverse function. The functional equation (6.8), in particular, implies that a C^∞ mapping φ representing an element from the manifold M has a real analytic extension $h \in \mathcal{F}$ which is C^∞ on S^1.

[3] Arne Karl-August Beurling, 1905–1986, Lars Valerian Ahlfors, 1907–1996; this extension appeared in a 1956 paper [37] with the nonalphabetic listing of the authors as Ahlfors had insisted because of the contribution made by Beurling. Ahlfors "... felt mostly like a secretary; the main ideas of the paper were due to Beurling" (see [28]).

Let $f \in \tilde{\Sigma}_0$ represent an element from $\varphi \in M$. Let $g \in \tilde{S}$ be the adjoint map, $g^{-1} \circ f\big|_{S^1} = \varphi$. If h is the Douady-Earle extension of φ, then $g \circ h\big|_{S^1} \equiv f\big|_{S^1}$ and $g \circ h$ is a quasiconformal extension of $f \in \tilde{\Sigma}_0$. Given $\varphi \in M$ we construct the mapping f^μ that satisfies the normalization of the class $\tilde{\Sigma}_0$ and whose Beltrami coefficient is

$$\mu_f(\zeta) = \frac{\overline{F_\zeta} F_{\bar{w}} - F_{\bar{\zeta}} \overline{F_w}}{\overline{F_{\bar{\zeta}}} F_{\bar{w}} - F_\zeta \overline{F_w}}, \quad w = h(\zeta), \ \zeta \in U, \tag{6.9}$$

with $\mu_f(\zeta) = 0$ for $\zeta \in U^*$. The equivalence class $[f^\mu]$ is a point of the universal Teichmüller space T. So the Douady-Earle extension defines an explicit embedding of M into T.

6.3.3 Semi-flows on T and M

As was mentioned in Section 6.2, the Weil-Petersson metric defines a Lie algebra of vector fields on T by the Poisson bracket $[\nu_1, \nu_2] = \{\nu_2, \nu_1\} - \{\nu_1, \nu_2\}$, where $\nu_1, \nu_2 \in H$. One can define the Poisson bracket at all other points of T by left translations of the universal modular group.

We proceed restricting ourselves to M embedded into T. The complex form of Green's formula implies that (6.7) for $f(\zeta) \equiv \zeta$ is equivalent to

$$\delta_d \zeta = \frac{-1}{\pi} \iint\limits_U \frac{\partial_{\bar{w}}(w d(w)) d\sigma_w}{w - \zeta}, \tag{6.10}$$

where the distributional derivative $\partial_{\bar{w}}$ is given in the unit disk U, $d(w)$ is a continuous extension of the C^∞ function $d(e^{i\theta}) \in \text{Vect } S^1$ into U that has $L^s(U)$ distributional derivatives in U, $s > 2$, and $d\sigma_w$ is the area element in U. Thus, one can extract the elements from H that are of the form $\nu(\zeta) = \zeta \partial_{\bar{\zeta}} d(\zeta)$, where $\partial_{\bar{\zeta}}$ means $\partial / \partial \bar{\zeta}$.

We are going to deduce an exact form of ν using the Douady-Earle extension. For this we start with the variation of the element

$$\varphi(e^{i\theta}, \tau) = e^{i\theta}(1 + \tau i d(e^{i\theta})) + o(\tau), \quad \varphi \in M, \ d \in \text{Vect } S^1,$$

and τ is small. The Beltrami coefficient of the extended quasiconformal map h has its variation as $\mu_h(\zeta) = \tau \nu(\zeta) + o(\tau)$, where

$$\nu(\zeta) = \left. \frac{\frac{\partial}{\partial \tau}\left(\overline{F^\tau_\zeta} F^\tau_{\bar{w}} - F^\tau_{\bar{\zeta}} \overline{F^\tau_w}\right)}{\overline{F^\tau_{\bar{\zeta}}} F^\tau_{\bar{w}} - F^\tau_\zeta \overline{F^\tau_w}} \right|_{\tau=0,\, w=\zeta}, \zeta \in U, \tag{6.11}$$

where

$$F^\tau(\zeta, w) = \frac{1}{2\pi} \int\limits_{S^1} \left(\frac{\varphi(z,\tau) - w}{1 - \overline{w}\varphi(z,\tau)} \right) \frac{1 - |\zeta|^2}{|\zeta - z|^2} |dz| = 0. \tag{6.12}$$

Thus, $\nu(\zeta)$ depends only on $d(e^{i\theta})$. We will give explicit formulae in the next section. They can be obtained substituting $\varphi(e^{i\theta},0)=e^{i\theta}$, and taking into account that

$$\overline{F^\tau_\zeta}F^\tau_{\bar w}-F^\tau_{\bar\zeta}\overline{F^\tau_w}\Big|_{\tau=0,\,w=\zeta}=0.$$

The Lie algebra Vect S^1 is embedded into the Lie algebra of H by (6.11), (6.12). Hence, a flow given on M corresponding to a vector $d\in$ Vect S^1 is represented as a flow on the universal Teichmüller space T corresponding to the vector $\nu\in H$ given by (6.11).

6.4 Infinitesimal descriptions of semi-flows

First of all we give an explicit formula that connects the vectors $d(e^{i\theta})$ from Vect S^1 with corresponding tangent vectors $\nu(\zeta)\in H$ to the universal Teichmüller space T, making use of the Douady-Earle extension. These vectors give the infinitesimal description of semi-flows on M and T respectively.

Theorem 6.4.1. *Let $d(e^{i\theta})\in$ Vect S^1 be the infinitesimal description of a flow φ in M. Then, the corresponding infinitesimal description $\nu(\zeta)\in H$ of this flow embedded into T is given by the function*

$$\nu(\zeta)=\frac{3}{2\pi}\int_0^{2\pi}\left(\frac{1-|\zeta|^2}{(1-e^{i\theta}\bar\zeta)^2}\right)^2 e^{2i\theta}d(e^{i\theta})d\theta. \tag{6.13}$$

Proof. Let $\varphi(\zeta,\tau)=e^{i(\theta+\tau d(e^{i\theta}))}$ and $h(\zeta,\tau)$ be the Douady-Earle extension of φ into the unit disk U, $\zeta\in U$ by means of (6.12). If $\tau=0$, then $h(\zeta,0)\equiv\zeta$. We calculate

$$\begin{aligned}
\partial_\zeta F^\tau(\zeta,w) &= \frac{1}{2\pi}\int_0^{2\pi}\left(\frac{\varphi(e^{i\theta},\tau)-w}{1-\overline{w}\varphi(e^{i\theta},\tau)}\right)\frac{e^{i\theta}(\bar\zeta-e^{-i\theta})^2}{|\zeta-e^{i\theta}|^4}d\theta,\\
\partial_{\bar\zeta} F^\tau(\zeta,w) &= \frac{1}{2\pi}\int_0^{2\pi}\left(\frac{\varphi(e^{i\theta},\tau)-w}{1-\overline{w}\varphi(e^{i\theta},\tau)}\right)\frac{e^{-i\theta}(\zeta-e^{i\theta})^2}{|\zeta-e^{i\theta}|^4}d\theta,\\
\partial_w F^\tau(\zeta,w) &= \frac{1}{2\pi}\int_0^{2\pi}\left(\frac{-1}{1-\overline{w}\varphi(e^{i\theta},\tau)}\right)\frac{1-|\zeta|^2}{|\zeta-e^{i\theta}|^2}d\theta,\\
\partial_{\bar w} F^\tau(\zeta,w) &= \frac{1}{2\pi}\int_0^{2\pi}\left(\frac{\varphi(e^{i\theta},\tau)(\varphi(e^{i\theta},\tau)-w)}{(1-\overline{w}\varphi(e^{i\theta},\tau))^2}\right)\frac{1-|\zeta|^2}{|\zeta-e^{i\theta}|^2}d\theta.
\end{aligned}$$

Substituting $\tau = 0$ and $w = \zeta$ we have

$$\begin{aligned}
\partial_\zeta F^\tau(\zeta, w)\Big|_{\tau=0,w=\zeta} &= \frac{1}{1-|\zeta|^2}, \\
\partial_{\bar\zeta} F^\tau(\zeta, w)\Big|_{\tau=0,w=\zeta} &= 0, \\
\partial_w F^\tau(\zeta, w)\Big|_{\tau=0,w=\zeta} &= \frac{-1}{1-|\zeta|^2}, \\
\partial_{\bar w} F^\tau(\zeta, w)\Big|_{\tau=0,w=\zeta} &= 0.
\end{aligned}$$

We will use the properties of the Douady-Earle extension. Let us fix a point $\zeta_0 \in U$ and choose two Möbius transformations σ, δ of U such that $\delta(0) = \zeta_0$ and $\sigma(0) = h(\zeta_0, \tau)$. We set $g = \sigma^{-1} \circ h \circ \delta$. Then, $g(0,\tau) = 0$, $\dot g(0,\tau) = 0$ and

$$\begin{aligned}
\partial_\zeta g(0,\tau) &= \partial_\zeta h(\zeta_0, \tau)\frac{\delta'(0)}{\sigma'(0)}, \\
\partial_{\bar\zeta} g(0,\tau) &= \partial_{\bar\zeta} h(\zeta_0, \tau)\frac{\overline{\delta'(0)}}{\sigma'(0)}.
\end{aligned}$$

So we see that

$$\frac{\partial_{\bar\zeta} h(\zeta_0,\tau)}{\partial_\zeta h(\zeta_0,\tau)} = \frac{\partial_{\bar\zeta} g(0,\tau)}{\partial_\zeta g(0,\tau)}\frac{\delta'(0)}{\overline{\delta'(0)}}.$$

By the property of the Douady-Earle extension we have that the function $g(\zeta, \tau)$, $\zeta \in U$ is the extension of $g(e^{i\theta}, \tau)$ by means of (6.12). If $\tau = 0$, then $g(\zeta, 0) \equiv \zeta$. Now we put $\psi(e^{i\theta}, \tau) = g(e^{i\theta}, \tau)$ in (6.12) and calculate variations in τ,

$$\begin{aligned}
\frac{\partial}{\partial\tau}\partial_{\bar\zeta} F^\tau(\zeta, w)\Big|_{\tau=0,w=\zeta=0} &= \frac{1}{2\pi}\int_0^{2\pi}\left(\frac{e^{i\theta}-\zeta}{(1-\overline{\zeta}e^{i\theta})^3}\right)\frac{1-|\zeta|^2}{|\zeta-e^{i\theta}|^2}\frac{\partial\psi(e^{i\theta},\tau)}{\partial\tau}\Big|_{\tau=0,\zeta=0} d\theta \\
&= \frac{1}{2\pi}\int_0^{2\pi} e^{i\theta}\dot\psi(e^{i\theta}, 0)d\theta, \\
\frac{\partial}{\partial\tau}\partial_{\bar w} F^\tau(\zeta, w)\Big|_{\tau=0,w=\zeta=0} &= \frac{1}{2\pi}\int_0^{2\pi}\left(\frac{2e^{i\theta}-\zeta-|\zeta|^2e^{i\theta}}{(1-\overline{\zeta}e^{i\theta})^3}\right)\frac{1-|\zeta|^2}{|\zeta-e^{i\theta}|^2}\frac{\partial\psi(e^{i\theta},\tau)}{\partial\tau}\Big|_{\tau=0,\zeta=0} d\theta \\
&= \frac{1}{2\pi}\int_0^{2\pi} 2e^{i\theta}\dot\psi(e^{i\theta}, 0)d\theta.
\end{aligned}$$

Then, we can obtain the explicit form of the variation of the Beltrami coefficient by (6.11) as

$$\frac{\partial}{\partial \tau} \frac{\partial_{\bar{\zeta}} g(0,\tau)}{\partial_{\zeta} g(0,\tau)}\bigg|_{\tau=0} = \frac{3}{2\pi} \int_0^{2\pi} e^{i\theta} \dot{\psi}(e^{i\theta},0) d\theta. \tag{6.14}$$

The Möbius transformation δ does not depend on τ whereas σ does. Explicitly, we put

$$\sigma^{-1} \circ h \circ \delta(\zeta) = \frac{h(\delta(\zeta),\tau) - h(\zeta_0,\tau)}{1 - h(\delta(\zeta),\tau)\overline{h(\zeta_0,\tau)}},$$

where $\delta(\zeta) = (\zeta + \zeta_0)(1 + \zeta\bar{\zeta}_0)^{-1}$. We denote $e^{i\alpha} = \delta(e^{i\theta})$. Therefore, denoting

$$e^{i\alpha} = \delta(e^{i\theta}) = \frac{e^{i\theta} + \zeta_0}{1 + \bar{\zeta}_0 e^{i\theta}},$$

we have

$$\dot{g}(e^{i\theta},0) = \frac{\dot{h}(e^{i\alpha},0)(1 - |\zeta_0|^2) - \dot{h}(\zeta_0,0)(1 - \bar{\zeta}_0 e^{i\alpha}) + \overline{\dot{h}(\zeta_0,0)} e^{i\alpha}(e^{i\alpha} - \zeta_0)}{(1 - \bar{\zeta}_0 e^{i\alpha})^2}.$$

Then,

$$e^{i\theta} d\theta = \frac{1 - |\zeta_0|^2}{(1 - e^{i\alpha}\bar{\zeta}_0)^2} e^{i\alpha} d\alpha,$$

and changing variables in (6.14), we obtain

$$\frac{\partial}{\partial \tau} \frac{\partial_{\bar{\zeta}} g(0,\tau)}{\partial_{\zeta} g(0,\tau)}\bigg|_{\tau=0} = \frac{3}{2\pi} \int_0^{2\pi} \left(\frac{1 - |\zeta_0|^2}{(1 - e^{i\alpha}\bar{\zeta}_0)^2} \right)^2 e^{2i\alpha} d(e^{i\alpha}) d\alpha.$$

Taking into account that $\delta'(0) = 1$ we come to the statement of the theorem. □

Corollary 6.4.2. *If* $q = \max_{\theta \in [0,2\pi]} |d(e^{i\theta})|$, *then*

$$|\nu(\zeta)| \leq 3 \frac{1 + |\zeta|^2}{1 - |\zeta|^2} q.$$

Proof. The formula given in the preceding theorem implies

$$\nu(\zeta) = \frac{3}{2\pi} \int_0^{2\pi} \frac{1 - |\zeta|^2}{(1 - e^{i\alpha}\bar{\zeta})^2} e^{i\alpha} d(\delta(e^{i\theta})) e^{i\theta} d\theta.$$

Changing variables $\alpha \to \theta$ we obtain

$$\nu(\zeta) = \frac{3}{2\pi} \int_0^{2\pi} \frac{e^{i\theta} + \zeta}{1 + e^{i\theta}\bar{\zeta}} \frac{(1 + e^{i\theta}\bar{\zeta})^2}{1 - |\zeta|^2} e^{i\theta} d(\delta(e^{i\theta})) d\theta. \tag{6.15}$$

Next, we obviously estimate $|\nu|$ as in the statement of the corollary. □

As we see, the given estimate is good enough when $|\zeta|$ is not close to 1. Let us now give an asymptotic estimate for $|\nu(\zeta)|$ in the case $|\zeta| \sim 1$.

Corollary 6.4.3. *There exists a constant M independent of ζ such that*

$$|\nu(\zeta)| \leq M\frac{1-|\zeta|^2}{|\zeta|^2}.$$

In particular, $|\nu(\zeta)| = O(1-|\zeta|^2)$ as $|\zeta| \sim 1$.

Proof. We integrate by parts the right-hand side in the formula (6.13) twice and come to the expression

$$\nu(\zeta) = -\frac{(1-|\zeta|^2)}{4\pi\bar{\zeta}^2}\int\limits_0^{2\pi}\frac{1-|\zeta|^2}{(1-e^{i\theta}\bar{\zeta})^2}\left(i\frac{\partial[e^{i\theta}d(e^{i\theta})]}{\partial\theta}+\frac{\partial^2[e^{i\theta}d(e^{i\theta})]}{\partial\theta^2}\right)d\theta. \tag{6.16}$$

The absolute value of the above integral is bounded because of the Poisson kernel in it and due to the smoothness of the function d. □

6.5 Parametric representation of univalent maps with quasiconformal extensions

6.5.1 Semigroups of conformal maps

The basic ideas that we use in this section come from Goryainov's works [134], [135] and the monograph by Shoikhet [309].

We consider the semigroup $\mathcal{G}$ of conformal univalent maps from U^* into itself with composition as the semigroup operation. This makes $\mathcal{G}$ a topological semigroup with respect to the topology of local uniform convergence on U^*. We impose the natural normalization for such conformal maps: $\Phi(\zeta) = \beta\zeta + b_0 + \frac{b_1}{\zeta} + \cdots$, $\zeta \in U^*$, $\beta > 0$. The unit of the semigroup is the identity. Let us construct on $\mathcal{G}$ a one-parameter semi-flow Φ^τ, that is, a continuous homomorphism from $\mathbb{R}^+$ into $\mathcal{G}$, with the parameter $\tau \geq 0$. For any fixed $\tau \geq 0$ the element Φ^τ is from $\mathcal{G}$ and is represented by a conformal map $\Phi(\zeta,\tau) = \beta(\tau)\zeta + b_0(\tau) + \frac{b_1(\tau)}{\zeta} + \cdots$ from U^* onto the domain $\Phi(U^*,\tau) \subset U^*$. The element Φ^τ satisfies the following properties:

- $\Phi^0 = id$;
- $\Phi^{\tau+s} = \Phi(\Phi(\zeta,\tau),s)$, for $\tau, s \geq 0$;
- $\Phi(\zeta,\tau) \to \zeta$ locally uniformly in U^* as $\tau \to 0$.

In particular, $\beta(0) = 1$. This semi-flow is generated by a vector field $v(\zeta)$ if for each $\zeta \in U^*$ the function $w = \Phi(\zeta, \tau)$, $\tau \geq 0$ is a solution of an autonomous differential equation $dw/d\tau = v(w)$ with the initial condition $w|_{\tau=0} = \zeta$. The semi-flow can be extended to a symmetric interval $(-t, t)$ by putting $\Phi^{-\tau} = \Phi^{-1}(\zeta, \tau)$. Certainly, the latter function is defined on the set $\Phi(U^*, \tau)$. Admitting this restriction for negative τ we have a one-parameter family Φ^τ for $\tau \in (-t, t)$.

For a semi-flow Φ^τ, $\tau \geq 0$ on $\mathcal{G}$ there is an infinitesimal generator at $\tau = 0$ constructed by the following procedure. Any element Φ^τ is represented by a conformal map $\Phi(\zeta, \tau)$ that satisfies the Schwarz Lemma for the maps $U^* \to U^*$. Hence,

$$\mathrm{Re}\, \frac{\zeta}{\Phi(\zeta, \tau)} \leq \left| \frac{\zeta}{\Phi(\zeta, \tau)} \right| \leq 1, \quad \zeta \in U^*,$$

where the equality sign is attained only for $\Phi^0 = id \simeq \Phi(\zeta, 0) \equiv \zeta$. Therefore, the following limit exists (see, e.g., [134], [135], [309])

$$\lim_{\tau \to 0} \mathrm{Re}\, \frac{\zeta - \Phi(\zeta, \tau)}{\tau \Phi(\zeta, \tau)} = -\mathrm{Re}\, \frac{\left.\frac{\partial \Phi(\zeta, \tau)}{\partial \tau}\right|_{\tau=0}}{\zeta} \leq 0,$$

and the representation

$$\left.\frac{\partial \Phi(\zeta, \tau)}{\partial \tau}\right|_{\tau=0} = \zeta p(\zeta)$$

holds, where $p(\zeta) = p_0 + p_1/\zeta + \cdots$ is an analytic function in U^* with positive real part, and

$$\left.\frac{\partial \beta(\tau)}{\partial \tau}\right|_{\tau=0} = p_0. \tag{6.17}$$

In [136] it was shown that Φ^τ is even C^∞ with respect to τ. The function $\zeta p(\zeta)$ is an infinitesimal generator for Φ^τ at $\tau = 0$, and the following variational formula holds:

$$\Phi(\zeta, \tau) = \zeta + \tau\, \zeta p(\zeta) + o(\tau), \quad \beta(\tau) = 1 + \tau p_0 + o(\tau). \tag{6.18}$$

The convergence is thought of as local uniform. We rewrite (6.18) as

$$\Phi(\zeta, \tau) = (1 + \tau p_0)\zeta + \tau\, \zeta(p(\zeta) - p_0) + o(\tau) = \beta(\tau)\zeta + \tau\, \zeta(p(\zeta) - p_0) + o(\tau). \tag{6.19}$$

Now let us proceed with the semigroup $\mathcal{G}^{qc} \subset \mathcal{G}$ of quasiconformal automorphisms of $\overline{\mathbb{C}}$. A quasiconformal map Φ representing an element of $\mathcal{G}^{qc}$ satisfies the Beltrami equation in $\overline{\mathbb{C}}$,

$$\Phi_{\bar{\zeta}} = \mu_\Phi(\zeta) \Phi_\zeta,$$

with the distributional derivatives $\Phi_{\bar{\zeta}}$ and Φ_ζ, where $\mu_\Phi(\zeta)$ is a measurable function vanishing in U^* and essentially bounded in U by

$$\|\mu_\Phi\| = \operatorname{ess\,sup}_U |\mu_\Phi(\zeta)| \leq k < 1,$$

for some k. If k is sufficiently small, then the function $\dfrac{\Phi - b_0}{\beta}$ satisfies the variational formula (see, e.g., [221])

$$\frac{\Phi(\zeta) - b_0}{\beta} = \zeta - \frac{1}{\pi}\iint\limits_{U} \frac{\mu_\Phi(w)d\sigma_w}{w - \zeta} + o(k), \tag{6.20}$$

where $d\sigma_w$ stands for the area element in the w-plane.

Now for each τ small and $\Phi^\tau \in \mathcal{G}^{qc}$ the mapping $h(\zeta, \tau) = \frac{\Phi(\zeta,\tau)-b_0(\tau)}{\beta(\tau)}$ is from Σ_0^{qc} and represents an equivalence class $[h^\tau] \in T$. Consider the one-parameter curve $x^\tau \in T$ that corresponds to $[h^\tau]$ and a velocity vector $\nu(\zeta) \in H$ (that is not trivial), such that

$$\mu_h(\zeta, \tau) = \mu_\Phi(\zeta, \tau) = \tau\nu(\zeta) + o(\tau).$$

We take into account that $\Phi(\zeta, 0) \equiv \zeta$ in U^* and is extended up to the identity map of $\overline{\mathbb{C}}$.

The formula (6.20) can be rewritten for $\Phi(\zeta, \tau)$ as

$$\frac{\Phi(\zeta, \tau) - b_0(\tau)}{\beta(\tau)} = \zeta - \frac{\tau}{\pi}\iint\limits_{U} \frac{\nu(w)d\sigma_w}{w - \zeta} + o(\tau). \tag{6.21}$$

Comparing with (6.19) we come to the following conclusion about Φ:

$$\Phi(\zeta, \tau) = \beta(\tau)\zeta + \tau p_1 - \frac{\tau}{\pi}\iint\limits_{U} \frac{\nu(w)d\sigma_w}{w - \zeta} + o(\tau). \tag{6.22}$$

The relations (6.18), 6.19, 6.22 imply that

$$p(z) = p_0 + \frac{p_1}{\zeta} - \frac{1}{\pi}\iint\limits_{U} \frac{\nu(w)d\sigma_w}{\zeta(w - \zeta)}. \tag{6.23}$$

The constants p_0, p_1 and the function ν must be such that $\operatorname{Re} p(z) > 0$ for all $z \in U^*$.

We summarize these observations in the following theorem.

Theorem 6.5.1. *Let Φ^τ be a semi-flow in $\mathcal{G}^{qc}$. Then it is generated by the vector field $v(\zeta) = \zeta p(\zeta)$,*

$$p(z) = p_0 + \frac{p_1}{\zeta} - \frac{1}{\pi}\iint\limits_{U} \frac{\nu(w)d\sigma_w}{\zeta(w - \zeta)},$$

where $\nu(\zeta) \in H$ is a harmonic Beltrami differential and the holomorphic function $p(\zeta)$ has positive real part in U^*.

This theorem implies that at any point $\tau \geq 0$ we have

$$\frac{\partial \Phi(\zeta, \tau)}{\partial \tau} = \Phi(\zeta, \tau)p(\Phi(\zeta, \tau)).$$

6.5.2 Evolution families and differential equations

A subset $\Phi^{t,s}$ of $\mathcal{G}$, $0 \leq s \leq t$ is called an *evolution family* in $\mathcal{G}$ if

- $\Phi^{t,t} = id$;
- $\Phi^{t,s} = \Phi^{t,r} \circ \Phi^{r,s}$, for $0 \leq s \leq r \leq t$;
- $\Phi^{t,s} \to id$ locally uniformly in U^* as $t, s \to \tau$.

In particular, if Φ^{τ} is a one-parameter semi-flow, then Φ^{t-s} is an evolution family. We consider a subordination chain of mappings $f(\zeta, t)$, $\zeta \in U^*$, $t \in [0, t_0)$, where the function $f(\zeta, t) = \alpha(t)z + a_0(t) + a_1(t)/\zeta + \dots$ is a meromorphic univalent map $U^* \to \overline{\mathbb{C}}$ for each fixed t and $f(U^*, s) \subset f(U^*, t)$ for $s < t$. Let us assume that this subordination chain exists for t in an interval $[0, t_0)$.

Let us pass to the semigroup $\mathcal{G}^{qc}$. So $\Phi^{t,s}$ now has a quasiconformal extension to U and being restricted to U^* is from $\mathcal{G}$. Moreover, $\Phi^{t,s} \to id$ locally uniformly in $\mathbb{C}$ as $t, s \to \tau$.

For each t fixed in $[0, t_0)$ the map $f(\zeta, t)$ has a quasiconformal extension into U (that can be assumed even real analytic). An important presupposition is that $f(\zeta, t)$ generates a *non-trivial path* in the universal Teichmüller space T. This means that for any $t_1, t_2 \in [0, t_0)$, $t_1 \neq t_2$, the mapping $f(\zeta, t_2)$, $\zeta \in U^*$, can not be obtained from $f(\zeta, t_1)$ by a Möbius transform, or taking into account the normalization of f, by multiplying by a constant. We construct the superposition $f^{-1}(f(\zeta, s), t)$ for $t \in [0, t_0)$, $s \leq t$. Putting $s = t - \tau$ we denote this mapping by $\Phi(\zeta, t, \tau)$.

Now we suppose the following conditions for $f(\zeta, t)$.

(i) The maps $f(\zeta, t)$ form a subordination chain in U^*, $t \in [0, t_0)$.

(ii) The map $f(\zeta, t)$ is holomorphic in U^*, $f(\zeta, t) = \alpha(t)\zeta + a_0(t) + a_1(t)/\zeta + \cdots$, where $\alpha(t) > 0$ and differentiable with respect to t.

(iii) The map $f(\zeta, t)$ is a quasiconformal homeomorphism of $\overline{\mathbb{C}}$.

(iv) The chain of maps $f(\zeta, t)$ is not trivial.

(v) The Beltrami coefficient $\mu_f(\zeta, t)$ of this map is differentiable with respect to t locally uniformly in U, vanishes in some neighbourhood of U^* (independently of t).

The function $\Phi(\zeta, t, \tau)$ is embedded into an evolution family in $\mathcal{G}$. It is differentiable with regard to τ and t in $[0, t_0)$, and $\Phi(\zeta, t, 0) = \zeta$. Fix t and let $D_\tau = \Phi^{-1}(U^*, t, \tau) \setminus U^*$. Then, there exists $\nu \in H$ such that the Beltrami coefficient μ is of the form $\mu_\Phi(\zeta, t, \tau) = \tau\nu(\zeta, t) + o(\tau)$ in $U \setminus D_\tau$, $\mu_\Phi(\zeta, t, \tau) = \mu_f(\zeta, t-\tau)$ in D_τ, and vanishes in $\hat{U}^*$. We make τ sufficiently small such that $\mu_\Phi(\zeta, t, \tau)$ vanishes in D_τ too. Therefore, $\zeta = \lim_{\tau\to 0} \Phi(\zeta, t, \tau)$ locally uniformly in $\mathbb{C}$ and $\Phi(\zeta, t, \tau)$ is embedded now into an evolution family in $\mathcal{G}^{qc}$. The identity map is embedded into a semi-flow $\Phi^\tau \subset \mathcal{G}^{qc}$ (which is smooth) as the initial point with the same velocity vector

$$\left.\frac{\partial \Phi(\zeta, t, \tau)}{\partial \tau}\right|_{\tau=0} = \zeta p(\zeta, t), \quad \zeta \in U^*,$$

that leads to equation (6.1) (the semi-flow Φ^τ is tangent to the evolution family at the origin). Actually, the differentiable trajectory $f(\zeta, t)$ generates a pencil of tangent smooth semi-flows with starting tangent vectors $\zeta p(\zeta, t)$ (that may be only measurable with respect to t). The projection to the universal Teichmüller space is shown in Figure 6.3.

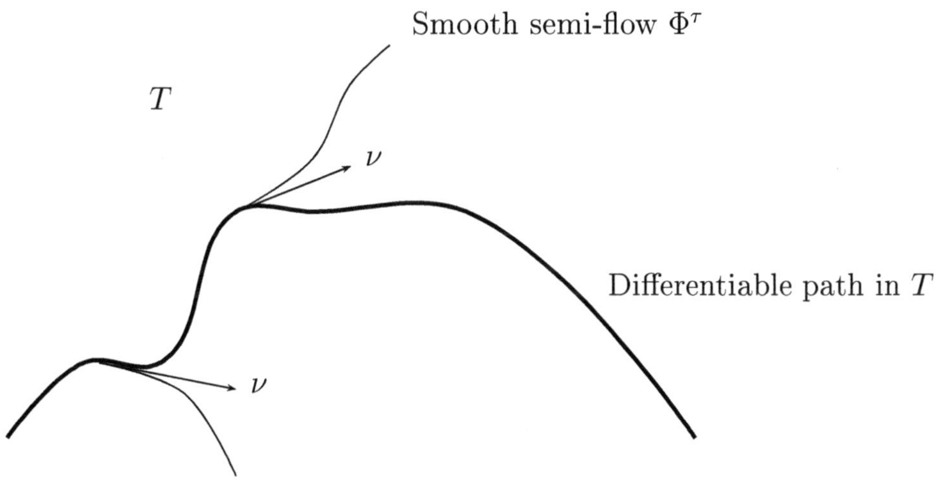

Figure 6.3: The pencil of tangent smooth semi-flows

The requirement of non-triviality makes it possible to use the variation (6.21). Therefore, the conclusion is that the function $f(\zeta, t)$ satisfies the equation (6.1) where the function $p(\zeta, t)$ is given by

$$p(\zeta, t) = p_0(t) + \frac{p_1(t)}{\zeta} - \frac{1}{\pi} \iint\limits_U \frac{\nu(w, t) d\sigma_w}{\zeta(w - \zeta)},$$

and has positive real part. The existence of $p_0(t), p_1(t)$ comes from the existence of the subordination chain. We can assign the normalization to $f(\zeta, t)$ controlling the change of the conformal radius of the subordination chain by e^{-t}. In this case, changing variables we obtain $p_0 = 1$, $p_1 = 0$.

Summarizing the conclusions about the function $p(\zeta, t)$ we come to the following result.

Theorem 6.5.2. *Let $f(\zeta, t)$ be a subordination chain of maps in U^* that exists for $t \in [0, t_0)$ and satisfies the conditions (i–v). Then, there are a real-valued function $p_0(t) > 0$, a complex-valued function $p_1(t)$, and a harmonic Beltrami differential $\nu(\zeta, t)$, such that* $\operatorname{Re} p(\zeta, t) > 0$ *for* $\zeta \in U^*$,

$$p(\zeta, t) = p_0(t) + \frac{p_1(t)}{\zeta} - \frac{1}{\pi} \iint\limits_U \frac{\nu(w,t) d\sigma_w}{\zeta(w - \zeta)}, \quad \zeta \in U^*,$$

and $f(\zeta, t)$ satisfies the differential equation

$$\frac{\partial f(\zeta, t)}{\partial t} = -\zeta \frac{\partial f(\zeta, t)}{\partial \zeta} p(\zeta, t), \quad \zeta \in U^*, \tag{6.24}$$

in $t \in [0, t_0)$.

In the above theorem the function $\nu(\zeta, t)$ belongs to the space of harmonic differentials. We ask now about another but equivalent form of ν as well as whether one can extend the equation (6.24) onto the whole complex plane.

Writing $w = f(\zeta, t - \tau)$, $\Phi(\zeta, t, \tau) = f^{-1}(w, t)$ we calculate the dilatation of the function $\Phi(\zeta, t, \tau)$ in U. Note that Φ is differentiable by t, τ.

$$\mu_\Phi = \frac{\Phi_{\bar\zeta}}{\Phi_\zeta} = \frac{f_w^{-1} w_{\bar\zeta} + f_{\bar w}^{-1} \bar w_{\bar\zeta}}{f_w^{-1} w_\zeta + f_{\bar w}^{-1} \bar w_\zeta} = \frac{w_{\bar\zeta} + \mu_{f^{-1}} \bar w_{\bar\zeta}}{w_\zeta + \mu_{f^{-1}} \bar w_\zeta} = \frac{\bar w_{\bar z}}{w_\zeta} \frac{\mu_w \dfrac{w_z}{\bar w_{\bar\zeta}} - \mu_f \dfrac{f_\zeta}{\bar f_{\bar\zeta}}}{1 - \mu_f \bar\mu_w \dfrac{f_\zeta \bar w_{\bar\zeta}}{w_\zeta \bar f_{\bar\zeta}}}.$$

We use that $\mu_{f^{-1}} \circ f = -\mu_f f_\zeta / \bar f_{\bar\zeta}$. Finally, μ_f, f_ζ, $f_{\bar\zeta}$ are differentiable by t almost everywhere in $t \in [0, t_0)$, locally uniformly in $\zeta \in U$, and

$$\nu_0(\zeta, t) = \lim_{\tau \to 0} \frac{\mu_\Phi}{\tau} = -\frac{\bar f_{\bar\zeta}}{f_\zeta} \frac{\dfrac{\partial}{\partial t}\left(\mu_f \dfrac{f_\zeta}{\bar f_{\bar\zeta}}\right)}{1 - |\mu_f|^2},$$

where the limit exists a.e. with respect to $t \in [0, t_0)$ locally uniformly in $\zeta \in U$, or in terms of the inverse function

$$\nu_0(\zeta, t) = \left(\frac{f_w^{-1}}{\bar f_{\bar w}^{-1}} \frac{\dfrac{\partial \mu_{f^{-1}}}{\partial t}}{1 - |\mu_{f^{-1}}|^2} \right) \circ f(\zeta, t).$$

Sometimes, it is much better to operate just with dilatations, avoiding functions, so we can rewrite the last expression as

$$\nu_0(z,t) = -\mu_f(z,t) \left[\frac{\frac{\partial \log \mu_{f^{-1}}}{\partial t}}{1-|\mu_{f^{-1}}|^2} \circ f(z,t) \right].$$

Remark 6.5.3. *The function $\nu(\zeta,t)$ in Theorem 6.5.2 may be replaced by the function $\nu_0(\zeta,t)$ that belongs to the same equivalence class in H.*

Let us consider one-parameter families of maps in U^* normalized by $f(\zeta,t) = e^{-t}\zeta + \dfrac{a_1(t)}{\zeta} + \cdots$. The inverse result to the Löwner-Kufarev equation states that given a holomorphic function $p(\zeta,t) = 1 + p_1(t)/\zeta + \cdots$ in $\zeta \in U^*$ with positive real part the solution of the equation (6.24) presents a subordination chain (see, e.g., [270]). This enables us to give a condition for ν_0 that guarantees a normalized one-parameter non-trivial family of maps $f(\zeta,t)$ to be a subordination chain.

Theorem 6.5.4. *Let $f(\zeta,t)$ be a normalized one-parameter non-trivial family of maps for $\zeta \in U^*$ which satisfies the conditions (ii–v) and is defined in an interval $[0,t_0)$. Let each $f(\zeta,t)$ be a homeomorphism of $\overline{\mathbb{C}}$ which is meromorphic in U^*, is normalized by $f(\zeta,t) = e^{-t}\zeta + \dfrac{a_1(t)}{\zeta} + \cdots$, and satisfies (6.24). Let the quasiconformal extension to U be given by a Beltrami coefficient $\mu_f = \mu(\zeta,t)$ which is differentiable with respect to t almost everywhere in $t \in [0,t_0)$. If*

$$\|\nu_0\|_\infty < \frac{\pi}{4\int_0^1 s\mathbf{K}(s)ds} \approx 0.706859\ldots,$$

where $\nu_0(\zeta,t)$ is as above and $\mathbf{K}(\cdot)$ is the complete elliptic integral, then $f(\zeta,t)$ is a normalized subordination chain.

Proof. Let $|\zeta| = \rho$, $w = re^{i\theta}$. We calculate

$$\begin{aligned}
\left|\frac{1}{\pi}\iint\limits_{U}\frac{\nu_0(w,t)d\sigma_w}{\zeta(w-\zeta)}\right| &\leq \frac{\|\nu_0\|_\infty}{\rho\pi}\iint\limits_{U}\frac{d\sigma_w}{|w-z|} = \frac{\|\nu_0\|_\infty}{\rho^2\pi}\iint\limits_{U}\frac{d\sigma_w}{|1-w/z|} \\
&= \frac{\|\nu_0\|_\infty}{\pi}\int\limits_0^1\int\limits_0^{2\pi}\frac{rdr\,d\theta}{\rho^2|1-re^{i\theta}/z|} \\
&= \frac{\|\nu_0\|_\infty}{\pi}\int\limits_0^1\int\limits_0^{2\pi}\frac{rdr\,d\theta}{\rho^2|1-re^{i\theta}/\rho|} \\
&= \frac{\|\nu_0\|_\infty}{\pi}\int\limits_0^1\int\limits_0^{2\pi}\frac{rdr\,d\theta}{\rho^2\sqrt{1+\frac{r^2}{\rho^2}-2\frac{r}{\rho}\cos\theta}} \\
&= \frac{\|\nu_0\|_\infty}{\pi}\int\limits_0^{1/\rho}\int\limits_0^{2\pi}\frac{sds\,d\theta}{\sqrt{1+s^2-2s\cos\theta}} \\
&\leq \frac{\|\nu_0\|_\infty}{\pi}\int\limits_0^1\int\limits_0^{2\pi}\frac{sds\,d\theta}{\sqrt{1+s^2-2s\cos\theta}} \\
&= \frac{4\|\nu_0\|_\infty}{\pi}\int\limits_0^1 s\mathbf{K}(s)ds < 1.
\end{aligned}$$

Then $\operatorname{Re} p(z,t) > 0$ that implies the statement of the theorem. □

Remark 6.5.5. *If* $\|\nu_0(\cdot,t)\|_\infty \leq q$, *then*

$$\frac{1+|\mu(\zeta,t)|}{1-|\mu(\zeta,t)|} \leq e^{2tq}\frac{1+|\mu(\zeta,0)|}{1-|\mu(\zeta,0)|}.$$

This obviously follows from the inequality

$$\frac{\partial|\mu_f|}{\partial t} = \frac{\partial|\mu_{f^{-1}}|}{\partial t} \leq |\dot{\mu}_{f^{-1}}|.$$

Remark 6.5.6. *Let us remark that the function* ν_0 *can be unilateraly discontinuous on* S^1 *in* $\overline{U}$, *therefore, it is not possible, in general, to use the Borel-Pompeiu formula to reduce the integral in* p *to a contour integral.*

The equation (6.24) is just the Löwner-Kufarev equation in partial derivatives with a special function $p(z,t)$ given in the above theorems.

Now we discuss the possibility of extending the equation (6.24) to all of $\mathbb{C}$. We differentiate the function $\Phi(\zeta,t,\tau)$ with respect to τ when $\zeta \in U \cup U^*$. It follows that

$$\left.\frac{\partial \Phi(\zeta,t,\tau)}{\partial \tau}\right|_{\tau=0} = \frac{-\overline{f_\zeta}}{|f_\zeta|^2-|f_{\bar\zeta}|^2}\dot f + \frac{f_{\bar\zeta}}{|f_\zeta|^2-|f_{\bar\zeta}|^2}\overline{\dot f} =: G(\zeta,t).$$

This formula can be rewritten in the form

$$\dot f(\zeta,t) = -(f_\zeta G(\zeta,t) + f_{\bar\zeta}\bar G(\zeta,t)).$$

Taking into account the equation (6.24) in U^* we have in the whole plane

$$\dot f(\zeta,t) = \begin{cases} -(f_\zeta G(\zeta,t) + f_{\bar\zeta}\bar G(\zeta,t)), & \text{for } \zeta\in\overline{U}, \\ -\zeta f_\zeta p(\zeta,t), & \text{for } \zeta\in U^*, \end{cases} \tag{6.25}$$

where $p(\zeta,t)$ is a holomorphic in U^* function with the positive real part by Theorem 6.5.2.

The variational formula (6.22) and differentiation of the singular integral imply that $G_{\bar\zeta}(\zeta,t)=\nu_0(\zeta,t)$, $\zeta\in U$. Now let us clarify what is G. Let us consider $\zeta\in U$. The Pompeiu formula leads to

$$G(\zeta,t) = h(\zeta,t) - \frac{1}{\pi}\iint\limits_U \frac{\nu_0(w,t)d\sigma_w}{w-\zeta}, \quad \zeta\in U$$

where $h(\zeta,t)$ is a holomorphic function with respect to ζ. The function G is continuous in $\overline{U}$ and by the Cauchy theorem

$$h(\zeta,t) = \frac{1}{2\pi i}\int\limits_{S^1}\frac{G(w,t)}{w-\zeta}dw.$$

To obtain the boundary values of the function $G(w,t)$, $|w|=1$, we shall use the second line in (6.25). Unfortunately, in general, it is not possible to use the same function f in both lines of (6.25) to obtain boundary values of G. Indeed, the mapping $f(\zeta,t)$ is differentiable regarding t a.e. in $t\in[0,t_0)$ locally uniformly in $\zeta\in\mathbb{C}$, and continuous in $\zeta\in\mathbb{C}$ for almost all $t\in[0,t_0)$. Therefore, the function $-(f_\zeta G(\zeta,t)+f_{\bar\zeta}\bar G(\zeta,t))$, $\zeta\in U$ is an extension of $-\zeta f'(\zeta,t)p(\zeta,t)$, $\zeta\in U^*$, whereas f_ζ, $\zeta\in U$ is not necessarily an extension of f', $\zeta\in U^*$.

A simple example of this situation is as follows. Let us consider the function

$$f(\zeta,t) = \begin{cases} e^{-t}\left(c\zeta + \dfrac{\bar\zeta}{c}\right), & \text{for } \zeta\in\overline{U}, \\ e^{-t}\left(c\zeta + \dfrac{1}{c\zeta}\right), & \text{for } \zeta\in U^*, \end{cases}$$

where $c>1$. This mapping forms a subordination chain with the dilatation $\mu(\zeta)$ that vanishes in U^* and is the constant $1/c^2$ in U. This chain is trivial, but it is not important for our particular goal here because we do not use at this stage the crucial variation. Then,

$$G(\zeta,t) = \begin{cases} \zeta, & \text{for } \zeta \in U, \\ \zeta \dfrac{c^2\zeta^2+1}{c^2\zeta^2-1}, & \text{for } \zeta \in U^*, \end{cases}$$

and it splits into two parts that can not be glued on S^1. The same is true for the derivatives f_ζ in U and f' in U^*.

If $\mu(\zeta,t)$ satisfies the condition (v) in a neighbourhood of S^1 in $\overline{U}$, then the derivatives f_ζ, $f_{\bar\zeta}$, $\zeta \in U$ have a continuation onto S^1 and

$$F(\zeta,t) = \frac{\overline{f_\zeta}\zeta f'p(\zeta,t) - f_{\bar\zeta}\overline{\zeta f'p(\zeta,t)}}{|f_\zeta|^2 - |f_{\bar\zeta}|^2}, \quad \zeta \in S^1,$$

where $\zeta f'p(\zeta,t)$ is thought of as the angular limits that exist a.e. on S^1. Moreover, in a neighbourhood of S^1 the derivative $f_{\bar\zeta}$ vanishes and the function $F(\zeta,t)$ can be written on S^1 as $F(\zeta,t) = \zeta p(\zeta,t)$. In turn,

$$h(\zeta,t) = \frac{1}{2\pi i}\int\limits_{S^1} \frac{wp(w,t)}{w-\zeta}dw.$$

This information allows us to formulate the following theorem.

Theorem 6.5.7. *Let $f(\zeta,t)$ be a subordination non-trivial chain of maps in U^* that exists for $t \in [0,t_0)$ and satisfies the conditions (i–v).*

(i) *For $\zeta \in U^*$ there exists a holomorphic function $p(\zeta,t)$ given by Theorem 6.5.2 such that*

$$\dot f(\zeta,t) = -\zeta f'(\zeta,t)p(\zeta,t).$$

(ii) *For $\zeta \in U$ there exists a continuous in ζ function $F(\zeta,t)$ given by*

$$F(\zeta,t) = \frac{1}{2\pi i}\int\limits_{S^1} \frac{wp(w,t)}{w-\zeta}dw - \frac{1}{\pi}\iint\limits_{U} \frac{\nu_0(w,t)d\sigma_w}{w-\zeta},$$

$$\nu_0(\zeta,t) = \frac{f_w^{-1}}{\bar f_{\bar w}^{-1}} \frac{\frac{\partial \mu_{f^{-1}}}{\partial t}}{1-|\mu_{f^{-1}}|^2} \circ f(\zeta,t),$$

such that

$$\dot f(\zeta,t) = -f_\zeta F(\zeta,t) - f_{\bar\zeta}\bar F(\zeta,t).$$

6.5.3 The Löwner-Kufarev ordinary differential equation

Dually to the Löwner-Kufarev partial derivative equation there is the Löwner-Kufarev ordinary differential equation. A function $g \in \Sigma_0$ is represented as a limit

$$\lim_{t\to\infty} e^{-t}w(\zeta,t), \tag{6.26}$$

where the function $w = g(\zeta, t)$ is a solution of the equation

$$\frac{dw}{dt} = wp(w,t), \tag{6.27}$$

almost everywhere in $t \in [0,\infty)$, with the initial condition $g(\zeta, 0) = \zeta$. The function $p(\zeta, t) = 1 + p_1(t)/\zeta + \cdots$ is analytic in U^*, measurable with respect to $t \in [0,\infty)$, and its real part Re $p(\zeta, t)$ is positive for almost all $t \in [0,\infty)$. The equation (6.27) is known as the *Löwner-Kufarev ordinary differential equation.* The solutions to (6.27) form a retracting subordination chain $g(\zeta, t)$, i.e., it satisfies the condition $g(U^*, t) \subset U^*$, $g(U^*, t) \subset g(U^*, s)$ for $t > s$, and $g(\zeta, 0) \equiv \zeta$.

The connection between (6.24) and (6.27) can be thought of as follows. Solving (6.24) by the method of characteristics and assuming s as the parameter along the characteristics, we have

$$\frac{dt}{ds} = 1, \quad \frac{d\zeta}{ds} = \zeta p(\zeta, t), \quad \frac{df}{ds} = 0,$$

with the initial conditions $t(0) = 0$, $\zeta(0) = \zeta_0$, $f(\zeta, 0) = f_0(\zeta)$, where ζ_0 is in U^*. We see that the equation (6.27) is exactly the characteristic equation for (6.24).

Unfortunately, this approach requires the extension of $f_0(w^{-1}(\zeta,t))$ into U^* (w^{-1} means the inverse function) because the solution to (6.24) is the function $f(\zeta,t)$ given as $f_0(w^{-1}(\zeta,t))$, where $\zeta = w(z,s)$ is a solution of the initial value problem for the characteristic equation (6.27) that maps U^* into U^*. Therefore, the solution of the initial value problem for the equation (6.24) may be non-univalent.

Solutions to the equation (6.27) are univalent functions $w(z,t) = e^{-t}\zeta + a_1(t)/\zeta + \cdots$ in the exterior of the unit disk that map U^* into itself. Conversely, every function from the class Σ_0 can be represented by the limit (6.26) where there exists a function $p(\zeta, t)$ with positive real part for almost all $t \geq 0$, such that $w(\zeta, t)$ is a solution to the equation (6.27) (see [270, pages 159–163]). Each function $p(\zeta, t)$ generates a unique function from the class Σ_0. The reciprocal statement is not true. In general, a function $g \in \Sigma_0$ can be determined by different functions p.

From [270, page 163] it follows that we can guarantee the univalence of the solutions to the Löwner-Kufarev equation in partial derivatives (6.24) assuming the initial condition $f_0(\zeta)$ given by the limit (6.26) with the function $p(\cdot, t)$ chosen to be the same in the equations (6.24) and (6.27), see [275].

The Löwner partial derivative equation (6.24) and the ordinary derivative equation (6.27) admit the Hamiltonian interpretation. Let us rewrite equation (6.24) in the form

$$\frac{\partial f'(\zeta,t)}{\partial t} = \frac{\partial \mathfrak{H}(\zeta, f', t)}{\partial \zeta}, \tag{6.28}$$

where $\mathfrak{H}(\zeta, f', t) = -\zeta f'(\zeta, t)p(\zeta, t)$, and the derivative f' is taken with respect to the complex variable ζ. Interpreting the function $\mathfrak{H}(\zeta, f', t)$ as a Hamiltonian we must write

$$\frac{\partial \zeta}{\partial t} = -\frac{\partial \mathfrak{H}(\zeta, f', t)}{\partial f'} = \zeta p(\zeta, t). \tag{6.29}$$

Equations (6.28) and (6.29) constitute the conjugate pair of Hamilton's relations, assuming the initial conditions $f(\zeta, 0) = f_0(\zeta)$ and $\zeta(\omega, 0) = \omega$ and rewriting $\zeta = w(\omega, t)$, $\omega \in U^*$.

Our goal is to deduce a form of the function p on the case of the subclass Σ_0^{qc}. Let a one-parameter family of maps $w = g(\zeta, t)$, $g \in \Sigma_0^{qc}$, satisfy the following conditions.

(i) The maps $g(\zeta, t)$ form a retracting subordination chain $g(U^*, 0) \subset U^*$.

(ii) The map $g(\zeta, t)$ is meromorphic in U^*, $g(\zeta, t) = \alpha(t)\zeta + a_1(t)/\zeta + \cdots$, where $\alpha(t) > 0$ and differentiable with respect to t.

(iii) The map $g(\zeta, t)$ is a quasiconformal homeomorphism of $\overline{\mathbb{C}}$.

(iv) The chain of maps $g(\zeta, t)$ is not trivial.

(v) The Beltrami coefficient $\mu_g(\zeta, t)$ of this map is differentiable with respect to t locally uniformly in U.

Note that in this case we do not need a strong assumption (v) in Section 6.5.2.

Set

$$H(\zeta, t, \tau) = g(g(\zeta, t), \tau) = \beta(\tau)w + b_0(\tau) + \frac{b_1(\tau)}{w} + \cdots,$$

where $w = g(\zeta, t)$. For each fixed t the mapping $g(\zeta, t)$ generates a smooth semi-flow H^τ in $\mathcal{G}^{qc}$ which is tangent to the path $g(\zeta, t + \tau)$ at $\tau = 0$. Therefore, we use the velocity vector $wp(w, t)$ (that may be only measurable with regard to t) with $w = g(\zeta, t)$ and obtain

$$\left.\frac{\partial H(\zeta, t, \tau)}{\partial \tau}\right|_{\tau=0} = g(\zeta, t)p(g(\zeta, t), t).$$

As before, the trajectory $g(\zeta, t)$ generates a pencil of tangent smooth semi-flows with the tangent vectors $wp(w, t)$, $w = g(z, t)$. Since $g(U^*, t) \in U^*$ for any $t > 0$, we can consider the limit

$$\lim_{\tau \to 0} \frac{H(\zeta, t, \tau) - g(\zeta, t)}{\tau g(\zeta, t)}.$$

We have that

$$\left.\frac{\partial H(\zeta, t, \tau)}{\partial \tau}\right|_{\tau=0} = \frac{\partial g(\zeta, t)}{\partial t} = g(\zeta, t)p(g(\zeta, t), t), \tag{6.30}$$

where $p(\zeta,t) = p_0(t) + p_1(t)/\zeta + \cdots$ is an analytic function in U^* that has positive real part for almost all fixed t. The equation defined by (6.30) is an evolution equation for the path $g(\zeta,t)$ and the initial condition is given by $g(\zeta,0) = \zeta$.

We suppose that all $g(\zeta,t)$ admit real analytic quasiconformal extensions and the family is non-trivial in the above sense. The function $g(w,\tau) = (H(\zeta,t,\tau) - b_0(\tau))/\beta(\tau)$ can be extended to a function from Σ_0^{qc} and it represents an equivalence class $[g^\tau] \in T$. There is a one-parameter path $y^\tau \in T$ that corresponds to a tangent velocity vector $\nu(w,t)$ such that

$$\mu_g(w,\tau) = \tau\nu(w,t) + o(\tau), \quad w = g(z,t).$$

We calculate explicitly the velocity vector making use of the Beltrami coefficient for a superposition:

$$\nu(w,t) = \lim_{\tau\to 0} \frac{\mu_{g(w,\tau)} \circ g(\zeta,t)}{\tau} = \lim_{\tau\to 0} \frac{1}{\tau} \frac{\mu_{H(\zeta,t,\tau)} - \mu_{g(\zeta,t)}}{1 - \bar{\mu}_{g(\zeta,t)}\mu_{H(\zeta,t,\tau)}} \frac{g_\zeta(\zeta,t)}{\bar{g}_{\bar{\zeta}}(\zeta,t)},$$

or

$$\nu(w,t) = \frac{\frac{\partial \mu_{g(\zeta,t)}}{\partial t}}{1 - |\mu_{g(\zeta,t)}|^2} \frac{g_\zeta}{\bar{g}_{\bar{\zeta}}} \circ g^{-1}(w,t), \quad \zeta \in U. \tag{6.31}$$

It is natural to implement an intrinsic parametrization using the Teichmüller distance $\tau_T(0,[g^t]) = t$, and assume the conformal radius to be $\beta(t) = e^t$ that implies $p_0 = 1$. The assumption of non-triviality allows us to use the variational formula (6.22) to state the following theorem.

Theorem 6.5.8. *Let $g(\zeta,t)$ be a retracting subordination chain of maps defined in $t \in [0,t_0)$ and $\zeta \in U^*$. Each $g(\zeta,t)$ is a homeomorphism of $\overline{\mathbb{C}}$ which is meromorphic in U^*, $g(\zeta,t) = e^t\zeta + b_1/\zeta + \cdots$, with an e^{2t}-quasiconformal extension to U given by a Beltrami coefficient $\mu(\zeta,t)$ that is differentiable with regard to t a.e. in $[0,t_0)$. The initial condition is $g(\zeta,0) \equiv \zeta$. Then, there is a function $p(\zeta,t)$ such that $\operatorname{Re} p(\zeta,t) > 0$ for $\zeta \in U^*$, and*

$$p(w,t) = 1 - \frac{1}{\pi} \iint\limits_{g(U,t)} \frac{\nu(u,t)d\sigma_u}{w(u-w)}, \quad w \in g(U^*,t),$$

where $\nu(u,t)$ is given by the formula (6.31), $\|\nu\|_\infty < 1$, and $w = g(\zeta,t)$ is a solution to the differential equation

$$\frac{dw}{dt} = wp(w,t), \quad w \in g(U^*,t), \tag{6.32}$$

with the initial condition $g(\zeta,0) = \zeta$.

Remark 6.5.9. *Taking into account the superposition we have*

$$p(g(\zeta,t),t) = 1 - \frac{1}{\pi} \iint\limits_U \frac{\dot{\mu}_g g_u^2(u,t)d\sigma_u}{g(\zeta,t)(g(u,t) - g(\zeta,t))},$$

where $u \in U$, $\zeta \in U^*$.

Remark 6.5.10. *The function $wp(w,t)$ has a continuation into $g(U,t)$ given by*

$$\frac{dw}{dt} = F(w,t),$$

where the function $F(w,t)$ is a solution to the equation

$$\frac{\partial F}{\partial \bar{w}} = \frac{g_\zeta^2 \dot{\mu}_g}{|g_\zeta|^2 - |g_{\bar{\zeta}}|^2} \circ g^{-1}(w,t).$$

In contrary to the Löwner-Kufarev equation in partial derivatives, the function F is the continuation of p in U through S^1. The solution exists by the Pompeiu integral and can be written as

$$\begin{aligned} F(w,t) &= h(w,t) - \frac{1}{\pi}\iint\limits_{g(U,t)} \frac{g_\zeta^2 \dot{\mu}_g}{|g_\zeta|^2 - |g_{\bar{\zeta}}|^2} \circ g^{-1}(u,t) \frac{d\sigma_u}{u-w} \\ &= h(w,t) - \frac{1}{\pi}\iint\limits_{g(U,t)} \frac{\nu(u,t)d\sigma_u}{u-w}, \end{aligned}$$

where $w \in g(U,t)$, $h(w,t)$ is a holomorphic function with respect to w, that can be written as

$$h(w,t) = \frac{1}{2\pi i} \int\limits_{\partial g(U,t)} \frac{up(u,t)}{u-w} du.$$

Reciprocally, given a function $F(u,t)$, $u \in g(U,t)$, we can write the function $p(w,t)$ as

$$p(w,t) = 1 - \frac{1}{\pi}\iint\limits_{g(U,t)} \frac{F_{\bar{u}}(u,t)d\sigma_u}{w(u-w)},$$

where $w \in g(U^*,t)$.

6.5.4 Univalent functions smooth on the boundary

Let us consider the class $\tilde{\Sigma}$ of functions $f(\zeta) = \alpha\zeta + a_0 + a_1/\zeta + \cdots$, $\zeta \in U^*$, such that being extended onto S^1 they are C^∞ on S^1. Repeating considerations of the preceding subsection for the embedding of M into the Teichmüller space T we come to the following theorem.

Theorem 6.5.11. *Let $f(\zeta,t)$ be a non-trivial subordination chain of maps that exists for $t \in [0,t_0)$ and $\zeta \in U^*$. Each $f(\zeta,t)$ is a homeomorphism $U^* \to \overline{\mathbb{C}}$ and belongs to $\tilde{\Sigma}$ for every fixed t. All these maps have quasiconformal extensions to U and there are a real-valued function $p_0(t) > 0$, complex-valued functions $p_1(t)$, and real-valued C^∞ functions $d(e^{i\theta},t)$ such that $\operatorname{Re} p(\zeta,t) > 0$ for $\zeta \in U^*$,*

$$p(\zeta,t) = p_0(t) + \frac{p_1(t)}{\zeta} - \frac{1}{2\pi}\int_0^{2\pi} \frac{e^{i2\theta} d(e^{i\theta},t)d\theta}{\zeta(e^{i\theta}-\zeta)}, \quad \zeta \in U^*,$$

and $f(\zeta,t)$ satisfies the differential equation

$$\frac{\partial f(\zeta,t)}{\partial t} = -\zeta \frac{\partial f(\zeta,t)}{\partial \zeta} p(\zeta,t), \quad \zeta \in U^*.$$

Theorems 6.5.2 and 6.5.11 are linked as follows. For a given subordination chain of maps $f(\zeta,t) \in \tilde{\Sigma}$, that exists for $t \in [0,t_0)$ and $\zeta \in U^*$, there is a C^∞ function $d(e^{i\theta},t)$ by Theorem 6.5.11 and we can construct the function $\nu(\zeta,t)$ by the Douady-Earle extension and the formula (6.11). Then, the function $f(\zeta,t)$ satisfies the equation of Theorem 6.5.2 with $p(\zeta,t)$ defined by such $\nu(\zeta,t)$.

Let us consider the ordinary Löwner-Kufarev equation for the functions smooth on S^1. If the retracting chain $g(\zeta,t)$ is smooth on S^1, then we use again the embedding of M into T and reach a similar result.

Theorem 6.5.12. *Let $g(\zeta,t)$ be a retracting non-trivial subordination chain of normalized maps that exists for $t \in [0,t_0)$ and $\zeta \in U^*$. Each $g(\zeta,t)$ is meromorphic in U^*, smooth on S^1, and $g(\zeta,t) = \beta(t)\zeta + b_0(t) + \frac{b_1(t)}{\zeta} + \cdots$, $\beta(t) > 0$. An additional assumption is that $g : U^* \to U^*$ for each fixed t. Then, there are a real-valued function $p_0(t)$, a complex-valued function $p_1(t)$, and a smooth real-valued function $d(e^{i\theta},t)$, such that $\operatorname{Re} p(\zeta,t) > 0$ for $\zeta \in U^*$,*

$$p(\zeta,t) = p_0(t) + \frac{p_1(t)}{\zeta} - \frac{1}{2\pi i} \int_{S^1} \left(\frac{zg'(z,t)}{g(z,t)} \right)^2 \frac{d(z,t)dz}{g(z,t) - \zeta}, \quad \zeta \in U^*,$$

and $w = g(\zeta,t)$ is a solution to the differential equation

$$\frac{dw}{dt} = wp(w,t), \quad w \in g(U^*,t)$$

with the initial condition $g(\zeta,0) = \zeta$.

Remark 6.5.13. *If we work with normalized functions*

$$g(\zeta,t) = e^t \zeta + \frac{b_1(t)}{\zeta} + \cdots,$$

then $p_0(t) \equiv 1$, $p_1(t) \equiv 0$.

6.5.5 An application to Hele-Shaw flows

Theorem 6.5.11 is linked to the Hele-Shaw free boundary problem as follows. Starting with a smooth boundary Γ_0, the one-parameter family $\Gamma(t)$ consists of smooth curves as long as the solutions exist. Let us consider the equation (6.3). Under injection we have a subordination chain of domains $\Omega(t)$. The Schwarz kernel can be developed as

$$\frac{\zeta + e^{i\theta}}{\zeta - e^{i\theta}} = 1 + \frac{2e^{i\theta}}{\zeta} + \frac{2e^{2i\theta}}{\zeta(\zeta - e^{i\theta})}.$$

Therefore, in Theorem 6.5.11 we can put

$$p_0(t) = \frac{1}{2\pi}\int_0^{2\pi} \frac{1}{|f'(e^{i\theta},t)|^2}d\theta, \quad p_1(t) = \frac{1}{\pi}\int_0^{2\pi} \frac{e^{i\theta}}{|f'(e^{i\theta},t)|^2}d\theta,$$

and

$$d(e^{i\theta},t) = \frac{-2}{|f'(e^{i\theta},t)|^2}.$$

Apart from the trivial case of growing ellipses there are no self-similar solutions, and therefore the Hele-Shaw dynamics $f(\zeta,t)$ generates a non-trivial path in T. Thus, given a Hele-Shaw evolution $\Gamma(t) = f(S^1,t)$ we observe a differentiable non-trivial path on T, such that at any time t the tangent vector ν is a harmonic Beltrami differential given by

$$\nu(\zeta,t) = \frac{-3}{\pi}\int_0^{2\pi} \frac{(1-|\zeta|^2)^2}{(1-e^{i\theta}\bar{\zeta})^4}\frac{e^{2i\theta}}{|f'(e^{i\theta},t)|^2}d\theta.$$

The corresponding cotangent vector is

$$\varphi(\zeta,t) = \frac{6}{\pi}\int_0^{2\pi} \frac{e^{-2i\theta}d\theta}{(1-e^{-i\theta}\zeta)^4|f'(e^{i\theta},t)|^2}.$$

6.6 Fractal growth

Benoit Mandelbrot (b. 1924, Warsaw) in 1977, 1983 brought to the world's attention that many natural objects simply do not have a preconceived form determined by a characteristic scale. Many of the structures in space and processes reveal new features when magnified beyond their usual scale in a wide variety of natural and industrial processes, such as crystal growth, vapor deposition, chemical dissolution, corrosion, erosion, fluid flow in porous media and biological growth a surface or an interface, biological processes. A fractal ("fractal" from Latin "fractus") is a rough or fragmented geometric shape that can be subdivided in parts, each of which is (at least approximately) a reduced-size copy of the whole. Fractals are generally self-similar, independent of scale, and have (by Mandelbrot's own definition) a Hausdorff dimension strictly greater than the topological dimension. There are many mathematical structures that are fractals, e.g., the Sierpinski triangle, the Koch snowflake, the Peano curve, the Mandelbrot set, and the Lorenz attractor. One of the ways to model a fractal is the process of fractal growth that can be either stochastic or deterministic. A nice overview of fractal growth phenomena is found in [341].

Many models of fractal growth patterns combine complex geometry with randomness. A typical and important model for pattern formation is *Diffusion-Limited Aggregation* (DLA) (see a survey in [161]). Considering colloidal particles

Figure 6.4: DLA clusters

undergoing Brownian motion in some fluid and letting them adhere irreversibly on contact with another one bring us to the basics of DLA. Fix a seed particle at the origin and start another one from infinity letting it perform a random walk. Ultimately, that second particle will either escape to infinity or contact the seed, to which it will stick irreversibly. Next another particle starts at infinity to walk randomly until it either sticks to the two-particle cluster or escapes to infinity. This process is repeated to an extent limited only by the modeler's patience. The clusters generated by this process are highly branched and fractal (see Figure 6.4).

The DLA model was introduced in 1981 by Witten and Sander [345], [346]. It was shown to have relations to dielectric breakdown [248], one-phase fluid flow in porous media [57], electro-chemical deposition [130], medical sciences [303], etc. A new conformal mapping language to study DLA was proposed by Hastings and Levitov [162], [163]. They showed that two-dimensional DLA can be grown by iterating stochastic conformal maps. Later this method was thoroughly handled in [72].

For a continuous random walk in 2-D the diffusion equation provides the law for the probability $u(z,t)$ that the walk reaches a point z at the time t,

$$\frac{\partial u}{\partial t} = \eta \Delta u,$$

where η is the diffusion coefficient. When the cluster growth rate per surface site is negligible compared to the diffusive relaxation time, the time dependence of the relaxation may be neglected (see, e.g., [346]). With a steady flux from infinity and the slow growth of the cluster, the left-hand side derivative can be neglected and we have just the Laplacian equation for u. If $K(t)$ is the closed aggregate at

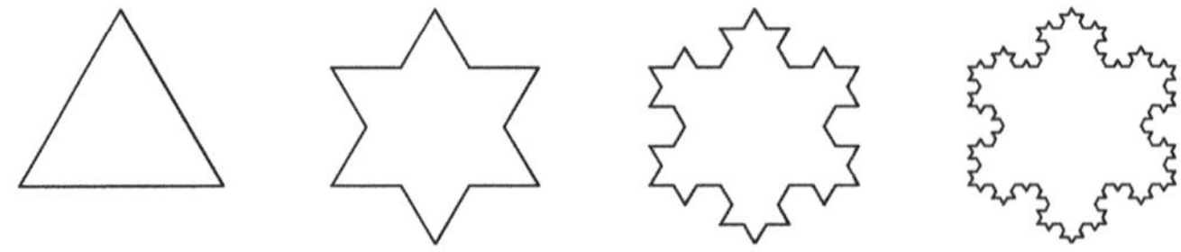

Figure 6.5: Koch's snowflake

the time t and $\Omega(t)$ is the connected part of the complement of $K(t)$ containing infinity, then the probability of the appearance of the random walker in $\mathbb{C}\setminus\Omega(t)$ is zero. Thus, the boundary condition $u(z,t)\big|_{\Gamma(t)} = 0$, $\Gamma(t) = \partial\Omega(t)$ is set. The only source of time dependence of u is the motion of $\Gamma(t)$. The problem resembles the classical Hele-Shaw problem, but the complex structure of $\Gamma(t)$ does not allow us to define the normal velocity in a good way although it is possible to do this in the discrete models.

Now let us construct a Riemann conformal map $f : U^* \to \overline{\mathbb{C}}$ which is meromorphic in U^*, $f(\zeta,t) = \alpha(t)\zeta + a_0(t) + \dfrac{a_1(t)}{\zeta} + \cdots$, $\alpha(t) > 0$, and maps U^* onto $\Omega(t)$. The boundary $\Gamma(t)$ need not even be a quasidisk, as considered earlier. While we are not able to construct a differential equation analogous to the Polubarinova-Galin one on the unit circle, the retracting Löwner subordination chain still exists, and the function $f(\zeta,t)$ satisfies the equation

$$\dot{f}(\zeta,t) = \zeta f'(\zeta,t)p_f(\zeta,t), \quad \zeta \in U^*, \tag{6.33}$$

where $p_f(\zeta,t) = p_0(t) + p_1(t)/\zeta + \cdots$ is a Carathéodory function: $\operatorname{Re} p(z,t) > 0$ for all $\zeta \in U^*$ and for almost all $t \in [0,\infty)$. A difference from the Hele-Shaw problem is that the DLA problem is well-posed on each level of discreteness by construction. An analogue of DLA model was treated by means of Löwner chains by Carleson and Makarov in [52]. In this section we follow their ideas as well as those from [170].

Of course, the fractal growth phenomena can be seen without randomness. A simplest example of such growth is the Koch snowflake (**Helge von Koch**, 1870–1924) (see Figure 6.5). DLA-like fractal growth without randomness can be found, e.g., in [73].

A new development, called Schramm-Löwner Evolution (previously called "Stochastic Löwner Evolution" by O. Schramm [306]) provides another probabilistic view on the Löwner chains and a new interpretation of the traditional conformal field theory approach. The basic idea is to replace the control function p in the classical Löwner equation by a function with driving parameter a one-dimensional Brownian motion (see [289], [306]).

Returning to the fractal growth we want to study a rather wide class of models with complex growing structure. We note that $\alpha(t) = \operatorname{cap} K(t) = \operatorname{cap}\Gamma(t)$. Let $M(0,2\pi)$ be the class of positive measures γ on $[0,2\pi]$. The control function

$p_f(\zeta, t)$ in (6.33) can be represented by the Riesz-Herglotz formula

$$p_f(\zeta, t) = \int_0^{2\pi} \frac{e^{i\theta} + \zeta}{e^{i\theta} - \zeta} d\gamma_t(\theta),$$

and $p_0(t) = \|\gamma_t\|$, where $\gamma_t(\theta) \in M(0, 2\pi)$ for almost all $t \geq 0$ and absolutely continuous in $t \geq 0$. Consequently, $\dot{\alpha}(t) = \alpha(t)\|\gamma_t\|$. There is a one-to-one correspondence between one-parameter (t) families of measures γ_t and Löwner chains $\Omega(t)$ (in our case of growing domains $\mathbb{C} \setminus \Omega(t)$ we have only surjective correspondence).

Example 1. Suppose we have an initial domain $\Omega(0)$. If the derivative of the measure γ_t with respect to the Lebesgue measure is the Dirac measure $d\gamma_t(\theta) \equiv \delta_{\theta_0}(\theta)d\theta$, then

$$p_f(\zeta, t) \equiv \frac{e^{i\theta_0} + \zeta}{e^{i\theta_0} - \zeta},$$

and $\Omega(t)$ is obtained by cutting $\Omega(0)$ along a geodesic arc. The preimage of the endpoint of this slit is exactly $e^{i\theta_0}$. In particular, if $\Omega(0)$ is a complement of a disk, then $\Omega(t)$ is $\Omega(0)$ minus a radial slit.

Example 2. Let $\Omega(0)$ be a domain bounded by an analytic curve $\Gamma(t)$. If the derivative of the measure γ_t with respect to the Lebesgue measure is

$$\frac{d\gamma_t(\theta)}{d\theta} = \frac{1}{2\pi|f'(e^{i\theta}, t)|^2},$$

then

$$p_f(\zeta, t) = \frac{1}{2\pi} \int_0^{2\pi} \frac{1}{|f'(e^{i\theta}, t)|^2} \frac{e^{i\theta} + \zeta}{e^{i\theta} - \zeta} d\theta,$$

and letting ζ tend to the unit circle we obtain $\mathrm{Re}\,[\dot{f}\ \overline{\zeta f'}] = 1$, which corresponds to the classical Hele-Shaw case, for which the solution exists locally in time.

In the classical Hele-Shaw process the boundary develops by fluid particles moving in the normal direction. In the discrete DLA models, either lattice or with circular patterns, the attaching particles are developed in the normal direction too. However, in the continuous limit it is usually impossible to speak of any normal direction because of the irregularity of $\Gamma(t)$.

In [52, Section 2.3] this difficulty was circumvented by evaluating the derivative of f occurring in γ_t in the above Löwner model slightly outside the boundary of the unit disk.

Let $\Omega(0)$ be any simply connected domain, $\infty \in \Omega(0)$, $0 \notin \Omega(0)$. The derivative of the measure γ_t with respect to the Lebesgue measure is

$$\frac{d\gamma_t(\theta)}{d\theta} = \frac{1}{2\pi|f'((1+\varepsilon)e^{i\theta}, t)|^2},$$

with sufficiently small positive ε. In this case the derivative is well defined.

It is worth mentioning that the estimate

$$\frac{\partial \operatorname{cap}\Gamma(t)}{\partial t} = \dot{\alpha}(t) \lesssim \frac{1}{\varepsilon}$$

would be equivalent to the Brennan conjecture (see [271, Chapter 8]) which is still unproved. However, Theorem 2.1 [52] states that if

$$R(t) = \max_{\theta\in[0,2\pi)} |f((1+\varepsilon)e^{i\theta}, t)|,$$

then

$$\limsup_{\Delta t\to 0} \frac{R(t+\Delta t) - R(t)}{\Delta t} \leq \frac{C}{\varepsilon},$$

for some absolute constant C. Carleson and Makarov [52] were, with the above model, able to establish an estimate for the growth of the cluster or aggregate given as a lower bound for the time needed to multiply the capacity of the aggregate by a suitable constant. This is an analogue of the upper bound for the size of the cluster in two-dimensional stochastic DLA given by [193].

Chapter 7

Hele-Shaw Evolution and Strings

In this section we shall give a short introduction to the classical string theory and apply it to the Hele-Shaw subordination evolution, in particular. In our setup the phase domain plays the role of the worldsheet of a bosonic string.

The Greek version of archetypal classical elements to explain patterns in nature, which dates from pre-Socratic times, consists of four elements Fire, Earth, Air, and Water (Latin derivatives are **ignis**, **terra**, **aer**, and **aqua**). They represent in Greek philosophy, science, and medicine the realms of the cosmos wherein all things exist and whereof all things consist. The ancients considered also the fifth element, **Ether or quinta essentia**, as a philosophical, heavenly, element in addition to the common four elements, as a unification substance, the prime element latent in all things. Sometimes Ether is associated with the Spirit.

In modern Physics the current understanding of matter is that there are four fundamental forces that have been identified and that govern all interactions. In our present Universe they have rather different properties. They are **gravity**, **the electromagnetic force**, **the strong force**, and **the weak force**. The first two are rather familiar in everyday life because of their long range scale. Strong and weak forces act at the atomic level at the range of $\sim 10^{-14}m$ and $\sim 10^{-15}m$ respectively.

- The *strong interaction* is responsible for binding the nuclei of atoms together. It is basically attractive, but can be effectively repulsive in some circumstances. Sometimes it is thought of as a shadow force of that binding three quarks together in a proton.
- The *weak force* is responsible for radioactive β-decay and neutrino interactions. An unstable neutron may decay into a proton (+), an electron (-) and an antineutrino.
- The *electromagnetic force* causes electric and magnetic effects such as the

repulsion between like electrical charges or the interaction of bar magnets. It can be attractive or repulsive, and acts only between pieces of matter carrying electrical charge.

- The *gravitational force* is weak, but very long ranged. It is always attractive, and acts between any two pieces of matter in the Universe since mass is its source. It binds together galaxies in clusters and governs the planet's motion.

Einstein's great dream (**Albert Einstein**, 1879–1955) was to come up with a unified theory of all forces. He believed that all forces were different aspects of the same force. Modern research shows that for objects at very high energy, the weak force and the electromagnetic force behave similarly. At the same time, the strong force appears to grow weaker. This caused the appearance of the so-called Theory of Everything and the Big Bang Theory on the global structure and history of the Universe. They suggest that the Universe originated from a single location of extremely high energy and matter densities. All forces were united in one superforce and perhaps it was acting on only one type of particles. After a huge explosion the Universe expanded, the energy decreased, unstable particles decayed, stable matter formed.

Theories that postulate the unification of the strong, weak, and electromagnetic forces are called Grand Unified Theories. Theories that add gravity and try to unify all four fundamental forces into a single force are called Superunified Theories. The theory that describes the unified electromagnetic and weak interactions is called the Standard Electroweak Theory, or sometimes just the Standard Model.

Grand Unified and Superunified Theories remain theoretical speculations so far as yet unproved, but there is strong experimental evidence for unification of the electromagnetic and weak interactions in the Standard Electroweak Theory. Furthermore, although Grand Unified Theories are not proved experimentally, there is strong circumstantial evidence to suggest that a theory at least like a Grand Unified Theory is required to make sense of the Universe.

The central paradox is the apparent incompatibility of two main theoretical foundations of particle physics. The first foundation is Einstein's general theory of relativity, which relates the force of gravity to the structure of space and time. The second theoretical foundation is quantum mechanics formulated for three of the four known forces of nature: the strong, weak and electromagnetic interactions, which can account for the atomic and subatomic world. Moreover, there have been discovered force-carrying particles for these three forces. They are for example, photons (the electromagnetic force), mesons and gluons (the strong force), and intermediate vector bosons or 'W' and 'Z' bosons (the weak force). No elementary particle that carries gravity (hypothetic "graviton") have been observed yet. The concept of what is "elementary", i.e. not built of anything else, changes as science progresses. Thirty years ago, it was believed that the proton was an elementary particle, but it has long since been known that it is built from three quarks.

One of the pretenders to be a unified theory is a so-called Higgs field. In the standard model of particle physics all the quarks and leptons turn out to be massless. **Peter Ware Higgs** (born May 29, 1929), FRSE, FRS, until recently the holder of a personal chair in theoretical physics at the University of Edinburgh and now an emeritus professor, proposed in 1964 a field that (like "Ether"!) permeates all space. All particles interact with this field and it gives them mass. The Higgs particle responsible for such a field has not yet been observed although attempts will be launched to try to observe the Higgs particle directly.

Until recently there seemed to be little hope that Einstein's theory of gravity, the fourth fundamental force, could be united with the precepts of quantum mechanics. In the last decade elementary particle physicists have become optimistic [74], [129], [160] that the theoretical impasse might be resolved. The optimism is based on striking developments in a new kind of theory known as **string** and **superstring** theory. The far-reaching claims made for this theory would, if correct, provide the much sought-after Theory of Everything, the unification of physics. The familiar elementary particles are understood as different modes of a single string. Superstring theory combines string theory with a mathematical structure called supersymmetry. It may be considered as the fifth element of modern physics.

7.1 Bosonic strings

7.1.1 Initial definitions

There are many sources for String Theory. We would recommend some modern more mathematical general reviews of Conformal Field Theory and String Theory as well as of adjacent algebraic structures [12], [78], [186], [189], [187], [225], [261]. Some of them are used in this section. Corresponding differential geometry background may be found in [88] or its Springer three-volume translation. A good source for the basics of Lagrangian and Hamiltonian mechanics is [18].

All elementary particles are either bosons or fermions. The spin-statistics theorem identifies the structural properties that differentiate fermions and bosons. Fermions are particles with semi-integer spin whereas bosons have integer spin. As a consequence of this fact, these two types of elementary particles represent wave-corpuscular duality. *Bosons can occupy the same relative position within space-time whereas fermions can not.* A typical boson is a photon whereas an electron is a typical fermion. Any composite particle is a number of bosons and fermions in interaction.

In classical Newtonian mechanics, in the absence of external forces, a point particle moves along the shortest curve (geodesic) between the initial and final positions. This is no longer so in quantum mechanics. In 1932, P. A. Dirac, a British theoretical physicist (**Paul Adrien Maurice Dirac** 1902–1984), laid the foundation stone [83] of what was destined to become, in the hands of Feynman, a new formulation of quantum mechanics. Within Feynman's interpretation (1948), the elementary particle can move along any path between the observed initial and final positions (**Richard Phillips Feynman**, 1918–1988). Thus, intermediate positions can not be determined any longer. However, we may assign a probability density to each path which is the higher, the "shorter" is the path, or in the setup of Lagrangian mechanics, the smaller the action is. All elementary particles are topological points but in Feynman's interpretation of quantum mechanics they possess some new internal structure. This can be expressed in the following definitions.

A *string* is a 1-dimensional object moving in D-dimensional space-time $\mathfrak{M}$. The first coordinate is treated as time. As the string moves in time, it sweeps a 2-dimensional surface in $\mathfrak{M}$ which is called the *worldsheet* (WS) of the string (analogously with the worldline of a Newtonian particle). We need to construct the probability density over the space of all surfaces connecting the initial and final "position" of the string. The density must be the higher the smaller is the action (least action or least energy principle). If we denote by $S[\phi]$ the action where ϕ is a quantity characterizing the surface (the metric further on), the idea of Gaussian distribution yields the simplest Feynman (path) integral[1]

$$\int_{\text{ini}}^{\text{fin}} D\phi \, e^{-\frac{1}{\hbar}S[\phi]}, \tag{7.1}$$

or the *partition function*, or the *correlator*, where $\hbar$ stands for the Planck constant and $D\phi$ is the most sophisticated part of this integral, which is roughly speaking a local product $\Pi_{x\in\mathfrak{M}} d\mu(\phi(x))$ of measures on $\mathfrak{M}$. The symbol $\int D\phi$ here is a concise way to represent the infinite-dimensional integral over all possible field configurations on all of space-time. The problem with the definition of $D\phi$ is that there is no translation-invariant measure on an infinite-dimensional vector space. The path integral is rather formal and may be given strict mathematical sense only in some simple situations. Up to now, only in dimension $D = 26$ is the theory consistent and the integral (7.1) mathematically well defined as an integral over the Teichmüller space with respect to the Weil-Petersson measure.

Since Feynman formally wrote down [113] the first path integral formulation of quantum mechanics, several definitions have been proposed from different perspectives. For example, Albeverio and Hoegh-Krohn [11] develop a general theory of oscillatory integrals on Hilbert spaces which contains as a particular case a definition of Feynman path integrals for nonrelativistic quantum mechanics, quantum statistical mechanics, and quantum field theory. Their formalism can be used for computing path integrals by computers more readily than for computing path integrals analytically. A good mathematical and historical treatment of Feynman's integral is found in [185]. The partition function (7.1) represents the transition probability between two observed points.

We shall parameterize WS by means of two parameters (τ, σ). Let x^μ be phase coordinates $\mu = 0, 1, \ldots, D-1$ of $\mathfrak{M}$, and let $\mathfrak{M}$ be flat. The relative space-time position of a point on the string WS is given by the functions $x^\mu(\tau, \sigma)$. The

[1]Throughout this chapter we shall use $\int$ for all integrals to avoid bulky notation.

parameters (τ, σ) have no physical significance, however the functions $x^\mu(\tau, \sigma)$ tell us how the string moves in $\mathfrak{M}$, and hence, they are the *dynamical variables.*

Strings may be *closed* if the dynamical variables are periodic with regard to σ or *open* otherwise. The parameter τ is normally taken in the interval $(-\infty, \infty)$ or $[0, \infty)$. We shall consider only closed strings.

Of course, 2-dimensional objects, worldsheets, may be generalized to any dimension. Edward Witten proposed the existence of so-called M-theory at a conference at the University of Southern California in 1995. According to Witten himself, 'M' stands for 'magic,' 'mystery' or 'membrane', depending on ones taste. He proposed a multidimensional object usually called p-brane (derivative of 'membrane') referring to its spatial dimension p (for example, a string is a 1-brane and a 'membrane' is a 2-brane). M-theory is a solution proposed for the unknown theory of everything which would combine all five superstring theories and 11-dimensional supergravity together. According to Witten, mathematical tools which have yet to be invented are needed in order to fully understand it.

7.1.2 Metrics

We shall use:

- the ambient Euclidean (+,+) or Lorenzian (–,+) metric $ds_{\mathfrak{M}}$ of $\mathfrak{M}$,

$$ds^2_{\mathfrak{M}} = \eta_{\mu\nu} dx^\mu dx^\nu, \quad \eta_{\mu\nu} = \begin{pmatrix} \pm 1 & 0 & \dots & 0 \\ 0 & 1 & \dots & 0 \\ \dots & \dots & \dots & \dots \\ 0 & 0 & \dots & 1 \end{pmatrix};$$

- the intrinsic metric ds_c of the canonical cylinder, which is also Euclidean or Lorenzian

$$ds^2_c = \eta_{\alpha\beta}\, d\sigma^\alpha d\sigma^\beta, \quad \begin{cases} \sigma^0 = \tau \\ \sigma^1 = \sigma \end{cases}, \quad \alpha, \beta = 0, 1.$$

  The intrinsic metric does not depend on the embedding of the worldsheet into $\mathfrak{M}$;

- the induced metric ds_{ws} on the worldsheet

$$ds^2_{ws} = \eta_{\mu\nu} \frac{\partial x^\mu}{\partial \sigma^\alpha} \frac{\partial x^\nu}{\partial \sigma^\beta}\, d\sigma^\alpha d\sigma^\beta = g_{\alpha\beta}\, d\sigma^\alpha d\sigma^\beta.$$

- an unknown apriori (variable) metric ds on the worldsheet as on a 2-D manifold, such that $ds^2 = h_{\alpha\beta}(\tau, \sigma)\, d\sigma^\alpha d\sigma^\beta$.

7.1.3 Dynamics

Dynamical variables $x^\mu(\tau,\sigma)$ satisfy the equations of motion derived by making use of the *least action principle.* In Lagrangian mechanics, a system with a configuration space $\mathfrak{N}$ is characterized by its *Lagrangian* L – a smooth, real-valued functional on the direct product of the tangent bundle $T\mathfrak{N}$ and the time axis. If $P\mathfrak{N}$ stands for the path space (which is an infinite-dimensional Fréchet manifold), the *action functional* S is the real-valued integral functional defined by $S=\int_{\text{ini}}^{\text{fin}} L dt$, where t is the time parameter. Turning to the space-time configuration space $\mathfrak{M}$ and a worldsheet in $\mathfrak{M}$ we work with the action functional $S=\int_{ws}\Phi(x)dx$ where the Lagrangian is included in to this spatial integral. So defined action must have some physical or geometrical relevance. In classical mechanics, action is the difference between kinetic and potential energies. Yôichirô Nambu (1921–present, from the University of Chicago, 1994/95 Wolf prize in Physics) in 1970 [243] and Tetsuo Gotô (from the Nihon University Tokyo) in 1971 [128] proposed to choose the area of the worldsheet as the simplest action. This action, known as the Nambu-Goto action, admits the form

$$S_{NG}=T\int\limits_{ws} dx = T\int\limits_{C} d\sigma^2\sqrt{|\det g_{\alpha\beta}|},$$

where C is the canonical cylinder and the quantity T, the string tension, has dimension mass per unit length. The string undergoes geodesic motion, i.e., the dynamical variables satisfy the Euler-Lagrange equation

$$\frac{1}{\sqrt{|\det g_{\alpha\beta}|}}\frac{\partial}{\partial\sigma^\alpha}\left(\sqrt{|\det g_{\alpha\beta}|}\, g^{\alpha\beta}\frac{\partial x^\mu}{\partial\sigma^\beta}\right)=0, \quad \mu=0,\dots,D-1,$$

where $g^{\alpha\beta}=g^{-1}_{\alpha\beta}$. The metric $g_{\alpha\beta}$ is defined on the worldsheet embedded into $\mathfrak{M}$, thus it depends on the mapping $x(\tau,\sigma)$, and as the equation for x is highly non linear. Moreover, the square root make it unpleasant to think of the path integrals as of Gaussian integrals. To overcome these problems Alexander Markovich Polyakov (1945–present, formerly at the Landau Institute in Moscow, currently at Princeton University, 1986 Dirac Medal, 1994 Lorentz Medal) proposed to introduce “Lagrange multipliers” in 1981 [267]. He considered the variable metric $h_{\alpha\beta}$ as an additional variable and the Dirichlet integral

$$S_P=\frac{T}{2}\int\limits_C d\sigma^2\sqrt{|\det h_{\alpha\beta}|}\, h^{\alpha\beta}\frac{\partial x^\mu}{\partial\sigma^\alpha}\frac{\partial x^\nu}{\partial\sigma^\beta}$$

as Polyakov action. The Euler-Lagrange equation (with regard to the variation of the worldsheet) for the Polyakov action S_P is formally the same as for the Nambu-Goto action S_{NG},

$$\frac{1}{\sqrt{|\det h_{\alpha\beta}|}}\frac{\partial}{\partial\sigma^\alpha}\left(\sqrt{|\det h_{\alpha\beta}|}\, h^{\alpha\beta}\frac{\partial x^\mu}{\partial\sigma^\beta}\right)=0, \quad \mu=0,\dots,D-1, \tag{7.2}$$

but h does not depend on x any more. So we have linear equations for $x(\tau, \sigma)$. The Euler-Lagrange equation must be supplemented by a boundary condition such that the mapping $x(\tau, \sigma)$ preserves the infinitesimal boundary structure (tangent vectors to ∂C go to tangent vectors to ∂WS).

There are three extra degrees of freedom in h but they can be removed by using the equation of motion for h,

$$\frac{\delta S_P}{\delta h^{\alpha\beta}} = 0,$$

where the left-hand side is a functional derivative. The two-dimensional energy-momentum tensor is defined as

$$T_{\alpha\beta} = \frac{-2}{T} \frac{1}{\sqrt{|\det h_{\alpha\beta}|}} \frac{\delta S_P}{\delta h^{\alpha\beta}},$$

and the equation of motion implies $T_{\alpha\beta} = 0$. Equation (7.2) yields that $x(\tau, \sigma)$ is conformal, i.e., that the metric $h_{\alpha\beta}$ on the canonical cylinder C is proportional to the induced metric on $WS = x(C)$. Moreover, $S_P = S_{NG}$ under $T_{\alpha\beta} = 0$, whereas in general, $S_P \geq S_{NG}$.

7.1.4 Symmetries

A guiding principle of modern physics is that there is a close connection between the standard conservation laws and the symmetries of space-time. There is no single theorem describing such a connection, however this principle is sometimes referred to as the "Noether theorem", although the Noether[2] 1918 theorem [250] itself contains only a very partial statement of it. By symmetries for the Polyakov action S_P we mean transformations keeping the action invariant.

- **Global symmetries**
  - Euclidean isometries or Poincaré invariance in the Lorenzian case (Poincaré group) applied to WS:

$$x^\mu \to x^\mu + b^\mu;$$

$$x^\mu \to x^\mu + \omega^\mu_\nu x^\nu;$$

where $\omega^\mu_\nu = -\omega^\nu_\mu$ are infinitesimal Euclidean or Lorenzian rotations.

[2]Emmy Noether, (born Nöther), 1882–1935, achieved the doctorate in 1907 under Paul Gordan. She was one of the most talented mathematicians of the early 20th century but she was refused permission to teach by the university of Göttingen, and her colleague, David Hilbert, had to advertise her courses in the university's prospectus under his own name. Finally, she was admitted to the faculty in 1919.

- **Local symmetries**
  - reparametrization invariance (diffeomorphisms in 2-D): $\sigma^\alpha \to \tilde{\sigma}^\alpha(\tau,\sigma)$ implies the invariance of the area element

$$d\tilde{\sigma}^2\sqrt{|\det \tilde{h}|} = d\sigma^2\sqrt{|\det h|}.$$

  - Weyl rescaling

$$h_{\alpha\beta}d\sigma^\alpha d\sigma^\beta \to e^{\rho(\tau,\sigma)}h_{\alpha\beta}d\sigma^\alpha d\sigma^\beta.$$

Weyl rescaling leaves $\sqrt{|\det h|}\ h_{\alpha\beta}$ invariant

Translations from the Euclidean isometries or the Poincaré group led to the conservation of energy and momentum, while infinitesimal Euclidean or Lorenz rotations led to the conservation of angular momentum. The Euclidean isometries and the Poincaré group appear as a global internal symmetry in $\mathfrak{M}$ but not in the 2-D space-time C.

The conformal or Weyl invariance implies that the energy-momentum tensor is always traceless: $h^{\alpha\beta}T_{\alpha\beta} = \operatorname{tr} T_{\alpha\beta} = T^\alpha_\alpha = 0$. The metric $h_{\alpha\beta}$ is defined up to a conformal factor $e^{\rho(\tau,\sigma)}$, so we can choose a *conformal gauge*, i.e., we can choose a general coordinate transformation that brings the metric to the form $h^{\alpha\beta} = e^{\rho(\tau,\sigma)}\eta_{\alpha\beta}$ with the metric η, and then we can apply Weyl scaling to eliminate the conformal factor. We consider the complex coordinates $d\zeta = id\tau + d\sigma$ and $d\bar{\zeta} = -id\tau + d\sigma$, $ds^2 = 4(d\zeta^2 + d\bar{\zeta}^2)$. In this gauge, the Polyakov action describes a free two-dimensional conformal field theory and the equation of motion for $x(\tau,\sigma) = x(\zeta,\bar{\zeta})$ is simply the free wave equation

$$\left(\frac{\partial^2}{\partial\zeta^2} + \frac{\partial^2}{\partial\bar{\zeta}^2}\right)x(\zeta,\bar{\zeta}) = 0.$$

The general solution to this equation is $x(\tau,\sigma) = \phi(\sigma+\tau) + \psi(\sigma-\tau)$. After the change of variables $\tau \to i\tau$ (Wick rotation) we get a complex form of it, $x(\zeta,\bar{\zeta}) = \phi(\zeta) + \psi(\bar{\zeta})$, where $\phi(\zeta)$ and $\psi(\bar{\zeta})$ describe the left-moving and the right-moving components of $x(\zeta,\bar{\zeta})$. For closed strings the worldsheet theory splits into two independent components. Wick rotation is a method of finding a solution to a problem in Minkowski space from a solution to a related problem in Euclidean space, by analytic continuation. This solution may then, under reverse substitution, yield a solution to the original problem.

Analogously, one may consider the Euclidean metric and space instead which is more familiar than the Minkowskii space. The equation of motion becomes the Laplacian

$$\left(\frac{\partial^2}{\partial\zeta\partial\bar{\zeta}}\right)x(\zeta,\bar{\zeta}) = 0.$$

In this case we get $x(\zeta,\bar{\zeta}) = \phi(\zeta) + \psi(\bar{\zeta})$ as a general solution. So in both cases the motion splits into the left-moving and the right-moving components of $x(\zeta,\bar{\zeta})$.

However, they are not completely equivalent from the mathematical point of view, because in the Laplacian case the solution must be harmonic, and therefore necessarily have strong regularity properties. In the Euclidean case one may interpret the above procedure in the following way. We consider the complex coordinates $d\zeta = id\tau + d\sigma$ and $d\bar{\zeta} = -id\tau + d\sigma$, the surface S diffeomorphic locally to the canonical cylinder C, and the conformally invariant metric $e^{\rho(\zeta,\bar{\zeta})}\delta_{\alpha\beta}$ with the Kronecker delta. Expressed in complex coordinates we have the invariant form

$$e^{\rho(\zeta,\bar{\zeta})}d\zeta^2 = 4\left(\frac{\partial x}{\partial \zeta}d\zeta\right)^2,$$

which is a holomorphic quadratic differential in terms of the local parameter ζ. The surface S is a critical point for S_P with respect to variations by families of diffeomorphisms. In general, the canonical surface C is a Riemann surface.

7.2 Correlator

7.2.1 Correlator and partition function

In probability theory and statistics, a *correlator*, also called a correlation coefficient, indicates the strength and direction of a linear relationship between two random variables. In general statistical usage, correlation or co-relation refers to the departure of two variables from independence. The correlator $\rho_{X,Y}$ between two random variables X and Y with expected values $E(X)$ and $E(Y)$ is

$$\rho_{X,Y} = \frac{E(XY) - E(X)E(Y)}{\sqrt{E(X^2) - E^2(X)}\,\sqrt{E(Y^2) - E^2(Y)}}.$$

The complete information on $\rho_{X,Y}$ is got by calculating $\langle X_k X_j\rangle = E(X_k X_j)$, $X_1 = X$, $X_2 = Y$, $j,k = 1,2$. Often, the quantity $\langle X_k X_j\rangle$ is called a correlator instead of $\rho_{X,Y}$.

Quantum particles, unlikely classical ones, can roughly speaking randomly deviate from their classical trajectories. Therefore, given the position and velocity of a particle at a given moment, we can not predict its position at a later time. The particle motion in $\mathfrak{M}$ in a potential field $\mathfrak{M} \to \mathbb{R}^D$ is described by a stochastic process $\phi(t) = (\phi_0(t), \ldots, \phi_{D-1}(t))$. By the Kolmogorov extension theorem [203], this process is completely described by the correlators

$$\langle \phi_{j_1}(t_1)\ldots\phi_{j_n}(t_n)\rangle, \quad n = 0, \ldots, D-1,$$

which by definition are the expected values of the products of random variables $\phi_{j_1}(t_1)\ldots\phi_{j_n}(t_n)$. A typical field of application of this idea is Brownian motion. A quantum particle is much less intuitive and much more difficult to visualize. But Feynman pointed out that the behaviour of a quantum particle in a potential field is correctly described by the same model by appropriate choosing of the weight, or in other words, of the correct measure in terms of integration. So the complete information on the dynamics of a quantum particle can be expressed by the correlator

$$\langle \phi_{j_1}(t_1,x_1)\dots\phi_{j_n}(t_n,x_n)\rangle = \int D\phi\, \phi_{j_1}(t_1,x_1)\dots\phi_{j_n}(t_n,x_n)e^{-\frac{1}{\hbar}S[\phi]},$$

normalized as

$$\int D\phi\, e^{-\frac{1}{\hbar}S[\phi]} = 1,$$

where $D\phi$ is a measure on the space of all possible paths in $\mathfrak{M}$, and $S[\phi]$ is an action. It is clear that such a definition is not satisfactory yet from the mathematical point of view because of the infinite dimensional integration. In the case of Brownian motion a justification is possible within the standard Lebesgue measure theory (theory of Wiener integrals). In quantum mechanics this setting is less satisfactory. However, physicists made several interesting conclusions using this path integral formalism, confirmed to be true, which gives certain expectations for a corresponding mathematical justification to be developed.

One of the important goals of any field theory, including conformal field theory, is to calculate the full set of partition functions. In the Euclidean setup of the bosonic string theory the heuristic idea is to make sense of the path integral

$$\int Dx\, Dh\, e^{-S_P[h]},$$

where we have put $\hbar = 1$ for simplicity. The integration is taken over the space of all mappings $x : C \to \mathfrak{M}$ (with prescribed boundary conditions) and over all metrics h on the canonical differentiable surface C with boundary (which we took previously to be a cylinder). Considering equivalent metrics we arrive at the space of deformations, the Teichmüller space. Finally, the measure is decomposed into a measure on the space of conformal factors (due to Weyl[3] symmetry) and the Weil-Petersson measure on the finite Teichmüller space. The details of this construction exceed the scope of our book so we refer the interested reader to [186].

7.2.2 Conformal anomaly. Liouville theory

The problem of the construction of correlators and a consistent theory of measure is crucial. As has been mentioned, the integration splits in two bits: integration with respect to equivalent coordinates (the Teichmüller space over C) and with

[3] Hermann Klaus Hugo Weyl, Peter to his close friends, 1885–1955. From 1913 to 1930 Weyl held the chair of mathematics at Zürich Technische Hochschule. In his first academic year in this new post he was a colleague of Einstein who was at this time working out the details of the theory of general relativity. André Weil, 1906–1998. He was a founding member, and *de facto* the early leader, of the influential Bourbaki group. Hans Petersson, 1902–1984.

respect to infinite dimensional metric factors from the Weyl symmetry. The latter locus is called the *conformal anomaly*. In string theory, conformal symmetry on the worldsheet, the local Weyl symmetry, must cancel if the theory is to be consistent. The required cancellation implies that the spacetime dimensionality must be equal to the critical dimension, which is either 26 in the case of bosonic string theory or 10 in the case of superstring theory. The classical theory has no mass scale so there is a conformal symmetry, but this is broken by a conformal anomaly.

In classical $2D$ gravity, the metric is purely gauge, because it has 3 components and there are 3 gauge symmetries, 2 diffeomorphisms and 1 Weyl rescaling of the metric. But after quantization, the trace of the metric becomes physical.

In lightcone quantization, the physical Hilbert space becomes unexpectedly large, because you cannot factor out the conformal symmetry unless the number of scalar fields equals $D = 26$, due to the conformal anomaly. Provided that $D < 26$, the theory is consistent in spite of the anomaly, because the Hilbert space has a positive-definite inner product (the spectrum is "ghost-free").

Alternatively, one can introduce the trace of the metric as a dynamical field already at the classical level; it is then known as the Liouville field (**Joseph Liouville**, 1809–1882). The parameters can be fine-tuned so that the Liouville field has central charge $c = 26 - D$, canceling the matter and ghost contributions, and proper conformal invariance is restored. However, the physical Hilbert space is still larger than expected classically, because it now includes the Liouville field.

Thus, the associated Liouville action is that of a soluble QFT, although the integration measure is not the usual one in the functional approach to quantum field theory. It is a theory of conformal anomaly. There is no particular reason for the length parameter to drop out of the problem, i.e., to apply Weyl invariance. Therefore, the string is non-critical in the sense that we use the metric $e^{\phi(\zeta,\bar{\zeta})}\eta_{\alpha\beta}$ instead of $\eta_{\alpha\beta}$. The correlator is the path integral

$$\langle A\rangle = \int\limits_{C(S)} D\phi\, A(\phi)\, e^{-\frac{1}{\hbar}S[\phi]}.$$

We take S to be an n-punctured spherec and $C(S)$ consists of all smooth conformal metrics $ds^2 = e^{\phi(\zeta,\bar{\zeta})}|d\zeta^2|$ with asymptotics $e^{\phi(\zeta,\bar{\zeta})} \sim (|\zeta - \zeta_j| \log|\zeta - \zeta_j|)^{-2}$ near the punctures ζ_j. This concrete situation was thoroughly treated by Takhtajan and Zograf [318], [348], [349]. The correct Liouville action for the non-critical strings becomes

$$S[\phi] = \lim_{\varepsilon\to 0}\left(\int\limits_{S(\varepsilon)} (|\phi_\zeta|^2 + e^\phi)d\sigma_\zeta + 2\pi n\log\varepsilon + 4\pi(n-2)\log|\log\varepsilon|\right), \tag{7.3}$$

where $S(\varepsilon)$ is obtained by removing ε-disks centered on the punctures. The least action principle leads to the Liouville equation

$$\frac{\partial^2}{\partial\zeta\partial\bar{\zeta}}\phi = \frac{1}{2}e^\phi.$$

The geometric meaning of this equation is that the conformal metric $e^{\phi(\zeta,\bar{\zeta})}|d\zeta|^2$ on S has constant negative curvature -1. The asymptotics guarantee that the metric is complete and the area of the surface S is finite due to the Gauss-Bonnet formula. The additional terms in the formula for $S[\phi]$ are present because the integrand (the first term) does not keep the local 2-form $|\phi_\zeta|^2 \, d\zeta \wedge d\bar{\zeta}$ invariant under the change of the local parameters on charts of S.

7.2.3 Superstrings

We already mentioned classical symmetries. Supersymmetry is a hypothetical symmetry that relates bosons and fermions. In supersymmetric theories, every fundamental fermion has a bosonic superpartner and vice versa. There is no direct experimental evidence so far that supersymmetry exists in the real world. However, there is some indirect evidence which suggests that supersymmetry may be found at energies not too far above those accessible by today's particle accelerators. When we include fermions in the worldsheet theory of the string, we automatically get a new type of symmetry called supersymmetry which relates bosons and fermions. Fermions and bosons are grouped together into supermultiplets which are related under the symmetry. This is the reason for the "super" in "superstrings". A consistent quantum field theory of superstrings exists only in 10 spacetime dimensions. We do not enter this topic and refer an interested reader to [189].

7.3 Virasoro algebra

7.3.1 Dirac and Kronecker δ-functions

These two special functions are of crucial importance in the present formalism. Whereas Dirac's δ is a distribution defined as a linear functional on the space of test functions and $\delta[\phi] = \phi(0)$ for every test function ϕ, Kronecker's δ_{kj} is a tensor; it can be written δ_k^j with a contravariant index j. This is a more accurate way to notate the identity matrix, considered as a linear mapping. The Kronecker delta has the so-called sifting property that for $j \in \mathbb{Z}$:

$$\sum_{k=-\infty}^{\infty} \delta_{kj} a_k = a_j.$$

This property is similar to one of the main properties of the Dirac delta function:

$$\int_{-\infty}^{\infty} \delta(x-y) f(x) dx = f(y),$$

and in fact Dirac's delta was named after the Kronecker delta because of this analogous property. For real x and y we have

$$\delta_{kj} = \frac{2}{\pi}\int_0^\pi \cos jx \cos kx \, dx,$$

and for $|x-y| < \pi$ we have

$$\begin{aligned}\delta(x-y) &= \frac{1}{\pi} + \frac{2}{\pi}\sum_{n=1}^{\infty} \cos nx \cos ny = \frac{2}{\pi}\sum_{n=1}^{\infty} \sin nx \sin ny \\ &= \frac{1}{2\pi}\sum_{n=-\infty}^{\infty} \exp(in(x-y)).\end{aligned}$$

For complex variables z and w the formal expansion is written as

$$\delta(z-w) = \sum_{n=-\infty}^{\infty} w^n z^{-1-n} = \cdots + \frac{z}{w^2} + \frac{1}{w} + \frac{1}{z} + \frac{w}{z^2} + \ldots$$

For an analytic function $f(z)$, Dirac's delta exhibits the standard properties

$$\delta(z-w) = \delta(w-z), \quad f(z)\delta(z-w) = f(w)\delta(z-w),$$

$$(z-w)\delta(z-w) = 0, \quad \mathrm{Res}_{\,z=0} f(z)\delta(z-w) = f(w),$$

$$\mathrm{Res}_{\,z=0}\frac{(z-w)^m}{n!}\frac{\partial^n}{\partial w^n}\delta(z-w) = \delta_{m,n}, \; m,n \geq 0,$$

$$\frac{(z-w)^m}{n!}\frac{\partial^n}{\partial w^n}\delta(z-w) = \frac{1}{(n-m)!}\frac{\partial^{n-m}}{\partial w^{n-m}}\delta(z-w), \;\; 0 \geq m \geq n.$$

7.3.2 Energy-momentum tensor

The main starting point is that the correlator of a total derivative is zero. The simplest application of this postulate is

$$\begin{aligned}0 = \int DxDh \frac{\delta}{\delta x^\mu} e^{-S_P[h]} &= -\int DxDh e^{-S_P[h]} \frac{\delta}{\delta x^\mu} S_P[h] \\ &= -\langle \frac{\delta}{\delta x^\mu} S_P[h] \rangle = \langle \frac{\partial^2}{\partial\zeta\partial\bar\zeta} x^\mu \rangle.\end{aligned}$$

To analyse this equation in a distributional sense we insert arbitrary test functions defined in a punctured neighbourhood of ζ and analogously get

$$\langle \frac{\partial^2}{\partial\zeta\partial\bar\zeta} x^\mu \cdot \text{insertion} \rangle = 0,$$

or that $x^\mu(\zeta,\bar\zeta)$ is harmonic. For the partition function $\langle x\rangle$ we obtain

$$0 = \int DxDh \frac{\delta}{\delta x^\mu} x^\nu e^{-S_P[h]}$$
$$= \int DxDhe^{-S_P[h]} \left(\delta_{\mu\nu}\delta^2(\zeta - w, \bar\zeta - \bar w) + \frac{\partial^2}{\partial\zeta\partial\bar\zeta} x^\mu(\zeta,\bar\zeta)x^\nu(w,\bar w) \right),$$

or

$$\frac{\partial^2}{\partial\zeta\partial\bar\zeta} x^\mu(\zeta,\bar\zeta)x^\nu(w,\bar w) = -\delta^2(\zeta - w, \bar\zeta - \bar w),$$

in a distributional sense, where $\delta^2(\zeta,\eta) = \delta(\zeta)\delta(\eta)$.

Let us consider the infinitesimal infinite dimensional conformal coordinate transformation $d\zeta\, d\bar\zeta \to (1 + \frac{\partial}{\partial\zeta}\varepsilon(\zeta))(1 + \frac{\partial}{\partial\bar\zeta}\bar\varepsilon(\zeta))d\zeta\, d\bar\zeta$ with $\frac{\partial}{\partial\bar\zeta}\varepsilon(\zeta) = 0$. The condition of the conformal symmetry implies that the energy-momentum tensor of the worldsheet theory $T_{\alpha\beta} = T(\zeta,\bar\zeta)$ is traceless and symmetric. There are only two independent components which can be chosen to be $T(\zeta) = T_{\zeta\zeta} = T_{11} - T_{22} - 2iT_{12}$ and $\bar T(\bar\zeta) = T_{\bar\zeta\bar\zeta} = T_{11} - T_{22} + 2iT_{12}$ (the components $T_{\zeta\bar\zeta}$ and $T_{\bar\zeta\zeta}$ vanish). So, the energy-momentum tensor $T(\zeta,\bar\zeta)$ splits into left-moving and right-moving parts $T(\zeta,\bar\zeta) = T(\zeta) + \bar T(\bar\zeta)$. The conservation of $T(\zeta,\bar\zeta)$ is expressed as

$$\frac{\partial}{\partial\bar\zeta} T(\zeta) = 0, \quad \frac{\partial}{\partial\zeta} \bar T(\bar\zeta) = 0,$$

Considering only the holomorphic component $T(\zeta)$ we have also $\frac{\partial}{\partial\bar\zeta} T\varepsilon(\zeta) = 0$ that corresponds to an infinite number of conserved quantities.

Now we continue working with the initial postulate. The 2-D coordinate invariance implies

$$\sum_{k=1}^{n} \langle \phi_{j_1}(\zeta_1)\dots\delta_\varepsilon\phi_{j_k}(\zeta_k)\dots\phi_{j_n}(\zeta_n)\rangle = -\frac{1}{2\pi i}\oint d\zeta \langle T(\zeta)\phi_{j_1}(\zeta_1)\dots\phi_{j_n}(\zeta_n)\rangle\varepsilon(\zeta).$$

This is the basic Ward identity of 2-D QFT (additional terms appear for higher-genus Riemann surfaces, see [101]). The contour integration is performed enclosing possible singularities of $T(\zeta)$. From the Ward identities, one may get useful *identities, operator product expansions* and *transformation laws.*

- **Identities:**
$$\frac{\partial}{\partial\bar\zeta} T(\zeta) = \frac{\partial}{\partial\zeta} \bar T(\bar\zeta) = 0, \quad T_{\zeta\bar\zeta} = T_{\bar\zeta\zeta} = 0.$$

- **Operator product expansions:**
$$T(\zeta)\phi_{j_k}(\zeta_k) \sim \frac{\phi'_{j_k}(\zeta_k)}{\zeta - \zeta_k} + \frac{\Delta_k}{(\zeta - \zeta_k)^2}\phi_{j_k}(\zeta_k),$$
$$T(\zeta)T(w) \sim \frac{c}{2(\zeta - w)^4} + \frac{2}{(\zeta - w)^2}T(w) + \frac{1}{\zeta - w}\partial T(w).$$

- **Transformation laws:**

$$T(\zeta')(d\zeta')^2 = (T(\zeta) - \frac{c}{12}S_{\zeta'}(\zeta))(d\zeta)^2,$$

under conformal change of variables $\zeta'(\zeta)$. With

$$\delta_\varepsilon \phi_{j_k}(\zeta) = -(\Delta_k \varepsilon' + \varepsilon\partial)\phi_{j_k}(\zeta),$$

$\phi_{j_k}(\zeta)$ transform as primary (conformal) holomorphic fields of conformal weight Δ_k,

$$\phi_{j_k}(\zeta) \to \left(\frac{\partial\zeta'}{\partial\zeta}\right)^{\Delta_k} \phi_{j_k}(\zeta').$$

Here $S_{\zeta'}(\zeta)$ is the Schwarzian derivative and c appears as a parameter of the theory and is called the *central charge*. It can not be determined from any general principle. The points ζ and w are taken in the same holomorphic chart. Only the holomorphic part of Dirac's δ participates.

Let us introduce the operators L_n, $n = 0, \pm 1, \pm 2, \ldots$ as the coefficients of the formal Laurent expansion

$$T(\zeta) = \sum_{n=-\infty}^{\infty} \frac{L_n}{\zeta^{n+2}}.$$

Then

$$L_n = \oint \frac{d\zeta}{2\pi i \zeta} \zeta^{n+2} T(\zeta),$$

where integration is performed over any contour encircling the origin. The operators L_n are called *Virasoro generators*[4] [342]. The operator product of the Virasoro generators $L_n \cdot L_m$ can be expressed as

$$L_n L_m = \oint_0 \frac{d\zeta}{2\pi i} \zeta^{n+1} T(\zeta) \oint_0 \frac{dw}{2\pi i} w^{n+1} T(w),$$

where $|w| > |\zeta|$. This product does not commute and the Poisson brackets are

$$[L_n, L_m] = L_n L_m - L_m L_n = (n-m)L_{n+m} + \frac{c}{12} n(n^2-1)\delta_{n,-m}.$$

Here the contours cancel except for a contribution as $\zeta \to w$, and we take into account the operator product $T(\zeta)T(w)$. This is the famous *Virasoro algebra* of string theory.

[4] Miguel Ángel Virasoro, born in Argentina in 1940. In 1995, Miguel Virasoro from the University of Rome was appointed to succeed the Nobel Laureate in Physics Abdus Salam (1926–1996) as the director of the International Centre for Theoretical Physics, Trieste, Italy. He was the Centre's director until the end of May 2002.

7.3.3 Central extension of the Witt algebra

Surprisingly, the same algebra was discovered earlier by Gelfand and Fuchs [123] as a non-trivial central extension of the group of diffeomorphisms of the unit circle.

A Killing vector field (**Wilhelm Karl Joseph Killing**, 1847–1923) is a vector field on a Riemannian manifold that preserves the metric. Killing fields are the infinitesimal generators of isometries; that is, flows generated by Killing fields are continuous isometries of the manifold. A Witt algebra is the Lie algebra of Killing vector fields defined on the Riemann sphere. The basis for these Killing fields is given by the holomorphic fields

$$L_n = -z^{n+1}\frac{\partial}{\partial z}.$$

The Poisson bracket of two Killing fields is

$$[L_m, L_n] = (n-m)z^{m+n+1}\frac{\partial}{\partial z} = (m-n)L_{m+n}.$$

The Virasoro algebra is the central extension of the Witt algebra.

To analyze and construct this central extension we first consider the real situation over the unit circle. Let us consider the Lie algebra Vect S^1 of smooth vector fields on S^1,

$$\phi(\theta)\frac{d}{d\theta},$$

where $\phi(\theta+2\pi)=\phi(\theta)$. The Poisson bracket in Vect S^1 is given as $[\phi_1,\phi_2] = \phi_1\phi_2' - \phi_1'\phi_2$.

Let I and G be Lie algebras. An exact sequence is a sequence of objects and morphisms between them such that the image of one morphism equals the kernel of the next. Let us consider the exact sequence of Lie algebras

$$0 \longrightarrow I \xrightarrow{f} E \xrightarrow{g} G \longrightarrow 0.$$

E is called the *central extension* of G by I if I belongs to the center of E. The central extension is given as $E \simeq G \oplus I$. A simple example is $[x+a]_E = [x,y]_G + [a,b]_I$. The Virasoro algebra is the unique (up to isomorphism) non-trivial central extension of Vect S^1 by $\mathbb{R}$ given by the *Gelfand-Fuchs cocycle*:

$$\omega(\phi_1,\phi_2) = \frac{1}{2\pi}\int\limits_0^{2\pi}(\phi_1'\phi_2'' - \phi_1''\phi_2')d\theta.$$

The Virasoro algebra Vir is a Lie algebra over the space Vect $S^1 \oplus \mathbb{R}$ defined by the commutator

$$[(\phi_1,a),(\phi_2,b)]_{Vir} = ([\phi_1,\phi_2]_{\text{Vect } S^1}, \frac{c}{12}\omega(\phi_1,\phi_2)),$$

where a and b are elements of the center, $ab - ba$ vanishes, and c is the central charge. Integration by parts leads to the 2-cocycle condition

$$\omega(\phi_1, [\phi_2, \phi_3]) + \omega(\phi_2, [\phi_3, \phi_1]) + \omega(\phi_3, [\phi_1, \phi_2]) = 0,$$

and

$$\omega(\phi_1, \phi_2) = -\frac{1}{4\pi}\int_0^{2\pi} (\phi_1' + \phi_1''')\phi_2 d\theta.$$

Correspondingly, we consider the group Diff S^1 of diffeomorphisms of the circle preserving its orientation $\theta \to f(\theta)$, where $\theta(\text{mod } 2\pi)$ is a parameter on S^1, $f(\theta + 2\pi) = f(\theta)$. The *Virasoro-Bott group* is the unique (up to isomorphism) non-trivial central extension of Diff S^1 given by the Thurston-Bott cocycle [40]

$$\Omega(f, g) = \frac{1}{2\pi}\int_0^{2\pi} \log((f \circ g)')d\log(g').$$

The Virasoro-Bott group is given by the following product on Diff $S^1 \times \mathbb{R}$,

$$(f, \alpha)(g, \beta) = (f \circ g, \alpha + \beta + \frac{c}{12}\Omega(f, g)).$$

We shall identify Vect S^1 with the functions with vanishing mean value over S^1. So we have

$$\phi(\theta) = \sum_{n=1}^{\infty} a_n \cos n\theta + b_n \sin n\theta.$$

Let us define a complex structure by the operator

$$J(\theta) = \sum_{n=1}^{\infty} -a_n \sin n\theta + b_n \cos n\theta.$$

Then $J^2 = -id$. On Vect $S^1 \oplus \mathbb{C}$, the operator J diagonalizes and we have

$$\phi \to \phi - iJ(\phi) = \sum_{n=1}^{\infty} (a_n - ib_n)e^{in\theta},$$

and the latter extends into the unit disk as a holomorphic function.

Taking the basis of Vect $S^1 \oplus \mathbb{C}$ as $e_n = -ie^{in\theta}\partial$ we get

$$[e_n, e_m] = (n - m)e_{n+m} + \frac{c}{12}n(n^2 - 1)\delta_{n,-m}.$$

So this shows that the Virasoro algebra is realizable both as a central extension of the Witt algebra and as an algebra of the Virasoro generators of CFT.

7.4 Hele-Shaw worldsheet

String theory is especially good to treat dynamical processes. There are two approaches to the Hele-Shaw evolution from the bosonic string point of view.

7.4.1 Integrable hierarchies

The first one considers worldsheets given by the time evolution of the boundary of phase domains. This approach, developed in, e.g., [209], [239], [344], leads to the construction of a solution of the dispersionless $2D$ Toda hierarchy (see [207], [208], [316] and the references therein). This solution satisfies the Polubarinova-Galin equation that becomes the string equation familiar in topological gravity and matrix models of $2D$ gravity. The main idea is that one can characterize a closed analytic curve by a set of harmonic Richardson's moments of the exterior of the domain surrounded by the curve together with its area:

$$C_k = -\int_{\text{exterior}} z^{-k} d\sigma_z\,, \qquad k \geq 1; \qquad C_0 = \int_{\text{interior}} d\sigma_z.$$

Here $z = x + iy$ and it is understood that the point $z = 0$ is inside the domain, whereas the point $z = \infty$ is outside. The integrals at $k = 1, 2$ are assumed to be properly regularized. In order to find the harmonic moments of the interior

$$C_{-k} = \int_{\text{interior}} z^k d\sigma_z\,, \qquad k \geq 1,$$

the set of moments of the exterior (and their complex conjugate) are identified with the times of the dispersionless 2D Toda hierarchy as follows:

$$t \equiv t_0 = \frac{C_0}{\pi}, \qquad t_k = \frac{C_k}{\pi k}, \qquad \bar{t}_k = \frac{\bar{C}_k}{\pi k}, \qquad k \geq 1.$$

These parameters are treated as independent variables. The derivatives of the harmonic moments of the interior C_{-k} with respect to the t_j, $\bar{t}_j$ satisfy the symmetric lattice equation

$$\frac{\partial C_{-k}}{\partial t_j} = \frac{\partial C_{-j}}{\partial t_k}, \qquad \frac{\partial C_{-k}}{\partial \bar{t}_j} = \frac{\partial \bar{C}_{-j}}{\partial t_k}.$$

Wiegmann and Zabrodin [344] constructed a real-valued τ-funcion, a solution of the so-called Hirota equation, such that

$$\frac{C_{-k}}{\pi} = \frac{\partial \log \tau}{\partial t_k}, \qquad \frac{\bar{C}_{-k}}{\pi} = \frac{\partial \log \tau}{\partial \bar{t}_k}, \qquad k \geq 1.$$

The Hirota equation reads as

$$S_{f^{-1}}(z) = \frac{6}{z^2} \sum_{k,n=1}^{\infty} \frac{1}{z^{n+k}} \frac{\partial^2 \log \tau}{\partial t_k \partial t_n},$$

where $z = f(\zeta)$ is the parametric map of the unit disk onto the exterior phase domain. The left-hand side is just the analytic component of the energy-momentum tensor.

7.4.2 Integrable structure of the Dirichlet problem

The above questions of integrable hierarchies of interior and exterior moments are closely related to the integrable structure of the Dirichlet problem [231], [210].

Laplacian growth is a special kind of domain variation. Following [231] we introduce the notation $\nabla(a)$ for the directional derivative corresponding to Hele-Shaw injection at the point a. This derivative can be regarded as a tangent vector in the infinite-dimensional space of smooth domains containing a, and it acts on functionals

$$F : \{\text{smooth bounded domains containing the point } a\} \to \mathbb{C}$$

as follows. Given a smooth domain D containing a, let $\{\Omega(t) : 0 \leq t < \varepsilon\}$ be the Hele-Shaw evolution with $Q = -1$ and $\Omega(0) = D$, so that

$$\frac{d}{dt}\int_{\Omega(t)} \varphi d\sigma = -\int_{\partial\Omega(t)} \varphi(z)\frac{\partial G(z,a)}{\partial n} ds$$

for every smooth function φ, and where $G(z,\zeta) := G_{\Omega(t)}(z,\zeta)$ is the Green function of the domain. Then, by definition,

$$(\nabla(a)F)(D) := \left.\frac{d}{dt}\right|_{t=0} F(\Omega(t)).$$

One can show that any infinitesimal domain variation of a smooth domain D can be approximated by linear combinations of $\nabla(a)$ with $a \in D$. See [301] for a general theorem in this direction. Taking a close to the boundary, the variation corresponding to $\nabla(a)$, i.e., Hele-Shaw injection at a, produces a "bump" on the boundary at a.

Example 7.4.1. *Let h be a fixed harmonic function defined in a neighbourhood of the closure of D and define F by*

$$F(D) = \int_D h d\sigma.$$

Then

$$(\nabla(a)F)(D) = \left.\frac{d}{dt}\right|_{t=0} \int_{\Omega(t)} h d\sigma = -\int_{\partial D} h(z)\frac{\partial G_D(z,a)}{\partial n} ds = h(a).$$

Taking, e.g., $h(z) = z^k$ we get moments such as in the previous subsection as special cases of the functional F, with

$$\nabla(a)\int_D z^k d\sigma = a^k$$

for $k \geq 0$.

Example 7.4.2. *As a second example, choose two points $b, c \in D$ and take F to be*

$$F(D) = G_D(b, c).$$

For a general domain variation, the change of this functional is given by the Hadamard variational formula: following classical tradition, let δn denote an infinitesimal change of the boundary in the normal direction and let δG_D denote the corresponding change of the Green function. Then the Hadamard formula is

$$\delta G_D(b, c) = -\int_{\partial D} \delta n \frac{\partial G(\cdot, b)}{\partial n} \frac{\partial G(\cdot, c)}{\partial n} ds.$$

On dividing with a corresponding infinitesimal time interval δt we may view $\delta n/\delta t$ as the normal velocity of the boundary. In our case (Hele-Shaw injection at a) we have

$$\frac{\delta n}{\delta t} = -\frac{\partial G(\cdot, a)}{\partial n}.$$

Thus, we get the beautiful formula

$$\nabla(a) G_D(b, c) = \int_{\partial D} \frac{\partial G(\cdot, a)}{\partial n} \frac{\partial G(\cdot, b)}{\partial n} \frac{\partial G(\cdot, c)}{\partial n} ds. \tag{7.4}$$

In particular we see that $\nabla(a) G_D(b, c)$ is completely symmetric in a, b and c.

The symmetry in the above formula (7.4) calls for an explanation. This will be provided in the next example, where an explicit version of the τ-function appears. We first remark that the Richardson conservation of moments for Hele-Shaw evolutions, which in a stronger form is built into the concept of weak solution and into the balayage formulation, contains an integrability statement: Hele-Shaw injection at a point a followed by injection at a point b gives the same result as if the injections had been performed in the opposite order. In infinitesimal form this statement becomes

$$\nabla(a)\nabla(b) = \nabla(b)\nabla(a).$$

Therefore, the above symmetry would be explained if we could find a functional E, such that

$$G(b, c) = \nabla(b)\nabla(c) E + \text{constant}.$$

Such a functional indeed exists and can be realized as a renormalized energy of the complementary domain, and which can also be identified with the logarithm of the τ-function.

Example 7.4.3. *Let B be a fixed bounded domain containing the closure of D and define*

$$E(D) = \frac{1}{4\pi} \int_{B \setminus D} \int_{B \setminus D} \log |z - \zeta| d\sigma_z d\sigma_\zeta.$$

In this functional the domain D occurs symmetrically at two places. Thus, the result of $\nabla(c)$ acting on D in $E(D)$ will be twice the result of its action on only one of the occurrences. This gives

$$\begin{aligned}\nabla(c)E(D) &= \frac{1}{2\pi}\frac{d}{dt}\bigg|_{t=0}\int_{B\setminus D}\int_{B\setminus\Omega(t)}\log|z-\zeta|d\sigma_\zeta d\sigma_z\\ &= \frac{1}{2\pi}\int_{B\setminus D}\int_{\partial D}\log|z-\zeta|\frac{\partial G_D(\zeta,c)}{\partial n}ds_\zeta d\sigma_z\\ &= -\frac{1}{2\pi}\int_{B\setminus D}\log|z-c|d\sigma_z = U^{B\setminus D}(c).\end{aligned}$$

Here we used that $\log|z-\zeta|$ is harmonic in D in the variable ζ when $z\in B\setminus D$. Recall that $U^{B\setminus D}$ denotes the potential of the set $B\setminus D$.

In order to continue we decompose the Green function $G(z,\zeta)$ as

$$G(z,\zeta) = -\frac{1}{2\pi}\log|z-\zeta| + H(z,\zeta)$$

with $H(z,\zeta)$ harmonic in each variable. Then, applying $\nabla(b)$ to the above and using that $G(z,c)=0$ on ∂D we obtain

$$\begin{aligned}\nabla(b)\nabla(c)E(D) &= -\frac{1}{2\pi}\int_{\partial D}\log|z-c|\frac{\partial G(\zeta,b)}{\partial n}ds_\zeta d\sigma_z\\ &= \frac{1}{2\pi}\int_{\partial D}(G(z,c)-H(z,c))\frac{\partial G(\zeta,b)}{\partial n}ds_\zeta d\sigma_z\\ &= H(b,c) = G(b,c) + \frac{1}{2\pi}\log|b-c|.\end{aligned}$$

In other words,

$$\nabla(b)\nabla(c)E(D) = G_D(b,c) + C(b,c),$$

where $C(b,c)$ is a constant with respect to variations of D. Applying finally $\nabla(a)$ we get

$$\nabla(a)\nabla(b)\nabla(c)E(D) = \nabla(a)G_D(b,c),$$

and the symmetry of $\nabla(a)G_D(b,c)$ is explained.

The τ-function [206], originally introduced in the context of integrable hierarchies, can in the present situation be chosen so that

$$\log\tau(D) = E(D).$$

One may also consider higher order derivatives, like defining $\nabla^{(k)}(a)$ by

$$\nabla^{(k)}(a)F(D) = \frac{\partial^k}{\partial a^k}\nabla(a)F(D).$$

This gives, e.g.,

$$\nabla^{(k)}(0)\int_D z^j d\sigma_z = k!\delta_{kj},$$

$$\nabla^{(k)}(0)E(D) = \text{constant}\int_{B\setminus D} z^{-k} d\sigma_z.$$

To connect to the results in the preceding subsection one should take D to be the exterior (unbounded) domain. Then the derivatives $\frac{\partial}{\partial t_k}$ will correspond to $\nabla^{(k)}(\infty)$ (suitably interpreted) and the statements there may be derived using the commutativity of the ∇-derivatives.

7.4.3 Hele-Shaw worldsheet

We propose another approach that treats the phase domain in the Hele-Shaw evolution as a worldsheet. Given the velocity field within the Hele-Shaw phase domain we consider the evolution of the level lines of the Green function (orthogonal to the stream lines) at each fixed moment t. Stream lines are solutions to the autonomous differential equation $dz/ds = v(z(s))$, where the conditions are given at the final point $s = T$, $z(T) \in \Gamma(t)$, where $\Gamma(t)$ is the boundary of the phase domain $\Omega(t)$ at the moment t. There is a problem at the initial point $s = 0$ because the origin is a singular point of the flow field v.

Let us consider the complex potential $W(z,t)$, $\operatorname{Re} W = p$. For each fixed t it is a multi-valued analytic function defined in $\Omega(t)$ whose real part solves the Dirichlet problem

$$\Delta p = -2\pi\delta_0(z), \quad z \in \Omega(t) \tag{7.5}$$

with the boundary condition

$$p\Big|_{\Gamma(t)} = 0, \tag{7.6}$$

so it is the Green function of the domain $\Omega(t)$ with a singular point at the origin. The non-trivial motion regarding time t is given by the kinematic condition on $\Gamma(t)$. We have

$$\frac{\partial W}{\partial z} = \frac{\partial p}{\partial x} - i\frac{\partial p}{\partial y}, \quad z = x + iy.$$

Since Green's function solves (7.5)–(7.6), we have the representation

$$W(z,t) = -\log z + w_0(z,t), \tag{7.7}$$

where $w_0(z,t)$ is an analytic regular function in $\Omega(t)$. Because of the conformal invariance of Green's function we have the superposition

$$(W \circ f)(\zeta, t) = -\log \zeta,$$

and the conformally invariant *complex velocity* is just $W'(z,t) = -\frac{f^{-1'}}{f^{-1}}(z,t)$, where $\zeta = f^{-1}(z,t)$ is the inverse to our parametric function f and prime means the complex derivative. Rewriting this relation we get

$$(W'(z,t)\,dz)^2 = \frac{d\zeta^2}{\zeta^2}. \tag{7.8}$$

The velocity field $(-\nabla p)$ is the conjugation of $(-W')$. In other words the velocity field is directed along the trajectories of the quadratic differential in the left-hand side of (7.8) for each fixed moment t. The equality (7.8) implies that the boundary $\partial\Omega(t)$ is the orthogonal trajectory of the differential $(W'(z,t)\,dz)^2$ with a double pole at the origin. The dependence on t yields that the trajectory structure of this differential changes in time, and in general, the stream lines are not inherited in time. These lines are geodesic in the conformal metric $|W'(z,t)||dz|$ generated by this differential.

We plan to construct a field theory for the Hele-Shaw evolution somehow analogous to the Liouville theory. However, the Liouville theory itself does not work directly any longer in our case. It is not directly connected with the classical Green function; symmetries given by the Poincaré group are not those relevant to the Hele-Shaw dynamics. Moreover, observe that being thought of as a Riemann surface, the phase domain $\Omega(t)\setminus\{0\}$ is hyperbolic and it admits the Poincaré metric with constant negative curvature

$$ds^2 = \frac{|f^{-1'}|^2}{|f^{-1}|^2 \log^2|f^{-1}|}|dz|^2.$$

The asymptotics about the origin and close to the hyperbolic boundary are such that the standard expression (7.3) (the principal term of it) for the Liouville action can not be used any more. Moreover, the action integral for the hyperbolic metric in the punctured unit disk has the following asymptotics:

$$\begin{aligned}\int\limits_{U^{(\varepsilon)}} (|\phi_z|^2 + e^{\phi})d\sigma_z \quad &\sim \quad -2\pi\log\varepsilon_1 - 4\pi\log|\log\varepsilon_1| \\ &+ \quad 4\pi\log|\log(1-\varepsilon_2)| - \frac{4\pi}{\log(1-\varepsilon_2)},\end{aligned}$$

where $U^{(\varepsilon)} = \{z : \varepsilon_1 < |z| < 1-\varepsilon_2\}$. Therefore, the corresponding integral for $\Omega(t)\setminus\{0\}$ will have a similar asymptotics plus terms containing $f'(0,t)$ and the boundary distortion (in our case the boundary derivatives) at the unit circle, which makes it difficult to operate with.

Let us instead use the flat *logarithmic metric* generated by (7.8) which is natural for the Laplacian growth

$$ds^2 = \frac{|f^{-1'}|^2}{|f^{-1}|^2}|dz|^2 = |W'|^2|dz|^2.$$

The hyperbolic boundary is not singular for this metric whereas the origin is. But it is a parabolic singularity which can be easily regularized.

The density of this metric satisfies the usual Laplacian equation $\varphi_{z\bar{z}} = 0$ in $\Omega(t) \setminus \{0\}$, where

$$ds^2 = e^{\varphi(z)}|dz|^2, \quad \varphi(z) = \log \frac{|f^{-1\prime}|^2}{|f^{-1}|^2}.$$

Obviously, the Laplacian equation is the Euler-Lagrange equation for the variational problem defined by the Dirichlet integral

$$\int_D |\phi_z|^2 d\sigma_z,$$

locally for any measurable set $D \subset \Omega(t) \setminus \{0\}$. However, this functional cannot be defined globally in $\Omega(t) \setminus \{0\}$ because of the parabolic singularity at the origin. To overcome this obstacle we define the classical action in the following way. Let $\Omega_\varepsilon(t) = \Omega(t) \setminus \{z : |z| \leq \varepsilon\}$ for a sufficiently small ε, $U^{(\varepsilon)} = \{\zeta : \varepsilon < |\zeta| < 1\}$. The function φ possesses the asymptotics

$$\varphi \sim \log \frac{1}{|z|^2}, \quad |\varphi_z|^2 \sim \frac{1}{|z|^2} \quad \text{as } z \to 0.$$

Therefore, the finite limit

$$\lim_{\varepsilon \to 0} \left\{ \int_{\Omega_\varepsilon(t)} |\varphi_z|^2 d\sigma_z + 2\pi \log \varepsilon \right\} =: \mathcal{S}[\varphi] \tag{7.9}$$

exists and we call it the *logarithmic action* for the Laplacian growth.

Remark 7.4.4. *As in the Liouville theory we have no reason to discard the conformal factor (Weyl rescaling) and we have the rotation invariance instead of the Poincaré group. Another important issue is that the infinitesimal structure at the boundary is invariant under the mapping f, i.e., the stream lines within the unit disk are radial, orthogonal to the unit circle and that they remain orthogonal under the conformal map f. This means that the boundary conditions are satisfied.*

Remark 7.4.5. *The logarithmic action represents the reduced area of the Hele-Shaw worldsheet in the Riemannian metric ds^2.*

Remark 7.4.6. *The potential $\varphi(z)$ possesses a particular geometric meaning. It gives rise to a vector field $\nabla\varphi$, whose projection onto each normal outward vector to the level line of Green's function is the curvature of this line, $\partial\varphi/\partial n = \varkappa$.*

Lemma 7.4.7. *The Euler-Lagrange equation for the variational problem for the logarithmic action $\mathcal{S}[\phi]$ is the Laplacian equation $\Delta\phi = -4\pi\delta_0(z)$, $z \in \Omega(t)$, where $\delta_0(z)$ is the Dirac distribution supported at the origin, where ϕ is taken from the class of twice differentiable functions in $\Omega(t) \setminus \{0\}$ with the asymptotics $\phi \sim -\log|z|^2$ as $z \to 0$.*

Proof. Let us consider first the integral

$$\mathcal{S}_\varepsilon[\phi] = \int\limits_{\Omega_\varepsilon(t)} |\phi_z|^2 d\sigma_z = \int\limits_{\mathbb{C}} \chi_{\Omega_\varepsilon(t)} |\phi_z|^2 d\sigma_z,$$

where $\chi_{\Omega_\varepsilon(t)}$ is the characteristic function of $\Omega_\varepsilon(t)$. Then, due to Green's theorem,

$$\begin{aligned} \lim_{h\to 0} \frac{\mathcal{S}_\varepsilon[\phi + hu] - \mathcal{S}_\varepsilon[\phi]}{h} &= 2\int\limits_{\mathbb{C}} \chi_{\Omega_\varepsilon(t)} \mathrm{Re}\, \phi_z \overline{u_z}\, d\sigma_z \\ &= -\frac{1}{2} \int\limits_{\Omega_\varepsilon(t)} u\Delta\phi\, d\sigma_z + \frac{1}{2} \int\limits_{\partial\Omega_\varepsilon(t)} u\frac{\partial\phi}{\partial n}\, ds, \qquad (7.10) \end{aligned}$$

for every $C^\infty(\mathbb{C})$ test function u supported in $\Omega(t)$. On the other hand, we have $\partial\phi/\partial n \sim -2/\varepsilon$ as $\varepsilon \to 0$ and $u = 0$ on $\partial\Omega(t)$. Therefore, the expression (7.10) tends to

$$-\frac{1}{2} \int\limits_{\Omega(t)} u\Delta\phi d\sigma_z - 2\pi u(0),$$

as $\varepsilon \to 0$, and the latter must vanish, that is equivalent to the Laplacian equation mentioned in the statement of the lemma. Obviously, the logarithmic term in the definition of $\mathcal{S}[\phi]$ does not contribute to the variation. □

Remark 7.4.8. *The boundary condition for the potential ϕ is $\phi(e^{i\theta}) = \log v_n$, where v_n is the normal boundary velocity of the Hele-Shaw evolution.*

Remark 7.4.9. *The above Dirichlet problem is related to an interesting property of plane domains discussed in [305]. If $\omega(z)$ is harmonic in a domain Ω and satisfies on its boundary the relation*

$$\frac{\partial^2\omega}{\partial s^2} = \varkappa \frac{\partial \omega}{\partial n},$$

then

$$-i\frac{\partial^2\omega}{\partial z^2} dz^2$$

is a quadratic differential on Ω.

Returning to the calculation of φ we get

$$\varphi_z = \frac{-1}{f'}\left(\frac{f''}{f'} + \frac{1}{\zeta}\right) \circ f^{-1}(z,t).$$

Hence, the action $\mathcal{S}$ can be expressed in terms of the parametric function as

$$\mathcal{S}[\varphi] \equiv \mathcal{S}[f] = \lim_{\varepsilon\to 0} \left\{ \int\limits_{U_\varepsilon} \left| \frac{f''}{f'} + \frac{1}{\zeta} \right|^2 d\sigma_\zeta + 2\pi \log\varepsilon \right\} + 2\pi \log|f'(0,t)|, \qquad (7.11)$$

or adding the logarithmic term in the integral we obtain

$$\mathcal{S}[f] = \int\limits_U \left(\left| \frac{f''}{f'} + \frac{1}{\zeta} \right|^2 - \frac{1}{|\zeta|^2} \right) d\sigma_\zeta + 2\pi \log |f'(0,t)|. \tag{7.12}$$

The functional (7.12) resembles the *universal Liouville action* defined by Takhtajan and Teo in [319]–[321] for quasicircles, which is based on conformal maps from the unit disk and from its exterior. However, (7.12) can not be directly reduced to the Liouville action.

Observe that the classical kinetic energy $\mathcal{E}$ for the harmonic potential p is calculated by the Dirichlet integral

$$\int\limits_D |2p_z|^2 d\sigma_z = \int\limits_D |W'|^2 d\sigma_z,$$

locally for any measurable set $D \subset \Omega(t) \setminus \{0\}$. However, this integral again cannot be defined globally in $\Omega(t) \setminus \{0\}$. Treating $\mathcal{E}$ in the same way as $\mathcal{S}$ we come to the finite limit

$$\mathcal{E} := \lim_{\varepsilon \to 0} \left\{ \int\limits_{\Omega_\varepsilon(t)} |W'|^2 d\sigma_z + 2\pi \log \varepsilon \right\},$$

or in terms of the parametric function f,

$$\mathcal{E} = \mathcal{E}[f] = 2\pi \log |f'(0,t)|.$$

The latter expression allows us to think of $\mathcal{E}$ as a capacity which exactly corresponds to the physical sense of $\mathcal{E}$ as minimal energy.

7.4.4 Variation of the logarithmic action

According to the construction the metric potential satisfies the Laplacian equation (singular at the origin) which is just the Euler-Lagrange equation for the logarithmic action. However, the non-trivial Hele-Shaw motion implies dependence of the entire structure on time. So now we aim to calculate the time-variation of the logarithmic action.

The Hele-Shaw problem being rewritten for the parametric time dependent function $f : U \to \Omega(t)$ admits the form of the Polubarinova-Galin equation, which is in principle, a reformulation of the kinematic condition. Namely, the relation

$$v_n = -\frac{\partial p}{\partial n}, \quad z \in \Gamma(t),$$

is equivalent to the fact that the function $f(\zeta, t)$ satisfies the non linear first-order partial differential equation

$$\operatorname{Re} \left(\dot{f}\, \overline{\zeta f'} \right) = 1, \quad |\zeta| = 1, \tag{7.13}$$

with the initial condition $f(\zeta,0) = f_0(\zeta)$. We denote by S_f the Schwarzian derivative

$$S_f = \frac{f'''}{f'} - \frac{3}{2}\left(\frac{f''}{f'}\right)^2,$$

and by

$$\varkappa(\theta,t) = \frac{\mathrm{Re}\left(1+\frac{e^{i\theta}f''}{f'}\right)}{|f'(e^{i\theta},t)|}$$

the curvature of the boundary $\partial\Omega(t)$ at the point $f(e^{i\theta},t)$.

Theorem 7.4.10. *Let $z = f(\zeta,t)$ be the parametric function for the Laplacian growth, $\mathcal{E}[f]$ the kinetic energy, and $\mathcal{S}[f]$ the logarithmic action. Then*

$$\frac{d}{dt}\mathcal{E}[f] = \int\limits_0^{2\pi} \frac{1}{|f'(e^{i\theta},t)|^2}d\theta,$$

and

$$\frac{d}{dt}(\mathcal{S}[f]+\mathcal{E}[f]) = 2\int\limits_0^{2\pi} \varkappa^2(\theta,t)\,d\theta + \int\limits_0^{2\pi} \frac{2}{|f'(e^{i\theta},t)|^2}\mathrm{Re}\,(e^{2i\theta}S_f)d\theta.$$

Proof. Making use of the Cauchy-Schwarz representation we extend equation (7.13) into the unit disk

$$\dot f = \zeta f' p(\zeta,t), \tag{7.14}$$

where

$$p(\zeta,t) = \frac{1}{2\pi}\int\limits_0^{2\pi} \frac{1}{|f'(e^{i\theta},t)|^2}\frac{e^{i\theta}+\zeta}{e^{i\theta}-\zeta}\,d\theta. \tag{7.15}$$

Immediately, we obtain that

$$\frac{d}{dt}\mathcal{E}[f] = \frac{d}{dt}2\pi\log|f'(0,t)| = \int\limits_0^{2\pi} \frac{1}{|f'(e^{i\theta},t)|^2}d\theta,$$

and hence,

$$\begin{aligned}\frac{d}{dt}\mathcal{S}[f] &= 2\mathrm{Re}\int\limits_U \left(\overline{\frac{f''}{f'}}+\frac{1}{\bar\zeta}\right)\left((1+\zeta\frac{f''}{f'})p(\zeta,t)+\zeta p'(\zeta,t)\right)' d\sigma_\zeta \\ &+ \int\limits_0^{2\pi} \frac{1}{|f'(e^{i\theta},t)|^2}d\theta. \end{aligned} \tag{7.16}$$

The integral in the first term of the right-hand side of (7.16) we rewrite by Green's theorem as

$$I=\frac{-1}{2i}\int\limits_{S^1}\left(\overline{\frac{f''}{f'}}+\frac{1}{\bar{\zeta}}\right)\left((1+\zeta\frac{f''}{f'})p(\zeta,t)+\zeta p'(\zeta,t)\right)d\bar{\zeta}-\frac{1}{2}\int\limits_0^{2\pi}\frac{1}{|f'(e^{i\theta},t)|^2}d\theta,$$

taking into account a singularity at the origin. Applying the Cauchy-Schwarz formula to the first term in I containing p we arrive at

$$\begin{aligned}2\operatorname{Re}I &= \int\limits_0^{2\pi}\left|1+e^{i\alpha}\frac{f''(e^{i\alpha},t)}{f'(e^{i\alpha},t)}\right|^2\frac{d\alpha}{|f'(e^{i\alpha},t)|^2}\\ &\quad+\operatorname{Re}\int\limits_0^{2\pi}\left(\overline{1+e^{i\alpha}\frac{f''(e^{i\alpha},t)}{f'(e^{i\alpha},t)}}\right)e^{i\alpha}p'(e^{i\alpha},t)d\alpha-\int\limits_0^{2\pi}\frac{1}{|f'(e^{i\theta},t)|^2}d\theta,\end{aligned}$$

or

$$\begin{aligned}\frac{d}{dt}S[f]=2\operatorname{Re}I+\int\limits_0^{2\pi}\frac{1}{|f'(e^{i\theta},t)|^2}d\theta &= \int\limits_0^{2\pi}\left|1+e^{i\alpha}\frac{f''}{f'}\right|^2\frac{d\alpha}{|f'(e^{i\alpha},t)|^2}\\ &\quad+\int\limits_0^{2\pi}\operatorname{Re}\left(1+e^{i\alpha}\frac{f''}{f'}\right)\operatorname{Re}e^{i\alpha}p'(e^{i\alpha},t)\,d\alpha\\ &\quad+\int\limits_0^{2\pi}\operatorname{Im}\left(1+e^{i\alpha}\frac{f''}{f'}\right)\operatorname{Im}e^{i\alpha}p'(e^{i\alpha},t)\,d\alpha.\end{aligned}$$

These equalities are thought of as limiting values making use of the analyticity of f on the boundary and $f'(\zeta,t)\neq 0$ for all ζ in the closure of the unit disk. Let us denote in the latter expression by J_2 the last integral and by J_1 the intermediate one. We have,

$$J_2=\int\limits_0^{2\pi}\operatorname{Im}\left(1+e^{i\alpha}\frac{f''}{f'}\right)\operatorname{Im}\frac{1}{2\pi}\int\limits_0^{2\pi}\frac{1}{|f'(e^{i\theta},t)|^2}\frac{2e^{i\theta}e^{i\alpha}}{(e^{i\theta}-e^{i\alpha})^2}\,d\theta\,d\alpha.$$

Obviously,

$$\frac{\partial}{\partial\theta}\left(\frac{e^{i\theta}+\zeta}{e^{i\theta}-\zeta}\right)=\frac{-2\,\zeta ie^{i\theta}}{(e^{i\theta}-\zeta)^2},\quad\text{and}\quad\frac{\partial}{\partial\theta}\left(\frac{1}{|f'(e^{i\theta},t)|^2}\right)=\frac{2}{|f'(e^{i\theta},t)|^2}\operatorname{Im}\frac{e^{i\theta}f''}{f'}.$$

Integrating by parts and applying the Cauchy-Schwarz formula again we obtain

$$J_2=-2\int\limits_0^{2\pi}\left(\operatorname{Im}\left(1+e^{i\theta}\frac{f''}{f'}\right)\right)^2\frac{d\theta}{|f'(e^{i\theta},t)|^2}.$$

Now we turn to J_1,

$$J_1 = \int_0^{2\pi} \mathrm{Re}\left(1 + e^{i\alpha}\frac{f''}{f'}\right) \mathrm{Re}\,\frac{1}{2\pi}\int_0^{2\pi} \frac{1}{|f'(e^{i\theta},t)|^2}\frac{2e^{i\theta}e^{i\alpha}}{(e^{i\theta}-e^{i\alpha})^2}\,d\theta\,d\alpha.$$

Here we change the order of integration and get

$$J_1 = \int_0^{2\pi} \frac{1}{|f'(e^{i\theta},t)|^2}\mathrm{Re}\,\frac{1}{2\pi}\int_0^{2\pi} \mathrm{Re}\left(1 + e^{i\alpha}\frac{f''}{f'}\right)\frac{2e^{i\theta}e^{i\alpha}}{(e^{i\theta}-e^{i\alpha})^2}\,d\alpha\,d\theta.$$

Integrating by parts we obtain

$$J_1 = \int_0^{2\pi} \frac{1}{|f'(e^{i\theta},t)|^2}\mathrm{Re}\,\frac{-i}{2\pi}\int_0^{2\pi} \frac{\partial}{\partial\alpha}\mathrm{Re}\left(1 + e^{i\alpha}\frac{f''}{f'}\right)\frac{e^{i\alpha}+e^{i\theta}}{e^{i\alpha}-e^{i\theta}}\,d\alpha\,d\theta.$$

The Cauchy formula gives

$$J_1 = \int_0^{2\pi} \frac{1}{|f'(e^{i\theta},t)|^2}\mathrm{Re}\left(\frac{e^{i\theta}f''}{f'} + \frac{e^{2i\theta}f'''}{f'} - \left(\frac{e^{i\theta}f''}{f'}\right)^2\right)d\theta,$$

or

$$J_1 = \int_0^{2\pi} \frac{1}{|f'(e^{i\theta},t)|^2}\mathrm{Re}\left(\frac{1}{2}\left(1+\frac{e^{i\theta}f''}{f'}\right)^2 + e^{2i\theta}S_f - \frac{1}{2}\right)d\theta,$$

where S_f is the Schwarzian derivative.

Summing up all these integrals we come to the conclusion that

$$\begin{aligned}\frac{d}{dt}\mathcal{S}[f] &= \int_0^{2\pi} \frac{1}{|f'(e^{i\theta},t)|^2}\mathrm{Re}\left(\frac{3}{2}\left(1+\frac{e^{i\theta}f''}{f'}\right)^2 + e^{2i\theta}S_f + \frac{1}{2}\right)d\theta \\ &\quad - \int_0^{2\pi} \frac{1}{|f'(e^{i\theta},t)|^2}\,d\theta.\end{aligned}$$

We observe that

$$\varkappa(\theta,t) = \frac{\mathrm{Re}\left(1+\frac{e^{i\theta}f''}{f'}\right)}{|f'(e^{i\theta},t)|}$$

is the curvature of the boundary $\partial\Omega(t)$ at the point $f(e^{i\theta},t)$. Integration by parts implies

$$\int_0^{2\pi} \frac{1}{|f'(e^{i\theta},t)|^2}\left(\operatorname{Im}\left(1+\frac{e^{i\theta}f''}{f'}\right)\right)^2 d\theta$$
$$= \frac{-1}{2}\int_0^{2\pi} \frac{1}{|f'(e^{i\theta},t)|^2}\operatorname{Re}\left(\frac{1}{2}\left(1+\frac{e^{i\theta}f''}{f'}\right)^2 + e^{2i\theta}S_f - \frac{1}{2}\right) d\theta.$$

So

$$\begin{aligned}\frac{d}{dt}\mathcal{S}[f] &= \frac{3}{2}\int_0^{2\pi} \varkappa^2(\theta,t)\,d\theta - \frac{3}{2}\int_0^{2\pi}\frac{1}{|f'(e^{i\theta},t)|^2}\left(\operatorname{Im}\left(1+\frac{e^{i\theta}f''}{f'}\right)\right)^2 d\theta \\ &+ \int_0^{2\pi}\frac{1}{|f'(e^{i\theta},t)|^2}\operatorname{Re}\left(e^{2i\theta}S_f+\frac{1}{2}\right)d\theta - \int_0^{2\pi}\frac{1}{|f'(e^{i\theta},t)|^2}d\theta \\ &= (\frac{3}{2}+\frac{3}{8})\int_0^{2\pi} \varkappa^2(\theta,t)\,d\theta - \frac{3}{8}\int_0^{2\pi}\frac{1}{|f'(e^{i\theta},t)|^2}\left(\operatorname{Im}\left(1+\frac{e^{i\theta}f''}{f'}\right)\right)^2 d\theta \\ &+ \int_0^{2\pi}\frac{1}{|f'(e^{i\theta},t)|^2}\operatorname{Re}\left((1+\frac{3}{4})e^{2i\theta}S_f+(\frac{1}{2}-\frac{3}{8})\right)d\theta - \int_0^{2\pi}\frac{1}{|f'(e^{i\theta},t)|^2}d\theta.\end{aligned}$$

Repeating this step we get

$$\begin{aligned}\frac{d}{dt}\mathcal{S}[f] &= \frac{3}{2}(1+\frac{1}{4}+\dots+\frac{1}{4^n})\int_0^{2\pi} \varkappa^2(\theta,t)\,d\theta \\ &-\frac{3}{2}\frac{1}{4^n}\int_0^{2\pi}\frac{1}{|f'(e^{i\theta},t)|^2}\left(\operatorname{Im}\left(1+\frac{e^{i\theta}f''}{f'}\right)\right)^2 d\theta \\ &+\int_0^{2\pi}\frac{1}{|f'(e^{i\theta},t)|^2}\operatorname{Re}\left((1+\frac{3}{4}+\dots+\frac{3}{4^n})e^{2i\theta}S_f+(\frac{1}{2}-\frac{3}{8}-\dots-\frac{3}{2}\frac{1}{4^n})\right)d\theta \\ &-\int_0^{2\pi}\frac{1}{|f'(e^{i\theta},t)|^2}d\theta,\end{aligned}$$

at n-th iteration. Taking the limit as $n\to\infty$, we finally obtain

$$\frac{d}{dt}\mathcal{S}[f] = 2\int_0^{2\pi} \varkappa^2(\theta,t)\,d\theta + \int_0^{2\pi}\frac{2}{|f'(e^{i\theta},t)|^2}\operatorname{Re}(e^{2i\theta}S_f)d\theta - \int_0^{2\pi}\frac{1}{|f'(e^{i\theta},t)|^2}d\theta,$$

as claimed in the statement of the theorem. □

In the simplest case of the circular evolution $f(\zeta,t)=\sqrt{2t}\zeta$ we have $\frac{d}{dt}\mathcal{S}[f]=\frac{d}{dt}\mathcal{E}[f]=\frac{\pi}{t}$.

7.4.5 Virasoro generators and Neretin polynomials

The entire necessary background for construction of the theory of unitary representations of Diff S^1 is found in the study of Kirillov's homogeneous Kählerian manifold $\mathcal{M}=$ Diff S^1/S^1, see [199]–[201]. The group Diff S^1 acts as a group of translations on the manifold $\mathcal{M}$ with the group S^1 as a stabilizer. The Kählerian geometry of $\mathcal{M}$ was described by Kirillov and Yuriev in [199]. The manifold $\mathcal{M}$ admits several representations, in particular, in the space of smooth probability measures, symplectic realization in the space of quadratic differentials. Let S stand for the usual class of all analytic univalent functions f in U normalized by $f(0)=0$, $f'(0)=1$. We shall use the analytic representation of $\mathcal{M}$ based on the class $\tilde{S}$ of functions from S which, being extended onto the closure $\overline{U}$ of U, are supposed to be smooth on S^1. The class $\tilde{S}$ is dense in S in the local uniform topology of U. There exists a canonical identification of $\tilde{S}$ with $\mathcal{M}$. As a consequence, $\tilde{S}$ is a homogeneous space under the left action of Diff S^1, see [8, Theorem 1.4.1] and [201]–[200].

The manifold $\mathcal{M}$ being considered as a realization $\tilde{S}$ admits affine coordinates $\{c_2,c_3,\dots\}$, where c_k is the k-th coefficient of a univalent function $f\in\tilde{S}$. Due to de Branges' theorem [43], $\mathcal{M}$ is a bounded open subset of $\{|c_k|<k+\varepsilon\}$.

The Goluzin-Schiffer variational formula lifts the actions from the Lie algebra Vect S^1 onto $\tilde{S}$. Let $f\in\tilde{S}$ and let $\nu(e^{i\theta})$ be a C^∞ real-valued function in $\theta\in(0,2\pi]$ from Vect S^1 making an infinitesimal action as $\theta\mapsto\theta+\tau\nu(e^{i\theta})$. Let us consider a variation of f given by

$$L_\nu[f](\zeta)=-\frac{f^2(\zeta)}{2\pi i}\int\limits_{S^1}\left(\frac{wf'(w)}{f(w)}\right)^2\frac{\nu(w)}{f(w)-f(\zeta)}\frac{dw}{w}. \tag{7.17}$$

Kirillov and Yuriev [199, 200] have established that the variations $L_\nu[f](\zeta)$ are closed with respect to the commutator and the induced Lie algebra is the same as Vect S^1. Moreover, Kirillov's result [201] states that there is the exponential map Vect $S^1\to$ Diff S^1 such that the subgroup S^1 coincides with the stabilizer of the map $f(\zeta)\equiv\zeta$ from $\tilde{S}$.

It is convenient [202] to extend (7.17) by complex linearity to $\mathbb{C}$Vect $S^1\to$ Vect $\tilde{S}$. Taking $\nu_k=-ie^{ik\theta}\partial$, $k\geq 0$ from the basis of $\mathbb{C}$Vect S^1, we obtain the expressions for $L_k=\delta_\nu f$, $f\in\tilde{S}\simeq\mathcal{M}$ (see formula (7.17)), as

$$L_0(f)=\zeta f'(\zeta)-f(\zeta),\quad L_k(f)=\zeta^{1+k}f'(\zeta).$$

The computation of L_k for $k<0$ is more difficult because poles of the integrand. For example,

$$L_{-1}(f)=f'(\zeta)-1-2c_2f(\zeta),\quad L_{-2}(f)=\frac{f'(\zeta)}{\zeta}-\frac{1}{f(\zeta)}-3c_2+(c_2^2-4c_3)f(\zeta),$$

(see, e.g., [202]). In terms of the coordinates $\{c_2,c_3,\dots\}$ on $\mathcal{M}$,

$$L_k=\partial_k+\sum_{n=1}^{\infty}(n+1)c_n\partial_{k+n},\quad L_0=\sum_{n=1}^{\infty}nc_{n+1}\partial_n,$$

for $k>0$, where $\partial_k=\partial/\partial c_{k+1}$.

Neretin [246] introduced the sequence of polynomials P_k, in the coordinates $\{c_2,c_3,\dots\}$ on $\mathcal{M}$ by the recurrent relations

$$L_m(P_n)=(n+m)P_{n-m}+\frac{c}{12}m(m^2-1)\delta_{n,m},\quad P_0=P_1\equiv 0,\quad P_k(0)=0,$$

where the central charge c is fixed. This gives, for example, $P_2=\frac{c}{2}(c_3-c_2^2)$, $P_3=2c(c_4-2c_2c_3+c_2^3)$. In general, the polynomials P_k are homogeneous with respect to rotations of the function f. It is worth mentioning that estimates of the absolute value of these polynomials have been a subject of investigations in the theory of univalent functions for a long time, e.g., for $|P_2|$ we have $|c_3-c_2^2|\leq 1$ (Bieberbach 1916 [38]), for estimates of $|P_3|$ see [138], [222], [322], [323]. For the Neretin polynomials one can construct the generatrix function

$$P(\zeta)=\sum_{k=1}^{\infty}P_k\zeta^k=\frac{c\zeta^2}{12}S_f(\zeta),$$

where $S_f(\zeta)$ is the Schwarzian derivative of f. Let $\nu\in\mathbb{C}\mathrm{Vect}\,S^1$ and ν^g be the associated right-invariant tangent vector field defined at $g\in\mathrm{Diff}\,S^1$. For the basis $\nu_k=-ie^{ik\theta}\partial$, one constructs the corresponding associated right-invariant basis ν_k^g. By $\{\psi_{-k}\}$ we denote the dual basis of 1-forms such that the value of each form on the vector ν_k^g is given as

$$(\psi_k,\nu_n^g)=\delta_{k+n,0}.$$

Let us construct the 1-form Ψ on $\mathrm{Diff}\,S^1$ by

$$\Psi=\sum_{k=1}^{\infty}(P_k\circ\pi)\psi_k,$$

where π means the natural projection $\mathrm{Diff}\,S^1\to\mathcal{M}$. This form appeared in [8], [9] in the context of the construction of a unitarizing probability measure for the

Neretin representation of $\mathcal{M}$. It is invariant under the left action of S^1. If $f \in \tilde{A}$ represents g and $\nu \in \text{Vect } S^1$, then the value of the form Ψ on the vector ν is

$$(\Psi, \nu)_f = \int_0^{2\pi} e^{2i\theta} \nu(e^{i\theta}) S_f \, d\theta,$$

see [8], [9]. So the variation of the logarithmic action becomes

$$\frac{d}{dt}(\mathcal{S}[f] + \mathcal{E}[f]) = 2 \int_0^{2\pi} \varkappa^2(\theta, t) \, d\theta + \text{Re}\,(\Omega, \nu)_f. \tag{7.18}$$

In this formula, we take into account the first coefficient, the conformal radius of the Laplacian evolution, that does not change the form Ω.

Problem 7.4.11. *The extended Toda hierarchy (see [53]) admits a nonabelian algebra of infinitesimal symmetries isomorphic to half of the Virasoro algebra [89]. It would be interesting to reveal the connections between the Hamiltonian approach through the Toda hierarchies and the action approach suggested in this section.*

Problem 7.4.12. *The multiply connected Hele-Shaw evolution was considered in [210] where its integrable structure was described by making use of a quasi-classical τ-function that generalizes the τ-function of the dispersionless Toda hierarchy. A natural problem is to try to blow up the process of collision of the components of the boundary of the phase domain and realize it as a worldsheet, see Fig. 7.1.*

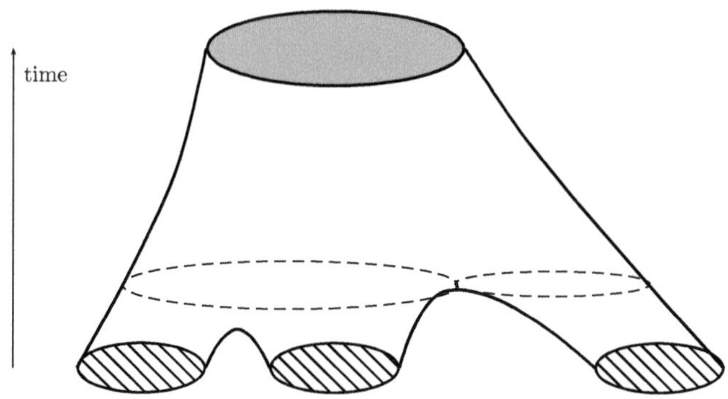

Figure 7.1: Hele-Shaw worldsheet

Bibliography

[1] F. Abergel, J. Mossino, *Caractérisation du problème de Muskat multidimensionnel et existence de solutions régulières*, C. R. Acad. Sci. Paris Sér. I Math. **319** (1994), no. 1, 35–40.

[2] B. Ablett, *An investigation of the angle of contact between paraffin wax and water*, Phil. Mag. **46** (1923), 244–256.

[3] O. Agam, E. Bettelheim, P. Wiegmann, A. Zabrodin, *Viscous fingering and a shape of an electronic droplet in the Quantum Hall regime*, arXiv: cond-mat/0111333, 2002.

[4] D. Aharonov, H. S. Shapiro, *Domains on which analytic functions satisfy quadrature identities*, J. Analyse Math. **30** (1976), 39–73.

[5] L. V. Ahlfors, *Lectures on quasiconformal mappings*, Van Nostrand Math. Stud., Princeton, N.J., 1966.

[6] L. V. Ahlfors, A. Beurling, *Conformal invariants and function-theoretic null-sets*, Acta Math., **83** (1950), no. 1-2, 101–129.

[7] L. V. Ahlfors, L. Bers, *Spaces of Riemann surfaces and quasiconformal mappings*, Collection of papers. Moscow: Inostrannaya Literatura., 1961

[8] H. Airault, P. Malliavin, *Unitarizing probability measures for representations of Virasoro algebra*, J. Math. Pures Appl. **80** (2001), no. 6, 627–667.

[9] H. Airault, P. Malliavin, A. Thalmaier, *Support of Virasoro unitarizing measures*, C. R. Acad. Sci. Paris, Ser. I **335** (2002), 621–626.

[10] J. M. Aitchison, S. D. Howison, *Computation of Hele-Shaw flows with free boundaries*, J. Comput. Phys. **60** (1985), no. 3, 376–390.

[11] S. Albeverio, R. Hoegh-Krohn, *Mathematical theory of Feynman path integral*, Lecture Notes in Mathematics 523, Springer-Verlag, 1976.

[12] S. Albeverio, J. Jost, S. Paycha, S. Scarlatti, *A mathematical introduction to string theory*, Variational problems, geometric and probabilistic methods. London Mathematical Society Lecture Note Series, 225. Cambridge University Press, Cambridge, 1997.

[13] I. A. Aleksandrov, *Parametric continuations in the theory of univalent functions*, Nauka, Moscow, 1976. (in Russian)

[14] M. Alimov, K. Kornev, G. Mukhamadullina, *Hysteretic effects in the problems of artificial freezing*, SIAM J. Appl. Math., **59** (1998), no. 2, 387–410.

[15] S. N. Antontsev, C. R. Gonçalves, A. M. Meirmanov, *Local existence of classical solutions to the well-posed Hele-Shaw problem*, Port. Math. (N.S.) **59** (2002), no. 4, 435–452.

[16] D. Armitage, S. Gardiner, *Classical potential theory*, Springer-Verlag, London, 2001.

[17] A. Arnéodo, Y. Couder, G. Grasseau, V. Hakim, M. Rabaud, *Uncovering the analytical Saffman-Taylor finger in unstable viscows fingering and diffusion-limited aggregation*, Phys. Rev. Lett. **63** (1989), no. 9. 984–987.

[18] V. I. Arnold, *Mathematical methods of classical mechanics*, Springer-Verlag, New York, 1989.

[19] F. G. Avkhadiev, L. A. Aksent'ev, M. A. Elizarov, *Sufficient conditions for the finite-valence of analytic functions, and their applications*, Itogi Nauki i Tekhniki, Mathematical analysis, Vol. 25, Akad. Nauk SSSR, Vsesoyuz. Inst. Nauchn. i Tekhn. Inform., Moscow, 1987, 3–121; English Transl.: J. Soviet Math., **49** (1990), no. 1, 715–799.

[20] C. Baiocchi, *Su alcuni problemi a frontiera libera connessi a questioni di idraulica*, Rend. Sem. Mat. Univ. e Politec. Torino **31** (1971/73), 69–80. (in Italian)

[21] J.-H. Bailly, *Local existence of classical solutions to first-order parabolic equations describing free boundaries*, Nonlinear Anal. **32** (1998), no. 5, 583–599.

[22] G. Baker, M. Siegel, S. Tanveer, *A well-posed numerical method to track isolated conformal map singularities in Hele-Shaw flow*, J. Comput. Phys. 120 (1995), no. 2, 348–364.

[23] C. M. Beaufoy, *Nautical and hydraulic experiments, with numerous scientific miscellanies*, London, at the private press of Henry Beaufoy, 1834.

[24] J. Becker, *Löwnersche Differentialgleichung und quasikonform fortsetzbare schlichte Funktionen*, J. Reine Angew. Math. **255** (1972), 23–43.

[25] J. Becker, *Löwnersche Differentialgleichung und Schlichtheitskriterien*, Math. Ann. **202** (1973), 321–335.

[26] J. Becker, *Conformal mappings with quasiconformal extensions.* Aspects of contemporary complex analysis (Proc. NATO Adv. Study Inst., Univ. Durham, Durham, 1979) Academic Press, London-New York, 1980, 37–77.

[27] J. Becker, Ch. Pommerenke, *On the Hausdorff dimension of quasicircles*, Ann. Acad. Sci. Fenn. Ser. A I Math. **12** (1987), no. 2, 329–333.

[28] B. Beckman, *Codebreakers. Arne Beurling and the Swedish Crypto program during World War II*, AMS, 2002.

[29] H. Begehr, R. P. Gilbert, *Hele-Shaw type flows in R^n*, Nonlinear Analysis **10** (1986), 65–85.

[30] H. Begehr, R. P. Gilbert, *Non-Newtonian Hele-Shaw flows in $n \geq 2$ dimensions*, Nonlinear Analysis **11** (1987), 17–47.

[31] P. P. Belinskii, *General properties of quasiconformal mappings*, Novosibirsk: Nauka., 1974 (Russian).

[32] M. Ben Amar, *Exact self-similar shapes in viscows fingering*, Phys. Review A **43** (1991), no. 10, 5724–5727.

[33] M. Ben Amar, *Viscous fingering in a wedge*, Phys. Review A **44** (1991), no. 6, 3673–3685.

[34] M. Ben Amar, V. Hakim, M. Mashaal, Y. Couder *Self-dilating viscows fingers in wedge-shaped Hele-Shaw cells*, Phys. Fluids A **3** (1991), no. 9, 2039–2042.

[35] D. Bensimon, *Stability of viscous fingering*, Phys. Rev. A bf 33 (1986), no. 2, 1302–1308.

[36] S. Bergman, M. Schiffer, *Kernel functions and elliptic differential equations in mathematical physics*, Academic Press, New York, 1953.

[37] A. Beurling, L. V. Ahlfors, *The boundary correspondence under quasiconformal mappings*, Acta Math. **96** (1956), 125–142.

[38] L. Bieberbach, *Über die Koeffizienten derjenigen Potenzreihen, welche eine schlichte Abbildung des Einheitskreises vermitteln*, S.-B. Preuss. Akad. Wiss. (1916), 940–955.

[39] J. Billingham, *Surface-tension-driven flow in fat fluit wedges and cones*, J. Fluid Mech., **397** (1999), 45-71.

[40] R. Bott, *On the characteristics classes of groups of diffeomorphisms*, Enseignment Math. (2) **23** (1977), 209–220.

[41] M. Boukrouche, *The quasiconformal mapping methods to solve a free boundary problem for generalized Hele-Shaw flows*, Complex Variables **28** (1996), 227–242.

[42] M. J. Boussinesq, *Calcul du pouvoir refroidissant des fluides*, J. Math. Pures et Appl., Sér. 6, **1** (1905), 285–332.

[43] L. de Branges, *A proof of the Bieberbach conjecture*, Acta Math. **154** (1985) no. 1–2, 137–152.

[44] D. A. Brannan, W. E. Kirwan, *On some classes of bounded univalent functions*, J. London Math. Soc. (2), **1** (1969), 431–443.

[45] L. Caffarelli, *The regularity of free boundaries in higher dimension*, Acta. Math. **139** (1977), 155–184.

[46] L. A. Caffarelli, *Compactness methods in free boundary problems*, Comm. Partial Diff. Equations **5** (1980), 427–448.

[47] L. A. Caffarelli, *A remark on the Hausdorff measure of a free boundary, and the convergence of coincedence sets*, Boll. Un. Mat. Ital. A(5) **18** (1981), no. 1, 109–113.

[48] L. A. Caffarelli, *The obstacle problem revisited, III. Existence theory, compactness and dependence on X*, J. Fourier Anal. Appl. **4** (1998), 383–402.

[49] L. Caffarelli, L. Karp, H. Shahgholian, *Regularity of the free boundary with application to the Pompeiu problem*, Ann. Math. **151** (2000), 269–292.

[50] L. A. Caffarelli, J. L. Vázquez, *Viscosity solutions for the porous medium equation*, Differential equations: La Pietra 1996 (Florence), Proc. Sympos. Pure Math., 65, Amer. Math. Soc., Providence, RI, 1999, 13–26.

[51] C. Carathéodory, *Untersuchungen über die konformen Abbildungen von festen und veränderlichen Gebieten*, Math. Ann. **72** (1912), 107–144.

[52] L. Carleson, N. Makarov, *Aggregation in the plane and Loewner's equation*, Comm. Math. Phys., **216** (2001), no. 3, 583–607.

[53] G. Carlet, B. Dubrovin, Y. Zhang, *The extended Toda hierarchy*, Mosc. Math. J. **4** (2004), no. 2, 313–332.

[54] H. D. Ceniceros, T. Y. Hou, H. Si, *Numerical study of Hele-Shaw flow with suction*, Physics of Fluids, **11** (1999), no. 9, 2471–2486.

[55] H. D. Ceniceros, T. Y. Hou, *The singular perturbation of surface tension in Hele-Shaw flows*, J. Fluid Mech. **409** (2000), 251–272.

[56] H. D. Ceniceros, J. M. Villalobos, *Topological reconfiguration in expanding Hele-Shaw flow*, J. Turbulence **3** (2002), no. 37, 1–8.

[57] D. Y. C. Chan, B. D. Hughes, L. Paterson, Ch. Sirakoff, *Simulating flow in porous media.* Phys. Rev. A (3) **38** (1988), no. 8, 4106–4120.

[58] P. Constantin, M. Pugh, *Global solution for small data to the Hele-Shaw problem*, Nonlinearity **6** (1993), 393–415.

[59] R. Courant, H. Robbins, *What Is Mathematics?* Oxford University Press, New York, 1941.

[60] R. Courant, D. Hilbert, *Methoden der mathematischen Physik. II*, Dritte Auflage. Heidelberger Taschenbücher, Band 30. Springer-Verlag, Berlin-New York, 1968.

[61] R. G. Cox, *Inertial and viscous effects on dynamic contact angles*, J. Fluid Mech. **357** (1998), 249–278.

[62] M. G. Crandall, P. L. Lions, *Viscosity solutions of Hamilton-Jacoby equations*, Trans. Amer. Math. Soc. **277** (1983), 1–42.

[63] D. Crowdy, *Theory of exact solutions for the evolution of a fluid annulus in a rotating Hele-Shaw cell*, Quart. Appl. Math. **60** (2002), no. 1, 11–36.

[64] D. Crowdy, *Constructing multiply-connected quadrature domains I: algebraic curves*, preprint 2003.

[65] D. Crowdy, *Quadrature domains and fluid dynamics*, Quadrature Domains and Applications, a Harold S. Shapiro Anniversary Volume. Birkhäuser, 2005.

[66] A. B. Crowley, *On the weak solution of moving boundary problems,* J. Inst. Math. Appl. **24** (1979), 43–57.

[67] L. J. Cummings, *Flow around a wedge of arbitrary angle in a Hele-Shaw cell*, European J. Appl. Math. **10** (1999), 547–560.

[68] L. J. Cummings, S. D. Howison, J. R. King, *Two-dimensional Stokes and Hele-Shaw flows with free surfaces*, European J. Appl. Math. **10** (1999), 635–680.

[69] L. J. Cummings, G. Richardson, M. Ben Amar, *Models of void electromigration,* European J. Appl. Math. **12** (2001), 97–134.

[70] L. M. Cummings, Yu. E. Hohlov, S. D. Howison, K. Kornev, *Two-dimensional solidification and melting in potential flows*, J. Fluid Mech., **378** (1999), 1–18.

[71] W.-S. Dai, M. Shelley, *A numerical study of the effect of surface tension and noise on an expanding Hele-Shaw bubble*, Phys. Fluids A **5** (1993), no. 9, 2131–2146.

[72] B. Davidovitch, H. G. E. Hentschel, Z. Olami, I. Procaccia, L. M. Sander, and E. Somfai, *Diffusion limited aggregation and iterated conformal maps*, Phys. Rev. E (3) **59** (1999), no. 2, part A, 1368–1378.

[73] B. Davidovitch, M. J. Feigenbaum, H. G. E. Hentschel, I. Procaccia, *Conformal dynamics of fractal growth patterns without randomness*, Phys. Rev. E **62** (2000), no. 2, 1706–1715.

[74] *ed.* P. C. W. Davies, J. R. Brown, *Superstrings. A theory of everything?*, Cambridge Univ. Press, 2006.

[75] S. H. Davis, *Theory of solidification*, Cambridge Univ. Press, 2001.

[76] P. Davis, *The Schwarz function and its applications*, Carus Math. Monographs 17, Math. Assoc. of America, 1974.

[77] A. J. DeGregoria, L. W. Schwartz, *A boundary-integral method for two-phase displacement in Hele-Shaw cells*, J. Fluid Mech. **164** (1986), 383–400.

[78] P. Deligne, P. Etingof, D. S. Freed, L. C. Jeffrey, D. Kazhdan, J. W. Morgan, D. R. Morrison, E. Witten (editors): *Quantum fields and strings: a course for mathematicians*, Vol. 1, 2 (Princeton, NJ, 1996/1997), Amer. Math. Soc., Providence, RI, 1999.

[79] A. S. Demidov, J.-P. Loheác, *The Stokes-Leibenson problem for Hele-Shaw flows*, Patterns and waves (Saint Petersburg, 2002), 103–124.

[80] A. S. Demidov, *Evolution of the perturbation of a circle in the Stokes-Leibenson problem for a Hele-Shaw flow*, Sovrem. Mat. Prilozh. No. 2, Differ. Uravn. Chast. Proizvod. (2003), 3–24; translation in J. Math. Sci. (N. Y.) **123** (2004), no. 5, 4381–4403.

[81] E. Di Benedetto, A. Friedman, *The ill-posed Hele-Shaw model and the Stefan problem for supercooled water*, Trans. Amer. Math. Soc., **282** (1984), no. 1, 183–204.

[82] E. Di Benedetto, A. Friedman, *Bubble growth in porous media*, Indiana Univ. Math. J. **35** (1986), no. 3, 573–606.

[83] P. A. Dirac, *Selected papers on quantum electrodynamics*, Dover Publications, New York, 1958, 312–320.

[84] J. Doob, *Classical potential theory and its probabilistic counterpart*, Springer-Verlag, Berlin, 1983.

[85] A. Douady, C. J. Earle, *Conformally natural extension of homeomorphisms of the circle*, Acta Math. **157** (1986), no. 1-2, 23–48.

[86] P. D. Dragnev, E. B. Saff, *Constrained energy problems with applications to orthogonal polynomials of a discrete variable*, J. Anal. Math. **72** (1997), 223–259.

[87] V. N. Dubinin, *Symmetrization in the geometric theory of functions of a complex variable*, Uspekhi Mat. Nauk **49** (1994), no. 1(295), 3–76 (Russian).

[88] B. A. Dubrovin, S. P. Novikov, A. T. Fomenko, *Modern geometry. Methods and applications*, Second edition. "Nauka", Moscow, 1986. (in Russian)

[89] B. Dubrovin, Y. Zhang, *Virasoro symmetries of the extended Toda hierarchy*, Comm. Math. Phys. **250** (2004), no. 1, 161–193.

[90] W. Dunham, *The mathematical universe*, John Wiley & Sons, 1994.

[91] J. Duchon, R. Robert, *Evolution d'une interface par capillarité et diffusion de volume, 1: existence locale en temps*, Ann. Inst. H. Poincaré, **1** (1984), 361–378.

[92] P. Duren, *Univalent functions*, Springer, New York, 1983.

[93] P. Duren, M. M. Schiffer, *Robin functions and energy functionals of multiply connected domains*, Pacific J. Math., **148** (1991), 251–273.

[94] P. Duren, J. Pfaltzgraff, *Robin capacity and extremal length*, J. Math. Anal. Appl., **179** (1993), 110–119.

[95] P. Duren, M. M. Schiffer, *Robin functions and distortion of capacity under conformal mapping*, Complex Variables Theory Appl., **21** (1993), 189–196.

[96] P. Duren, J. Pfaltzgraff, R. E. Thurman, *Physical interpretation and further properties of Robin capacity*, Algebra i Analiz, **9** (1997), no. 3, 211–219; English vers.: St. Petersburg Math. J., **9** (1998), no. 3, 607–614.

[97] P. Duren, J. Pfaltzgraff, *Hyperbolic capacity and its distortion under conformal mapping*, J. Anal. Math., **78** (1999), 205–218.

[98] P. Duren, *Robin capacity*, Computational Methods and Function Theory, N. Papamichael, St. Ruscheweyh, E. B. Saff (Eds), World Sci. Publ. (1999), 177–190.

[99] E. B. Dussan V., *On the spreading of liquids on solid surfaces: static and dynamic contact lines*, Ann. Rev. Fluid Mechanics **11** (1979), 371–400.

[100] P. Ebenfelt, B. Gustafsson, D. Khavinson, M. Putinar (editors): *Quadrature Domains and Applications*, a Harold S. Shapiro Anniversary Volume. Birkhäuser, 2005.

[101] T. Eguchi, H. Ooguri, *Conformal and current algebras on a general Riemann surface*, Nucl. Phys. B **282** (1987), no. 2, 308–328.

[102] C. M. Elliott, V. Janovský, *A variational inequality approach to Hele-Shaw flow with a moving boundary*, Proc. Roy. Soc. Edin., **A88** (1981), 93–107.

[103] C. M. Elliott, J. R. Ockendon, *Weak and variational methods for moving boundary problem*, Pitman, London, 1992.

[104] V. M. Entov, P. I. Ètingof, *Bubble contraction in Hele-Shaw cells*, Quart. J. Mech. Appl. Math., **44** (1991), no. 4, 507–535.

[105] V. M. Entov, P. I. Ètingof, D. Ya. Kleinbock, *On nonlinear inteface dynamics in Hele-Shaw flows*, Euro. J. Appl. Math., **6** (1995), no. 5, 399–420.

[106] J. Escher, G. Simonett, *On Hele-Shaw models with surface tension*, Math. Res. Lett., **3** (1996), no. 4, 467–474.

[107] J. Escher, G. Simonett, *Classical solutions for Hele-Shaw models with surface tension*, Adv. Differential Equations, **2** (1997), no. 4, 619–642.

[108] J. Escher, G. Simonett, *Classical solutions of multidimensional Hele-Shaw models*, SIAM J. Math. Anal. **28** (1997), no. 5, 1028–1047.

[109] J. Escher, G. Simonett, *A center manifold analysis for the Mullins-Sekerka model*, J. Differential Equations, **143** (1998), no. 2, 267–292.

[110] J. Escher, G. Simonett, *Moving surfaces and abstract parabolic evolution equations*, Topics in nonlinear analysis, 183–212, Progr. Nonlinear Differential Equations Appl. 35, Birkhäuser, Basel, 1999.

[111] L. Fejér, *Mechanische Quadraturen mit positiven Cotesschen Zahlen*, Math. Zeitschrift, **37** (1933), no. 2, 287–309.

[112] L. Fejér, G. Szegö, *Special conformal mappings*, Duke Math. J., **18** (1951), 535–548.

[113] R. P. Feynman, *The principle of least action in quantum mechanics*, Ph.D. Thesis, Princeton Univ., Princeton, N.J., 1942.

[114] J. M. Fitz-Gerald, J. A. McGeough, *Mathematical theory of electro-chemical machining: Anodic smoothing*, J. Inst. Math. Appl. **5** (1969), 387–408; *Anodic shaping*, J. Inst. Math. Appl. **5** (1969), 409–421; *Deburring and cavity forming*, J. Inst. Math. Appl. **6** (1970), 102–110.

[115] A. Friedman, *Variational principles and free boundary problems*, Wiley-Interscience, New York, 1982.

[116] F. D. Gakhov, *Boundary value problems*, Addison-Wesley, 1966.

[117] L. A. Galin, *Unsteady filtration with a free surface*, Dokl. Akad. Nauk USSR, **47** (1945), 246–249. (in Russian)

[118] F. P. Gardiner, *Teichmüller theory and quadratic differentials*, Wiley Interscience, N.Y., 1987

[119] F. Gardiner, N. Lakic, *Quasiconformal Teichmüller theory*, Mathematicsl Surveys and Monographs, vol. 76, Amer. Math. Soc., 2000.

[120] C. F. Gauss, *Allgemeine Lehrsätze in Beziehung auf die im verkehrten Verh"altnisse des Quadrats der Entfernung Wirkenden Anziehungs- und Abstossungs-Kräfte.* Gauss Werke **5**, pp. 197–242, 1840, Göttingen, 1867.

[121] F. W. Gehring, J. Väisälä, *The coefficients of quasiconformality of domains in the space*, Acta Math. **114** (1965), 60–120.

[122] F. W. Gehring, E. Reich, *Area distortion under quasiconformal mappings*, Ann. Acad. Sci. Fenn. Ser. A I, No. 388, 1966, 15 pp.

[123] I. M. Gel'fand, D. B. Fuchs, *Cohomology of the Lie algebra of vector fields on the circle*, Func. Anal. Appl. **2** (1968), no.4, 342–343.

[124] R. P. Gilbert, P. Shi, *Anisotropic Hele-Shaw flows*, Math. Methods Appl. Sci. **11** (1989), no. 4, 417–429.

[125] K. A. Gillow, S. D. Howison, *A bibliography of free and moving boundary problems for Hele-Shaw and Stokes flow*, Published electronically at URL>http://www.maths.ox.ac.uk/˜howison/Hele-Shaw.

[126] M. E. Goldstein, R. I. Reid, *Effect of flow on freezing and thawing of saturated porous media*, Proc. Royal Soc. London, Ser. A, **364** (1978), 45–73.

[127] G. M. Goluzin, *Geometric theory of functions of a complex variable*, Transl. Math. Monogr., Vol. 26, American Mathematical Society, Providence, R.I. 1969.

[128] T. Gotô, *Relativistic quantum mechanics of one-dimensional mechanical continuum and subsidiary conditon of dual resonance model*, Progr. Theor. Phys. **46** (1971), 1560–1569.

[129] M. B. Green, *Superstrings*, Scientific American (1986), Sept., page 44.

[130] D. Grier, E. Ben-Jacob, R. Clarke, and L. M. Sander, *Morphology and Microstructure in Electrochemical Deposition of Zinc*, Phys. Rev. Lett. **56** (1986), 1264-1267.

[131] A. W. Goodman, E. B. Saff, *On univalent functions convex in one direction*, Proc. Amer. Math. Soc., **73** (1979), no. 2, 183–187.

[132] A. W. Goodman, *Univalent functions, Vol I, II*, Mariner Publishing Company, Inc. U. South Florida, 1983.

[133] Y. N. Gordeev, K. G. Kornev, *Crystallization in forced flow: the Saffman-Taylor problem*, Euro. J. Appl. Math., **10** (1999), 535–545.

[134] V. V. Goryainov, *Fractional iterates of functions that are analytic in the unit disk with given fixed points*, Mat. Sb. **182** (1991), no. 9, 1281–1299; Engl. Transl. in Math. USSR-Sb. **74** (1993), no. 1, 29–46.

[135] V. V. Goryainov, *One-parameter semigroups of analytic functions*, Geometric function theory and applications of complex analysis to mechanics: studies in complex analysis and its applications to partial differential equations, 2 (Halle, 1988), Pitman Res. Notes Math. Ser., 257, Longman Sci. Tech., Harlow, 1991, 160–164.

[136] V. V. Goryainov, *One-parameter semigroups of analytic functions and a compositional analogue of infinite divisibility.* Proceedings of the Institute of Applied Mathematics and Mechanics, Vol. 5, Tr. Inst. Prikl. Mat. Mekh., 5, Nats. Akad. Nauk Ukrainy Inst. Prikl. Mat. Mekh., Donetsk, 2000, 44–57.(in Russian)

[137] D.-A. Grave, *Sur le problème de Dirichlet*, Assoc. Française pour l' Avancement des Sciences, Comptes-Rendus (Bordeaux), 111–136.

[138] L. Gromova, A. Vasil'ev, *On the estimate of the fourth-order homogeneous coefficient functional for univalent functions*, Ann. Polon. Math. **63** (1996), 7–12.

[139] H. Grötzsch, *Über einige Extremalprobleme der konformen Abbildungen. I–II*, Ber. Verh.- Sächs. Akad. Wiss. Leipzig, Math.–Phys. Kl., **80** (1928), 367–376.

[140] H. Grötzsch, *Über die Verzerrung bei nichtkonformen schlichten Abbildungen merfach zusammenhängender schlichten Bereiche*, Ber. Verh.- Sächs. Akad. Wiss. Leipzig, **82** (1930), 69–80.

[141] K. Gustafson, T. Abe, *The third boundary condition– was it Robin's?* Math. Intelligencer **20** (1998), no. 1, 63–71.

[142] B. Gustafsson, *Quadrature identities and the Schottky double*, Acta Appl. Math. **1** (1983), no. 3, 209–240.

[143] B. Gustafsson, *On a differential equation arising in a Hele-Shaw flow moving boundary problem*, Arkiv för Mat., **22** (1984), no. 1, 251–268.

[144] B. Gustafsson, *Applications of variational inequalities to a moving boundary problem for Hele-Shaw flows*, SIAM J. Math. Analysis, **16** (1985), no. 2, 279–300.

[145] B. Gustafsson, *Existence of weak backward solutions to a generalized Hele-Shaw flow moving boundary problem*, Nonlinear Analysis, **9** (1985), 203–215.

[146] B. Gustafsson, M. Sakai *Properties of some balayage operators, with applications to quadrature domains and moving boundary problems*, Nonlinear Analysis, **22** (1994), no. 10, 1221–1245.

[147] B. Gustafsson, M. Putinar, *An exponential transform and regularity of free boundaries in two dimensions*, Ann. Scuola Norm. Sup. Pisa Cl. Sci. (4) **26** (1998), 507–543.

[148] B. Gustafsson, M. Putinar, *Linear analysis of quadrature domains II*, Israel J.Math. **119** (2000), 187–216.

[149] B. Gustafsson, M. Sakai, *Sharp estimates of the curvature of some free boundaries in two dimensions*, Ann. Acad. Sci. Fenn. Math. **28** (2003), no. 1, 123–142.

[150] B. Gustafsson, M. Sakai, *On the curvature of the free boundary for the obstacle problem in two dimensions*, Monatsh. Math. **142** (2004), no. 1-2, 1–5.

[151] B. Gustafsson, D. Prokhorov, A. Vasil'ev, *Infinite lifetime for the starlike dynamics in Hele-Shaw cells*, Proc. Amer. Math. Soc. **132** (2004), no. 9, 2661–2669.

[152] B. Gustafsson, A. Vasil'ev, *Nonbranching weak and starshaped strong solutions for Hele-Shaw dynamics*, Arch. Math. (Basel) **84** (2005), no. 6, 551–558.

[153] B. Gustafsson, H.S Shapiro, *What is a quadrature domain?*, in [100]

[154] B. Gustafsson, *Lectures on balayage*, in Clifford algebras and potential theory, 17–63, Univ. Joensuu Dept. Math. Rep. Ser., 7, Univ. Joensuu, Joensuu, 2004.

[155] V. Ya. Gutlyanskiĭ, *Parametric representation of univalent functions*, Dokl. Akad. Nauk SSSR **194** (1970), 750–753; Engl. Transl. in Soviet Math. Dokl. **11** (1970), 1273–1276.

[156] V. Ya. Gutlyanskiĭ, *On some classes of univalent functions.* Theory of functions and mappings, "Naukova Dumka", Kiev, 1979, 85–97. (in Russian)

[157] V. Ya. Gutljanskiĭ, *The method of variations for univalent analytic functions with a quasiconformal extension*, Sibirsk. Mat. Zh. **21** (1980), no. 2, 61–78; Engl. Transl. in Siberian Math. J. **21** (1980), no. 2, 190–204.

[158] H. L. Guy, *H.S. Hele-Shaw 1854-1941*, Obituary Notices of Fellows of The Royal Society **3**(1941), no. 10. 790–811.

[159] Cheng Qi He, *A parametric representation of quasiconformal extensions*, Kexue Tongbao, **25** (1980), no. 9, 721–724.

[160] H. E. Haber, G. L. Kane, *Is nature supersymmetric?*, Scientific American (1986), June, page 42.

[161] T. C. Halsey, *Diffusion-Limited Aggregation: A Model for Pattern Formation*, Physics Today **53** (2000), no. 11, 36.

[162] M. B. Hastings, *Renormalization theory of stochastic growth*, Physical Review E **55** (1987), no. 1, 135-152.

[163] M. B. Hastings, L. S. Levitov, *Laplacian growth as one-dimensional turbulence*, Physica D, **116** (1998), no. 1-2 , 244-252.

[164] H. Hedenmalm, S. Shimorin, *Hele-Shaw flow on hyperbolic surfaces*, J. Math. Pures et Appl. **81** (2002), 187–222.

[165] H. Hedenmalm, A. Olofsson, *Hele-Shaw flow on weakly hyperbolic surfaces*, Indiana Univ. Math. J. 54 (2005), no. 4, 1161–1180.

[166] H. S. Hele-Shaw, *The flow of water*, Nature, **58** (1898), no. 1489, 34–36, 520.

[167] H. S. Hele-Shaw, *On the motion of a viscous fluid between two parallel plates*, Trans. Royal Inst. Nav. Archit., London, **40**(1898), 21.

[168] W. Hengartner, G. Schober, *A remark on level curves for domains convex in one direction*, Collection of articles dedicated to Eberhard Hopf on the occasion of his 70th birthday, Applicable Anal., **3** (1973), 101–106.

[169] G. Herglotz, *Über die analytische Fortsetzung des Potentials ins Innere der anziehenden Massen*, Gekrönte Preisschr. der Jablonowskischen Gesellsch. zu Leipzig, 56 pp.

[170] Yu. E. Hohlov, S. D. Howison, C. Huntingford, J. R. Ockendon, A. A. Lacey, *A model for non-smooth free boundaries in Hele-Shaw flows*, Quart. J. Mech. Appl. Math., **47** (1994), 107–128.

[171] Yu. E. Hohlov, S. D. Howison, *On the classification of solutions to the zero surface tension model for Hele-Shaw free boundary flows*, Quart. Appl. Math., **54** (1994), no. 4, 777–789.

[172] Yu. E. Hohlov, M. Reissig, *On classical solvability for the Hele-Shaw moving boundary problem with kinetic undercooling regularization*, Euro. J. Applied Math., **6** (1995), 421–439.

[173] Yu. E. Hohlov, D. V. Prokhorov, A. Vasil'ev, *On geometrical properties of free boundaries in the Hele-Shaw flow moving boundary problem*, Lobachevskiĭ J. Math., **1** (1998), 3–13. (electronic)

[174] T. Y. Hou, L. Lowengrub, M. Shelley, *Removing the stiffness from interfacial flow with surface tension*, J. Comput. Phys. **114** (1994), no. 2, 312–338.

[175] T. Y. Hou, *Numerical solutions to free boundary problems*, Acta numerica, 1995, 335–415.

[176] S. D. Howison, *Fingering in Hele-Shaw cells*, J. Fluid Mech. **167** (1986), 439-453.

[177] S. D. Howison, *Cusp development in Hele-Shaw flow with a free surface*, SIAM J. Appl. Math., **46** (1986), no. 1, 20–26.

[178] S. D. Howison, *Bubble growth in porous media and Hele-Shaw cells*, Proc. Roy. Soc. Edinburgh, Ser. A, **102** (1986), no. 1-2, 141–148.

[179] S. D. Howison, *Complex Variable methods in Hele-Shaw moving Boundary Problems*, European J. Appl. Math., **3** (1992), no. 3, 209–224.

[180] S. D. Howison, *A note on the two-phase Hele-Shaw problem*, J. Fluid Mech., **409** (2000), 243–249.

[181] S. D. Howison, J. King, *Explicit solutions to six free-boundary problems in fluid flow and diffusion*, IMA J. Appl. Math. **42** (1989), 155–175.

[182] G. P. Ivantzov, *The temperature field around a spherical, cylindrical, or point crystal growing in a cooling solution*, Dokl. Acad. Nauk USSR, **58** (1947), 567–569. (in Russian)

[183] J. Jenkins, *On the existence of certain general extremal metrics*, Ann. of Math., **66** (1957), no. 3, 440–453.

[184] J. Jenkins, *Univalent functions and conformal mapping*, Springer-Verlag, 1958.

[185] G. W. Johnson, M. L. Lapidus, *The Feynman integral and Feynman's operational calculus*, Oxford Mathematical Monographs. Oxford Science Publications. The Clarendon Press, Oxford University Press, New York, 2000.

[186] J. Jost, *Bosonic strings: a mathematical treatment*, Amer. Math. Soc., Providence, RI, 2001.

[187] V. Kac, *Vertex algebras for beginners*, Amer. Math. Soc., Providence, RI, 1998.

[188] L. P. Kadanoff, *Exact soutions for the Saffman-Taylor problem with surface tension*, Phys. Review Letters **65** (1990), no. 24, 2986–2988.

[189] M. Kaku, *Intraduction to superstrings*, Springer-Verlag, Berlin-New York, 1988.

[190] W. Kaplan, *Close-to-convex schlicht functions*, Michigan Math. J., **1** (1952), 169–185.

[191] J. B. Keller, M. Miksis, *Surface tension driven flows*, SIAM J. Appl. Math. **34** (1983), no. 2, 268–277.

[192] J. B. Keller, P. A. Milewski, J.-M. Vanden-Broeck, *Merging and wetting driven by surface tension*, European J. Mech. B-Fluids **19** (2000), 491–502.

[193] H. Kesten, *Hitting probabilities of random walks on* $\mathbb{Z}^d$, Stoch. Proc. Appl. **25** (1987), 165–184.

[194] I. C. Kim, *Uniqueness and existence results on the Hele-Shaw and the Stefan problems*, Arch. Ration. Mech. Anal. **168** (2003), no. 4, 299–328.

[195] M. Kimura, *Time local existence of a moving boundary of the Hele-Shaw flow with suction*, Europ. J. Appl. Math., **10** (1999), no.6, 581–605.

[196] D. Kinderlehrer, G. Stampacchia, *An introduction to variational inequalities and their applications*, Academic Press, New York-London, 1980.

[197] J. R. King, A. A. Lacey, J. L. Vázquez, *Persistence of corners in free boundaries in Hele-Shaw flows*, Euro. J. Appl. Math. **6** (1995), no. 5, 455-490.

[198] J. R. King, *Development of singularities in some moving boundary problems*, Euro. J. Appl. Math. **6** (1995), no. 5, 491–507.

[199] A. A. Kirillov, D. V. Yuriev, *Kähler geometry of the infinite-dimensional homogeneous space* $M = \mathrm{Diff}_+(S^1)/\mathrm{Rot}(S^1)$, Funktsional. Anal. i Prilozhen. **21** (1987), no. 4, 35–46. (in Russian)

[200] A. A. Kirillov, D. V. Yuriev, *Representations of the Virasoro algebra by the orbit method*, J. Geom. Phys. **5** (1988), no. 3, 351–363.

[201] A. A. Kirillov, *Kähler structure on the K-orbits of a group of diffeomorphisms of the circle*, Funktsional. Anal. i Prilozhen. **21** (1987), no. 2, 42–45.

[202] A. A. Kirillov, *Geometric approach to discrete series of unirreps for Vir*, J. Math. Pures Appl. **77** (1998), 735–746.

[203] A. N. Kolmogorov, *Foundations of probability theory*, Chelsea, New York, 1950.

[204] K. Kornev, G. Mukhamadullina, *Mathematical theory of freezing in porous media*, Proc. Royal Soc. London, Ser. A, **447** (1994), 281–297.

[205] K. Kornev, A. Vasil'ev, *Geometric properties of the solutions of a Hele-Shaw type equation*, Proc. Amer. Math. Soc., **128** (2000), no. 9, 2683–2685.

[206] I. K. Kostov, I. Krichever, M. Mineev-Weinstein, P. B. Wiegmann, A. Zabrodin, *The τ-function for analytic curves*, Random matrix models and their applications, Math. Sci. Res. Inst. Publ., 40, Cambridge Univ. Press, Cambridge, 2001, 285–299.

[207] I. M. Krichever, *The dispersionless Lax equations and topological minimal models*, Comm. Math. Phys. **143** (1992), no. 2, 415–429.

[208] I. M. Krichever, *The τ-function of the universal Whitham hierarchy, matrix models and topological field theories*, Comm. Pure Appl. Math. **47** (1994), no. 4, 437–475.

[209] I. Krichever, M. Mineev-Weinstein, P. Wiegmann, A. Zabrodin, *Laplacian growth and Whitham equations of soliton theory*, Phys. D **198** (2004), no. 1-2, 1–28.

[210] I. Krichever, A. Marshakov, A. Zabrodin, *Integrable structure of the Dirichlet boundary problem in multiply-connected domains*, Comm. Math. Phys. **259** (2005), no. 1, 1–44.

[211] S. L. Krushkal, *Quasiconformal mappings and Riemann surfaces*, A Halsted Press Book. Scripta Series in Mathematics. With a foreword by Lipman Bers. V. H. Winston & Sons, Washington, D.C.; John Wiley & Sons, New York-Toronto, Ont.-London, 1979

[212] P. P. Kufarev, *On one-parameter families of analytic functions*, Rec. Math. [Mat. Sbornik] N.S. **13(55)** (1943), 87–118.

[213] P. P. Kufarev, *A solution of the boundary problem for an oil well in a circle*, Doklady Akad. Nauk SSSR (N. S.) **60** (1948), 1333–1334. (in Russian)

[214] P. P. Kufarev, *The problem of the contour of the oil-bearing region for a circle with an arbitrary number of gaps*, Doklady Akad. Nauk SSSR (N.S.) **75** (1950), 507–510. (in Russian)

[215] O. Kuznetsova, *On polynomial solutions of the Hele-Shaw problem*, Sibirsk. Mat. Zh. **42** (2001), no. 5, 1084–1093; Engl. transl.: Siberian Math. J. **42** (2001), no. 5, 907–915.

[216] O. Kuznetsova, *Invariant families in the Hele-Shaw problem*, Preprint TRITA-MAT-2003-07, Royal Institute of Technology, Stokholm, Sweden, 2003.

[217] H. Lamb, *Hydrodynamics*, 6-th edition, Dover Publ., New York, 1932.

[218] M. A. Lavrentiev, *A general problem of the theory of quasi-conformal representation of plane regions*, Mat. Sbornik (N.S) **21(63)** (1947), 85–320.

[219] J. Ławrynowicz, J. Krzyż, *Quasiconformal mappings in the plane: parametric methods*, Lecture Notes in Math., Vol. 978, Springer-Verlag, Berlin-New York, 1983.

[220] A. Lecko, D. Partyka, *An alternative proof of a result due to Douady and Earle*, Annales Univ. Mariae Curie-Skłodowska, Sect. A, **42** (1988), 59–68.

[221] O. Lehto, *Univalent functions and Teichmüller spaces*, Graduate Texts in Mathematics, 109. Springer-Verlag, New York, 1987.

[222] P. Lehto, *On fourth-order homogeneous functionals in the class of bounded univalent functions*, Ann. Acad. Sci. Fenn. Ser. A I Math. Dissertationes No. 48, (1984), 1–46.

[223] L. S. Leibenzon, *Oil producing mechanics, Part II*, Neftizdat, Moscow, 1934.

[224] K. Löwner, *Untersuchungen über schlichte konforme Abbildungen des Einheitskreises*, Math. Ann. **89** (1923), 103–121.

[225] D. Lüst, S. Theisen, *Lectures on string theory*, Lecture Notes in Physics, Springer-Verlag, 1989.

[226] I. Loutsenko, *The variable coefficient Hele-Shaw problem, integrability and quadrature identities*, arXiv:math-ph/0510070.

[227] V. A. Maksimov, *On the stable shape of bodies solidified around a could sourse in a stream fluid*, Izv. Acad. Nauk USSR, Mekhanica, **4** (1965), no. 41, 41–45. (in Russian)

[228] V. A. Maksimov, *On the determination of the shape of bodies formed by solidification of the fluid phase of the stream*, Prikl. Mat. Mekh. (J. Appl. Math. Mech.), **40** (1976), no. 264, 264–272.

[229] I. Markina, A. Vasil'ev, *Long-pin perturbations of the trivial solution for Hele-Shaw corner flows*, Scientia. Ser. A, Math. Sci. **9** (2003), 33-43.

[230] I. Markina, A. Vasil'ev, *Explicit solutions for Hele-Shaw corner flows*, European J. Appl. Math. **15** (2004), 1–9.

[231] A. Marshakov, P. Wiegmann, A. Zabrodin, *Integrable structure of the Dirichlet boundary problem in two dimensions*, Comm. Math. Phys. **227** (2002), no. 1, 131–153.

[232] O. Martio, B. Øksendal, *Fluid flow in a medium distorted by a quasiconformal map can produce fractal boundaries*, European J. Appl. Math. **7** (1996), no. 1, 1–10.

[233] A. Marx, *Untersuchungen über schlichte Abbildungen*, Math. Ann., **107** (1932/33), 40–67.

[234] J. W. McLean, P. G. Saffman, *The effect of surface tension on the shape of fingers in a Hele Shaw cell*, J. Fluid Mech., **102** (1981), 455-469.

[235] E. B. McLeod, *The explicit solution of a free boundary problem involving surface tension*, J. Rat. Mech. Analysis **4** (1955), 557–567.

[236] E. Meiburg, G. M. Homsy, *Nonlinear unstable viscous fingers in Hele-Shaw flows. 2. Numerical simulation*, Phys. Fluids **31** (1988), no. 3, 429–439.

[237] A. M. Meirmanov, B. Zaltzman, *Global in time solution to the Hele-Shaw problem with change of topology*, Eur. J. Appl. Math. **13** (2002), 431–447.

[238] R. Merks, A. Hoekstra, J. Kaandorp, P. Sloot, *Models of coral growth, sponatneous branching, compactification and Laplacian growth applications*, J. Theoret. Biol. **224** (2003), 153–166.

[239] M. Mineev-Weinstein, A. Zabrodin, *Whitham-Toda hierarchy in the Laplacian growth problem*, J. Nonlinear Math. Phys. **8** (2001), 212–218.

[240] J. Mossino, *Inégalités isopérimétriques et applications en physique*, Travaux en Cours. Hermann, Paris, 1984. (in French)

[241] W. W. Mullins, *Grain boundary grooving by volume diffusion*, Trans. Metallurgical Society of AIME, **218** (1960), 354–361.

[242] N. I. Muskhelishvili, *Singular integral equations*, P. Noordhoff, Groningen, the Netherlands, 1953.

[243] Y. Nambu, *Duality and hydrodynamics*, Lectures at the Copenhagen symposium, 1970.

[244] S. Nag, *The complex analytic theory of Teichmüller spaces*, Wiley-Interscience Publ. John Wiley & Sons, Inc., New York, 1988.

[245] Z. Nehari, *Schwarzian derivatives and schlicht functions*, Bull. Amer. Math. Soc. **55** (1949), no. 6, 545–551.

[246] Yu. A. Neretin, *Representations of Virasoro and affine Lie algebras*, Encyclopedia of Mathematical Sciences, Vol. 22, Springer-Verlag, 1994, pp. 157–225.

[247] Q. Nie, F. R. Tian, *Singularities in Hele-Shaw flows*, SIAM J. Appl. Math., **58** (1998), no. 1, 34–54.

[248] L. Niemeyer, L. Pietronero, and H. J. Wiessmann, *Fractal dimension of dielectric breakdown*, Phys. Rev. Lett. **52** (1984), no. 12, 1033–1036.

[249] T. Nishida, *A note on a theorem of Nirenberg*, J. Differential Geom. **12** (1977), no. 4, 629–633.

[250] E. Noether, *Invariante Varlationsprobleme*, Nachr. d. König. Gesellsch. d. Wiss. zu Göttingen, Math-phys. Klasse (1918), 235–257.

[251] P. S. Novikov, *On the uniqueness of the inverse problem of potential theory*, Dokl. Akad. Nauk SSSR **18** (1938), 165–168 (in Russian)

[252] H. Ockendon, J. R. Ockendon, *Viscous Flow*, Cambridge U.P., 1995.

[253] J. R. Ockendon, S. D. Howison, *P. Ya. Kochina and Hele-Shaw in modern mathematics, natural sciences, and technology.* Prikl. Mat. Mekh. **66** (2002), no. 3, 515–524; Engl. transl.: J. Appl. Math. Mech. **66** (2002), no. 3, 505–512.

[254] M. Ohtsuka, *Dirichlet problem, extremal length, and prime ends*, Van Nostrand, New York, 1970.

[255] N. C. Overgaard, *A moving-boundary problem related to a fluid model for biofilm growth*, Preprint, 2005.

[256] E. Ozugurlu, J.-M. Vanden-Broeck, *The distortion of a bubble in a corner flow*, Euro. J. Appl. Math. **11** (2000), 171–179.

[257] C.-W. Park, G. M. Homsy, *Two-phase displacement in Hele-Shaw cells: theory*, J. Fluid Mech., **139** (1984), 291-308.

[258] N. B. Pleshchinskii, M. Reissig, *Hele-Shaw flows with nonlinear kinetic undercooling regularization*, Nonlinear Anal. **50** (2002), no. 2, 191–203.

[259] H. Poincare, *Sur les équations aux dérivées partielles de la physique mathematique*, Amer. J. Math. **12** (1890), 211–294.

[260] H. Poincare, *Théorie du potentiel Newtonien*, Paris, Gauthier-Villars, 1899.

[261] J. Polchinski, *String theory*, Cambridge Univ. Press, 1998.

[262] P. Ya. Polubarinova-Kochina, *On a problem of the motion of the contour of a petroleum shell*, Dokl. Akad. Nauk USSR, **47** (1945), no. 4, 254–257. (in Russian)

[263] P. Ya. Polubarinova-Kochina, *Concerning unsteady motions in the theory of filtration*, Prikl. Matem. Mech., **9** (1945), no. 1, 79–90. (in Russian)

[264] P. Ya. Polubarinova-Kochina, *Theory of groundwater movement*, Princeton Univ. Press, 1962.

[265] G. Pólya, G. Szegö, *Isoperimetric Inequalities in Mathematical Physics*, Princeton, NJ: Princeton University Press, 1951.

[266] G. Pólya, M. Schiffer, *Sur la représentation conforme de l'extérieur d'une courbe fermée convexe*, C. R. Acad. Sci. Paris, **248** (1959), 2837–2839.

[267] A. M. Polyakov, *Quantum geometry of bosonic strings*, Phys. Lett. B **103** (1981), no. 3, 207–210.

[268] Ch. Pommerenke, *Über die Subordination analytischer Funktionen*, J. Reine Angew. Math. **218** (1965), 159–173.

[269] Ch. Pommerenke, *On the logarithmic capacity and conformal mapping*, Duke Math. J., **35** (1968), 321–325.

[270] Ch. Pommerenke, *Univalent functions, with a chapter on quadratic differentials by G. Jensen*, Vandenhoeck & Ruprecht, Göttingen, 1975.

[271] Ch. Pommerenke, *Boundary behaviour of conformal maps*, Springer, Berlin, 1992.

[272] Ch. Pommerenke, A. Vasil'ev, *Angular derivatives of bounded univalent functions and extremal partitions of the unit disk*, Pacific. J. Math. **206** (2002), no. 2, 425-450.

[273] G. Prokert, *Existence results for Hele-Shaw flow driven by surface tension*, European J. Appl. Math., **9** (1998), no. 2, 195–221.

[274] D. V. Prokhorov, *Level lines of functions that are convex in the direction of an axis.*, Mat. Zametki (Math. Notes), **44** (1988), no. 4, 523–527. (in Russian)

[275] D. Prokhorov, A. Vasil'ev, *Convex Dynamics in Hele-Shaw cells*, Intern. J. Math. and Math. Sci. **31** (2002), no. 11, 639–650.

[276] D. Prokhorov, A. Vasil'ev, *Univalent functions and integrable systems*, Comm. Math. Phys. **262** (2006), no. 2, 393–410.

[277] Qing Nie, Fei Ran Tian, *Singularities in Hele-Shaw flows*, SIAM J. Appl. Math., **58** (1998), no. 1, 34–54.

[278] T. Ransford, *Potential theory in the complex plane*, London Math. Soc. Student Texts, 28, Cambridge Univ. Press, 1995.

[279] Y. Reichelt, *Moving boundary problems for degenerate elliptic equations*, Nonlinear Analysis **27** (1996), 1207–1227.

[280] M. Reissig, L. von Wolfersdorf, *A simplified proof for a moving boundary problem for Hele-Shaw flows in the plane*, Ark. Mat., **31** (1993), no. 1, 101–116.

[281] M. Reissig, F. Hübner, *Analytical and numerical treatment of Hele-Shaw models with and without regularization*, Generalized analytic functions (Graz, 1997), Int. Soc. Anal. Appl. Comput., 1, Kluwer, 1998, 271–287

[282] M. Reissig, S. V. Rogosin, *Analytical and numerical treatment of a complex model for Hele-Shaw moving boundary value problems with kinetic undercooling regularization*, European J. Appl. Math. **10** (1999), no. 6, 561–579.

[283] S. Richardson, *Hele-Shaw flows with a free boundary produced by the injecton of fluid into a narrow channel*, J. Fluid Mech., **56** (1972), no. 4, 609–618.

[284] S. Richardson, *On the classification of solutions to the zero surface tension model for Hele-Shaw free boundary flows*. Quart. Appl. Math., **55** (1997), no. 2, 313–319.

[285] S. Richardson, *Hele-Shaw flows with time-dependent free boundaries involving a multiply-connected fluid region*, Europ. J. Appl. Math. **12** (2001), 677–688.

[286] S. Richardson, *Hele-Shaw flows with free boundaries in a corner or around a wedge. Part I: Liquid at the vertex*, Europ. J. Appl. Math. **12** (2001), 665–676.

[287] S. Richardson, *Hele-Shaw flows with free boundaries in a corner or around a wedge. Part II: Air at the vertex*, Europ. J. Appl. Math. **12** (2001), 677–688.

[288] M. S. Robertson, *Analytic functions starlike in one direction*, American J. Math., **58** (1936), 465-72.

[289] S. Rohde, O. Schramm, *Basic properties of SLE*, Ann. of Math. (2) 161 (2005), no. 2, 883–924.

[290] S. Rohde, M. Zinsmeister, *Some remarks on Laplacian growth*, Topology Appl. **152** (2005), no. 1-2, 26–43.

[291] S. Ruscheweyh, L. C. Salinas, *On the preservation of direction-convexity and the Goodman-Saff conjecture*, Ann. Acad. Sci. Fenn. Ser. A-I Math., **14** (1989), no. 1, 63–73.

[292] E. B. Saff, V. Totik, *Logarithmic potentials with external fields* (with Appendix B by Thomas Bloom), Springer-Verlag, Berlin.

[293] P. G. Saffman, G. I. Taylor, *The penetration of a fluid into a porous medium or Hele-Shaw cell containing a more viscous liquid*, Proc. Royal Soc. London, Ser. A, **245** (1958), no. 281, 312–329.

[294] P. G. Saffman, G. I. Taylor, *A note on the motion of bubbles in a Hele-Shaw cell and porous medium*, Quart. J. Mech. Appl. Math. **17** (1959), no. 3, 265–279.

[295] M. Sakai, *Quadrature domains*, Lecture Notes in Math. 934, Springer-Verlag, New York, 1982.

[296] M. Sakai, *Regularity of boundaries having a Schwarz function*, Acta Math. **166** (1991), 263–297.

[297] M. Sakai, *Regularity of boundaries in two dimensions*, Ann. Scuola Norm. Sup. Pisa Cl. Sci. (4) **20** (1993), 323–339.

[298] M. Sakai, *Application of variational inequalities to the existence theorem on quadrature domains*, Trans. Amer. Math. Soc. **276** (1993), 267–279.

[299] M. Sakai, *Regularity of boundaries of quadrature domains in two dimensions*, SIAM J. Math. Analysis **24** (1993), no. 2, 341–364.

[300] M. Sakai, *Sharp estimate of the distance from a fixed point to the frontier of a Hele-Shaw flow*, Potential Analysis **8** (1998), no. 3, 277–302.

[301] M. Sakai, *Linear combinations of harmonic measures and quadrature domains of signed measures with small support*, Proc. Edinburgh. Math. Soc. **42** (1999), 433-444.

[302] M. Sakai, *Restriction, localization, and microlocalization*, in [100]

[303] A. Savakis, S. Maggelakis, *Models of shrinking clusters with applications to epidermal wound healing*, Math. Comp. Modeling **25** (1997), no. 6, 1-6.

[304] D. G. Schaeffer, *A stability theorem for the obstacle problem*, Advances in Math. **17** (1975), no. 1, 34–47.

[305] M. Schiffer, N. S. Hawley, *Connections and conformal mapping*, Acta Math. **107** (1962), 175–274.

[306] O. Schramm, *Scaling limits of loop-erased random walks and uniform spanninig trees*, Israel J. Math. **118** (2000), 221–288.

[307] Shah Dao-Shing, *Parametric representation of quasiconformal mappings*, Science Record **3** (1959), 400–407.

[308] H. S. Shapiro, *The Schwarz function and its generalizzation to higher dimensions*, University of Arkansas Lecture Notes in the Mathematical Sciences, 9. A Wiley-Interscience Publication. John Wiley & Sons, Inc., New York, 1992.

[309] D. Shoikhet, *Semigroups in geometrical function theory*, Kluwer Academic Publishers, Dordrecht, 2001.

[310] M. Siegel, S. Tanveer, W. S. Dai, *Singular effects of surface tension in evolving Hele-Shaw flows*, J. Fluid Mech., **323** (1996), 201–236.

[311] M. Siegel, R. E. Caflisch, S. D. Howison, *Global existence, singular solutions and illposedness for the Muskat problem*, Comm. Pure Appl. Math. 57 (2004), no. 10, 1374–1411.

[312] T. Sjödin, *Partial Balayage – classical and probabilistic approaches*, Nonlinear Analysis, to appear.

[313] A. Yu. Solynin, *Modules and extremal metric problems*, Algebra i Analiz **11** (1999) no. 1, 3–86; English transl.: St. Petersburg Math. J. **11** (2000), no. 1, 1–70.

[314] J. Stankiewicz, *Some remarks concerning starlike functions*, Bull. Acad. Polon. Sci., Sér. Sci. Math. Astronom. Phys., **18** (1970), 143–146.

[315] K. Strebel, *Quadratic differentials*, Springer-Verlag, 1984

[316] K. Takasaki, T. Takebe, *Integrable hierarchies and dispersionless limit*, Rev. Math. Phys. **7** (1995), no. 5, 743–808.

[317] K. Takasaki, T. Takebe, *Radial Löwner equation and dispersionless cmKP hierarcy*, arXiv: nlin.SI/0601063, 2006, 1–18.

[318] L. A. Takhtajan, *Liouville theory: quantum geometry of Riemann surfaces*, Modern Phys. Lett. A **8** (1993), no. 37, 3529–3535.

[319] L. A. Takhtajan, L.-P. Teo, *Weil-Petersson metric on the universal Teichmüller space I. Curvature properties and Chern forms*, arXiv: math. CV/0312172, 2004.

[320] L. A. Takhtajan, L.-P. Teo, *Weil-Petersson metric on the universal Teichmüller space II. Kähler potential and period mapping*, arXiv: math. CV/0406408, 2004.

[321] L. A. Takhtajan, L.-P. Teo, *Weil-Petersson geometry of the universal Teichmuller space*, Progress in Math. **237** (2005), 219–227.

[322] O. Tammi, *Extremum problems for bounded univalent functions*, Lecture Notes in Mathematics, 646. Springer-Verlag, Berlin–New York, 1978.

[323] O. Tammi, *Extremum problems for bounded univalent functions II*, Lecture Notes in Mathematics, 913. Springer-Verlag, Berlin–New York, 1982.

[324] S. Tanveer, *Evolution of Hele-Shaw interface for small surface tension*, Phil. Trans. R. Soc. Lond., A, **343** (1993), no. 1668, 155–204.

[325] S. Tanveer, X. Xie, *Rigorous results in steady finger selection in viscous fingering*, Arch. Rational Mech. Anal. **166** (2003), 219–286.

[326] S. Tanveer, X. Xie, *Analyticity and non-existence of classical steady Hele-Shaw fingers*, Comm. Pure Appl. Math. **54** (2003), 353–402.

[327] O. Teichmüller, *Extremale quasikonforme Abbildungen und quadratische Differentiale*, Abhandl. Preuss. Akad. Wiss., Math.-Naturwiss. Kl. **22** (1940), 3–197.

[328] H. Thomé, M. Rabaud, V. Hakim, Y. Couder, *The Saffman-Taylor Instability: From the Linear to the Circular Geometry* Phys. Fluids A **1** (1989), 224–240.

[329] F. R. Tian, *A Cauchy integral approach to Hele-Shaw problems with a free boundary*, Arch. Rational Mech. Anal. **135** (1996), no. 2, 175–196.

[330] F. R. Tian, *Hele-Shaw problems in multidimensional spaces*, J. Nonlinear Science **10** (2000), 275–290.

[331] Yuhai Tu, *Saffman-Taylor problem in sector geometry.* Asymptotics beyond all orders (La Jolla, CA, 1991), NATO Adv. Sci. Inst. Ser. B Phys., 284, Plenum, New York, 1991, 175–186.

[332] J.-M. Vanden-Broeck, J. B. Keller, *Deformation of a bubble or drop in a uniform flow*, J. Fluid. Mech. **101** (1980), 673–686.

[333] A. N. Varchenko, P. I. Ètingof, *Why the boundary of a round drop becomes a curve of order four*, University Lecture Series, vol. 3, AMS, 1992.

[334] A. Vasil'ev, *Univalent functions in the dynamics of viscows flows*, Comp. Methods and Func. Theory, **1** (2001), no.2, 311-337.

[335] A. Vasil'ev, *Moduli of families of curves for conformal and quasiconformal mappings.* Lecture Notes in Mathematics, vol. 1788, Springer-Verlag, Berlin-New York, 2002.

[336] A. Vasil'ev, *Univalent functions in two-dimensional free boundary problems*, Acta Applic. Math, **79** (2003), no. 3, 249–280.

[337] A. Vasil'ev, I. Markina, *On the geometry of Hele-Shaw flows with small surface tension*, Interfaces and Free Boundaries, **5** (2003), no.2, 183–192.

[338] A. Vasil'ev, *Evolution of conformal maps with quasiconformal extensions*, Bull. Sci. Math. **129** (2005), no. 10, 831–859.

[339] Yu. P. Vinogradov, P. P. Kufarev, *On a problem of filtration*, Akad. Nauk SSSR. Prikl. Mat. Meh., **12** (1948), 181–198. (in Russian)

[340] Yu. P. Vinogradov, P. P. Kufarev, *On some particular solutions of the problem of filtration*, Doklady Akad. Nauk SSSR (N.S.) **57** (1947), 335–338. (in Russian)

[341] T. Vicsek, *Fractal growth phenomena*, World Scientific Publishing Co, Singapore, 1989.

[342] M. A. Virasoro, *Subsitiary conditions and ghosts in dual resonance models*, Phys. Rev. **D1** (1970), 2933–2936.

[343] N. Whitaker, *Numerical solution of the Hele-Shaw equations*, J. Comput. Phys. **90** (1990), no. 1, 176–199.

[344] P. B. Wiegmann, A. Zabrodin, *Conformal maps and integrable hierarchies*, Comm. Math. Phys. **213** (2000), no. 3, 523–538.

[345] T. A. Witten, Jr., L. M. Sander, *Diffusion-Limited Aggregation, and kinetic critical phenomenon*, Phys. Rev. Letters, **47** (1981), no. 2, 1400–1403.

[346] T. A. Witten, L. M. Sander, *Diffusion-Limited Aggregation*, Phys. Rev. B **27** (1983), no. 9, 5686–5697.

[347] S. Wolpert, *Thurston's Riemannian metric for Teichmüller space*, J. Differential Geom., **23** (1986), no. 2, 143–174.

[348] P. G. Zograf, L. A. Takhtajan, *On the Liouville equation, accessory parameters and the geometry of Teichmüller space for Riemann surfaces of genus 0*, Mat. Sb. (N.S.) **132(174)** (1987), no. 2, 147–166; translation in Math. USSR-Sb. **60** (1988), no. 1, 143–161.

[349] P. Zograf, L. Takhtajan, *Hyperbolic 2-spheres with conical singularities, accessory parameters and Kähler metrics on* $M_{0,n}$, Trans. Amer. Math. Soc. **355** (2003), no. 5, 1857–1867.

List of Symbols

$\mathbb{C}$	complex plane
$\overline{\mathbb{C}}$	Riemann sphere
U	unit disk
$\partial U = S^1$	unit circle
U^*	exterior of the unit disk
U_r	disk of radius r
H^+	right half-plane
$\mathbb{R}$	real line
$\mathbb{R}^+$	positive real axis
$\mathbb{R}^-$	negative real axis
$\bar{D}$	closure of D
$\text{int}\, D$	interior of D
$f \circ g$	superposition of f and g
$\delta_a(z)$	Dirac's distribution in $z \in \mathbb{C}$ supported at a
$\mathbf{F} \equiv {}_2\mathbf{F}_1$	Gauss hypergeometric function
$\mathbf{B}$	Euler's Beta-function
$\mathbf{K}$	complete elliptic integral
$S(z)$	Schwarz function
χ_Ω	characteristic function of Ω
$d\sigma_z$	area element in z-plane
Bal	partial balayage
$\text{dist}\,(\Gamma, a)$	distance from set Γ to point a
S	class of univalent functions in U normalized by $f(\zeta) = \zeta + a_2\zeta^2 + \dots$
Σ	class of univalent functions in U^* normalized by $f(\zeta) = \zeta + a_0 + a_1/\zeta + \dots$
Σ_0	subclass of Σ of functions with $a_0 = 0$
$\tilde{\Sigma}$	subclass of Σ of functions smooth on the boundary
$\tilde{\Sigma}_0$	subclass of Σ_0 of functions smooth on the boundary
Σ^{qc}	subclass of Σ of functions that admit quasiconformal extension into U
Σ_0^{qc}	subclass of Σ_0 of functions that admit quasiconformal extension into U
S^*	class of starlike functions in U

S^*_α	class of starlike functions of order α in U
$S^*(\alpha)$	class of strongly starlike functions of order α in U
C	class of convex functions in U
$C_{\mathbb{R}}$	class of convex functions in U in the direction of $\mathbb{R}$
$H^-_{\mathbb{R}}$	class of convex functions in H^+ in the negative direction of $\mathbb{R}$
$H^-_{\mathbb{R}}(\alpha)$	class of convex functions in H^+ of order α in the negative direction of $\mathbb{R}$
$m(D,\Gamma)$	modulus of a family of curves Γ in D
$M(D)$	conformal modulus of a doubly connected domain D
$R(D,a)$	conformal radius of D with respect to a
$m(D,a)$	reduced modulus of D with respect to a
$\operatorname{cap} C$	capacity of a condenser C
$\operatorname{cap}^{(h)} C$	hyperbolic capacity of a continuum C
$m_\triangle(D,a)$	reduced modulus of a triangle D with respect to its vertex a
$\mu_f(\zeta)$	dilatation of a quasiconformal map f
$\mathcal{F}$	family of all quasiconformal automorphisms of U
$\mathcal{F}_0$	family of all quasiconformal automorphisms of U normalized by $f(\pm 1) = \pm 1$, $f(i) = i$
$L^\infty(U)$	essentially bounded functions in U
T	universal Teichmüller space
$S_f(z)$	Schwarzian derivative
$\mathcal{M}$	universal modular group
$\tau(x,y)$	Teichmüller distance
$B(U)$	Banach space of all functions holomorphic in U equipped with the norm $\|\varphi\|_{B(U)} = \sup_{\zeta\in U} \lvert\varphi(\zeta)\rvert(1-\lvert\zeta\rvert^2)^2$
$A(U)$	Banach space of analytic functions with the finite L^1 norm in the unit disk
$A^2(U)$	Banach space of analytic functions φ with the finite norm $\|\varphi\|_{A^2(U)} = \iint_U \lvert\varphi(\zeta)\rvert^2(1-\lvert\zeta\rvert^2)^2 d\sigma_\zeta$
H	tangent space to T at the initial point
$\operatorname{Diff} S^1$	Lie group of C^∞ sense preserving diffeomorphisms of the unit circle S^1
$\operatorname{Rot} S^1$	group of rotations of S^1
$\operatorname{Vect} S^1$	Lie algebra of smooth tangent vector fields to S^1

Index

Advances in Mathematical Fluid Mechanics

Edited by
Galdi, G.P., University of Pittsburgh, USA
Heywood, J.G., University of British Columbia, Vancouver, Canada
Rannacher, R., University of Heidelberg, Germany

Advances in Mathematical Fluid Mechanics is a forum for the publication of high quality monographs, or collections of works, on the mathematical theory of fluid mechanics, with special regards to the Navier-Stokes equations. Its mathematical aims and scope are similar to those of the Journal of Mathematial Fluid Mechanics. In particular, mathematical aspects of computational methods and of applications to science and engineering are welcome as an important part of the theory. So also are works in related areas of mathematics that have a direct bearing on fluid mechanics.

Your Specialized Publisher in Mathematics
Birkhäuser

For orders originating from all over the world except USA/Canada/Latin America:

Birkhäuser Verlag AG
c/o Springer GmbH & Co
Haberstrasse 7
D-69126 Heidelberg
Fax: +49 / 6221 / 345 4 229
e-mail: birkhauser@springer.de
http://www.birkhauser.ch

For orders originating in the USA/Canada/Latin America:

Birkhäuser
333 Meadowland Parkway
USA-Secaucus
NJ 07094-2491
Fax: +1 201 348 4505
e-mail: orders@birkhauser.com

Galdi, G.P. / Heywood, J.G. / Rannacher, R. (eds.)
Contributions to Current Challenges in Mathematical Fluid Mechanics (2004)
ISBN 3-7643-7104-8

The mathematical theory of the Navier–Stokes equations presents still fundamental open questions that represent as many challenges for the interested mathematicians. This volume collects a series of articles whose objective is to furnish new contributions and ideas to these questions, with particular regard to turbulence modelling, regularity of solutions to the initial-value problem, flow in region with an unbounded boundary and compressible flow.
Contributors: A. Biryuk, D. Chae and J. Lee, A. Dunca, V. John and W.J. Layton, T. Hishida, T. Leonaviciene and K. Pileckas.

Gustafsson, B. / Vasiliev, A.
Conformal and Potential Analysis in Hele–Shaw Cells (2006).
ISBN 3-7643-7703-8

Neustupa, J. / Penel, P. (eds.)
Mathematical Fluid Mechanics. Recent Results and Open Questions (2000)
ISBN 3-7643-6593-5

Mathematical modeling and numerical simulation in fluid mechanics are topics of great importance both in theory and technical applications. The present book attempts to describe the current status in various areas of research. The 10 chapters, mostly survey articles, are written by internationally renowned specialists and offer a range of approaches to and views of the essential questions and problems. In particular, the theories of incompressible and compressible Navier–Stokes equations are considered, as well as stability theory and numerical methods in fluid mechanics.
Although the book is primarily written for researchers in the field, it will also serve as a valuable source of information to graduate students.

MIX
Papier aus verantwortungsvollen Quellen
Paper from responsible sources
FSC® C105338

If you have any concerns about our products,
you can contact us on
ProductSafety@springernature.com

In case Publisher is established outside the EU,
the EU authorized representative is:
Springer Nature Customer Service Center GmbH
Europaplatz 3, 69115 Heidelberg, Germany

Printed by Libri Plureos GmbH
in Hamburg, Germany